Knut Bjørlykke

Sedimentology and Petroleum Geology

With 186 Figures

Springer-Verlag Berlin Heidelberg New York
London Paris Tokyo Hong Kong

Professor Dr. KNUT OLAV BJØRLYKKE
Department of Geology
University of Oslo
P.O. Box 1047 Blindern
0316 Oslo 3, Norway

Translated from Norwegian by:

BEVERLY WAHL
Frydensgate 4 b
0564 Oslo 5, Norway

Title of the original Norwegian edition:
Lærebok i Sedimentologi og Petroleums Geologi
© by Universitetsforlaget, Oslo 1984

ISBN-13: 978-3-540-17691-6   e-ISBN-13: 978-3-642-72592-0
DOI: 10.1007/978-3-642-72592-0

Library of Congress Cataloging-in-Publication Data. Bjørlykke, K.O. (Knut Olav), 1938-
[Lærebok i sedimentologi og petroleums geologi. English] Sedimentology and petroleum geology / K.O. Bjørlykke. p. cm. Translation of: Lærebok i sedimentologi og petroleums geologi.
Bibliography: p. Includes index. 1. Petroleum–Geology. 2. Sedimentology.   I. Title. TN
870.5.B5713   1989   553.2'82–dc 19   89-5965

© Springer-Verlag Berlin Heidelberg 1989

Typesetting: International Typesetters, Inc., Makati, Philippines.

2132/3145-543210 – Printed on acid-free paper

Preface

This book is intended to give an introduction to sedimentology and petroleum geology at undergraduate level. These two subjects have been treated together because of the close links between sedimentology as an academic dicipline, petroleum geology, which is the application of sedimentology, and a number of other aspects of petroleum exploration and production. The oil industry ist by far the most important employer of sedimentologists and the lively interaction that takes place between the academic community and the research laboratories and exploration departments of the oil industry has been very fruitful for both parties. Our knowledge of sedimentary basins now depends to a very large extent on data obtained by commercial petroleum exploration. Studies of actual rocks in outcrops, particularly if they are extensive, will always be important for sedimentologists, but subsurface data like seismic sections and well logs provide us with in much information on the three-dimensional distribution of facies that we could not otherwise obtain. Subsurface techniques are certainly important for petroleum geologists, but also other sedimentologists should be able to use subsurface data. I have therefore included elementary introductions to the use of well logs and seismic methods in this book, with fundamentals of external controls on sedimentation such as basin subsidence and sea level changes.

I have tried to present the state of knowledge at this level without referring to the original research papers except when specific data are quoted or used in illustrations. The list of references also includes other important papers, particularly review papers and books for further reading.

April 1989   KNUT BJØRLYKKE

Acknowledgement

I am greatly indebted to the following colleagues who have taken the time to critically review parts of this book:
J.D. Collinson, J.I. Faleide, W.L. Galloway, S.R. Larter, D. Leythaeuser, M.R. Talbot, T.P. Scoffin, and N. Spjeldnæs.

Beverly Wahl is thanked for translating an earlier Norwegian version of this book into English.

Many illustrations including the cover have been made by Masaoki Adachi. I would also like to thank Elisabet Middlethon at Universitetsforlaget for her assistance.

Permission to reproduce illustrations from journals or other books is acknowledged.

KNUT BJØRLYKKE

Contents

1 Introduction to Sedimentology and Petroleum Geology

Sedimentology is the study of sedimentary rocks and their formation. The subject covers processes which produce sediments, such as weathering and erosion, transport and deposition by water or air, and the changes which take place in sediments after their deposition (diagenesis). Changes which take place in sedimentary rocks at temperatures of over 200°C are called *metamorphic processes* and are not dealt with here.

Like all natural sciences, sedimentology has an important descriptive component. In order to be able to describe sedimentary rocks, or to understand such descriptions, it is necessary to familiarise oneself with quite an extensive nomenclature. There are specialised names for types of sedimentary structures, grain-size distributions and mineralogical composition of sediments. We also have a genetic nomenclature, which names rock types according to the particular way in which we think they have formed. Examples of these are fluvial sediments, (which are deposited by rivers) and eolian (air-borne) sediments. The descriptive nomenclature is used as a basis for an interpretation of how the rock was formed. When we are reasonably confident about their origin we may use the genetic nomenclature.

Sedimentology covers studies of both *recent* (modern) sediments and older sedimentary rocks. By studying how sediments form today in so-called recent environments, we can understand the conditions under which various sedimentological processes take place. From such observations we may be able to recognise older sediments which have been formed in the same way. This is called using the principle of uniformitarianism, which has been of great importance in all geological disciplines since the time of James Hutton (1726–1797).

Conditions on the surface of the Earth have fluctuated widely throughout geological history, however, so that the principle of uniformitarianism cannot be applied without reservations. One important aspect of sedimentological research is attempting to reconstruct changes in environments on the Earth's surface throughout geological time. This applies particularly to climate, vegetation and the composition of the atmosphere and the oceans.

The study of the development of plants and animals throughout geological time (palaeontology) is important to sedimentology, not only for dating beds, but also because organisms are an important component of many sedimentary rocks (certain limestones, for example), and organic processes affect the weathering process and the composition of sea-water. Many organisms make very specific demands of their environment, and fossils are consequently a great help in reconstructing the environment in which the sediments were deposited Palaeoecology is the study of ecological conditions as we are able to reconstruct them on the basis of remains or traces of plants and animals in rocks. Traces of animals in sediments have proven to be very useful environmental indicators.

Studies of recent and older sedimentary rocks provide a fruitful two-way exchange of information in sedimentology. From studies of recent environments we can learn what general conditions particular processes require. In older rocks, however, we can study sections which encompass many millions of years of sedimentation, offering us a completely different record of the way sediment-ological processes can vary as a function of time. As a result, studies of older rocks also contribute to our understanding of the recent environment and offer non-uniformitarian explanations.

When we study rocks, we should attempt to give *objective descriptions* of the composition, structure etc. of the rocks, and on the basis of these try to *interpret* how they were formed. However, it is impossible to give a completely exhaustive, objective description of a rock. We know from experience that we have a tendency to observe what we are looking for, or what we anticipate finding. Descriptions of sedimentary sequences from the 1940's and 1950's, for example, contain few observations about sedimentary structures which we would consider fairly con-spicuous and important today. We observe sedimentary structures because we have learned to recognise them and understand their genetic significance. Our obser-vations are therefore selective.

Many sedimentologists use a standard checklist for what they should observe in the field, so that their descriptions are as comparable as possible. It is, however, important that field observations do not become too much of a routine. Facies analysis should be a creative process and the various models should be kept in mind when making the observations. It is also desirable to quantify observations as far as possible. Field observations can be supplemented by laboratory analyses. These might be surveys, texture analyses (e.g. grain distribution), microscope analyses (perhaps using a scanning electron microscope) or chemical analyses. Pure de-scriptions of sedimentary rocks are useful because they increase the data base on which we can build our interpretations. Nevertheless it is often most fruitful to have a theory or hypothesis against which to test our observations. Data collection can then be focussed on observations and measurements which can support or disprove the hypothesis.

Although there were a number of scattered investigations earlier, it was not until the end of the 1950's and the 1960's that geologists began to undertake systematic studies of recent environments of sedimentation with a view to finding connections between the external environment and the sediments which ac-cumulate. The environment governs the sedimentological processes which deter-mine what sort of sediments are formed and deposited. The connection we are trying to understand in modern environments is thus: environment → process → sediment.

Today a large number of modern environments of sedimentation have been studied in great detail. These include eolian and fluvial environments, deltas, beach zones, tidal flats and carbonate banks. Deep sea environments have naturally not been so easy to study, but modern sampling and remote sensing equipment and underwater TV cameras have made it easier to gather observations from this environment, too. In recent years systematic drilling through sedimentary layers in the ocean depths (Deep Sea Drilling Programme — DSDP, and Ocean Drilling Programme — ODP) has provided an entirely new picture of the geology of the ocean depths. Specially constructed diving ships (e.g. ALBIN) make it possible for geologists to observe the ocean floor directly at depths of up to about 3500 m and take samples of surface sediments. In addition geophysical, particularly seismic surveys, provide one of the most important bases for understanding the stratigraphy and geometry of sedimentary basins.

In studying older rocks we base our approach on certain features which we can observe or measure, and attempt to interpret the processes that produced them. Particular variations in grain size and sedimentary structures in profiles can be interpreted as having been formed through particular processes, e.g. eolian or aqueous processes. These processes (e.g. eolian or aqueous transport) also help us to reconstruct the environments to a certain extent. The direction of interpretation in studies of older sedimentary rocks is thus: sedimentary rock → processes → environment.

Applied geology has always been important, even for purely scientific research. The interests of economic recovery of raw materials from sedimentary rocks create a demand for sedimentologists and sedimentological research. Exploration for and recovery of raw materials also provide important scientific information. Sedimentary rocks contain raw materials of considerably greater value than those we find in metamorphic and eruptive rocks. The most important are oil, gas and coal deposits, but a very large amount of the world's ore deposits is also found in sedimentary rocks. In recent years we have seen an increasing tendency to interpret ores, even in metamorphic rocks, as having been formed through sedimentary processes. Limestone, clay, sand and gravel are also important raw materials which require sedimentological expertise.

The petroleum industry employs a very high percentage of the world's professional geologists. This industry has a particular need for research, and also has the financial capacity to invest in it. Because oil and gas are found largely in sedimentary rocks, exploration for and recovery of hydrocarbons is based to a large extent on sedimentology. Much of what we now know about the world's sedimentary basins and their regional geology is derived from seismic profiles which have been shot in connection with oil exploration and drilling for oil and gas. The oil industry has also helped to stimulate pure sedimentological research, and significant contributions to research in this area are published by the research laboratories of the oil companies. Research based on economic interests is also useful from a purely scientific point of view, because it often focusses on particular questions which may be quite fundamental.

In this book I have attempted to demonstrate how the oil potential of sedimentary basins is dependent on sedimentological factors. An area of increasing importance is so-called *production geology*, i.e. geological description and interpretation of oil and gas reservoirs. Attempting to increase the fraction of oil or gas which can be recovered from the reservoirs is one major task. This requires close teamwork between reservoir engineers and geologists, to establish in great detail the geometry and distribution of porosity and permeability of reservoir rocks. We also need very much to know more about the physical and chemical properties of the reservoir rock, for reasons which are discussed at the end of the book.

2 Textures

Textures in Clastic Sediments

The *textures* of clastic sediments include external properties of sediment grains, such as the size, shape and orientation of the grains. These properties can be described relatively objectively, and say a great deal about the origin and conditions of transport and deposition of the sediments.

Grain Size

By grain size we normally mean grain diameter, but the two are only strictly synonymous in the case of completely spherical particles. Most grains are not spherical, however, and it is difficult to identify a representative diameter, particularly in the case of elongated or flat grains. For this reason we have adopted the concept "nominal" diameter (d_n), defined as the diameter of a spherical body which has the same volume as the grain. In practice we are seldom in a position to measure the volume of each grain, and we therefore use indirect methods to measure the size of a grain. If the grain resembles a tri-axial ellipsoid it is useful for some purposes to measure the long axis (a), the intermediate axis (b) and the short axis (c) of a grain.

Sand and gravel can be most simply analysed by means of *mechanical sieving*. A bank of sieves consists of sieves with mesh sizes which decrease downwards. A sample is put in the uppermost sieve and the bank of sieves is shaken (Fig. 2.3a). Grains which are larger than the mesh size will remain, while smaller grains will fall through and perhaps remain lying on the next sieve. by weighing the fraction of the sample which remains on each sieve, we can construct a grain-size distribution curve. However, what passes between the square holes of the sieve depends greatly on grain shape, and is not an exact expression of the nominal diameter. The lower limit for sieve analyses is 0.04–0.03 mm, because finer particles exhibit too great cohesion, which makes it difficult fo them to pass through the finer sieves.

Fine silt and clay fractions can be analysed in a number of ways. Most classic methods are based on measurements of settling velocity in liquids, and are based on Stokes' Law:

When the settling velocity of grains (falling through water, for example) is constant, the resistance to the movement (friction), which acts upwards, must be equal to the force of gravity, which acts downwards (Fig. 2.1).

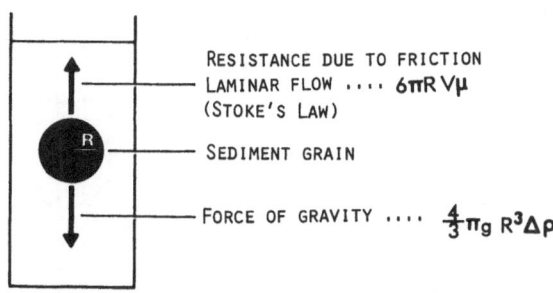

Fig. 2.1. The forces acting on a falling sediment grain. We can then work out the relationship between settling velocity of sediment grains in water and grain size (radius). See text

$6 \pi R v \mu = 4/3\pi gR^3 \Delta\rho.$
Friction   Gravity
$v = 2/9\mu \cdot gR^2 \cdot \Delta \rho.$

Here μ is viscosity (N.s.n^{-2}) and $\Delta\rho$ the difference in density between the grain and the liquid. R is the radius (cm) of the sediment grain. We see that the settling velocity (v) increases with the square of the radius (R). For settling through water one approaches:

$\log v = 2 \cdot \log R + c$ (const. approx. 4.5)
$v = cms^{-1}.$

This applies for small particles, where the flow of the liquid around the grain is laminar (see p. 17) and the concentration of grains is low.

When the grains are larger than about 0.1–0.5 mm, the settling velocity increases, and turbulence develops around the grain. The frictional resistance therefore increases, and in the case of larger grains (>1 mm) settling velocity increases approximately in proportion to the square root of the radius.

$\log v = \sqrt{\log 4/3 \, \Delta\rho \, R}$

or

$\log v = \frac{1}{2}\log R + c$ (Rubey's Law).

The settling velocity is sensitive to temperature variations, which affect the viscosity of the water.

However, it is not practicable to measure the settling velocity of each grain, particularly because in most cases we want to analyse a representative sample. We can, however, measure the settling velocities of sediment grains indirectly. One commonly used method employs a hydrometer (Fig. 2.2). We suspend the sample in a cylinder with a mixer so that at a time T_0 we have an even distribution of all grain sizes and therefore of density throughout the cylinder. A hydrometer floats in the suspension and registers the density of the suspension on a scale on the upper part of the tube. If we let the suspension stand for a period of time (T_1) the grains will sink towards the bottom. At the level at which the bulb of the hydrometer is floating, all the largest particles will already have passed, so that the density at that level is reduced. This reduction in density is registered on the hydrometer scale. After a further period of time (T_2) all those grains which are slightly smaller will also have

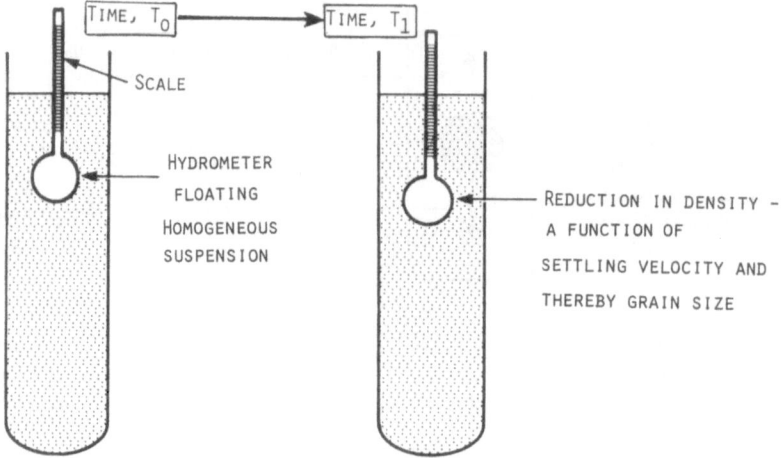

Fig. 2.2. Diagram showing principles of hydrometer analysis. A sample is dispersed into a homogeneous suspension. A hydrometer floating in the suspension records the change in density of the suspension as a function of time. The density of the suspension is a function of grain size. See text

passed this level, and the density will be further reduced. By taking successive readings of the density we may plot a curve which expresses density reduction as a function of time. Since density reduction is a function of settling velocity, this curve can be calibrated to give a grain-size distribution curve. We can also determine the density of the suspension as a function of time by pipetting out small amounts of suspension from a particular level in the cylinder at particular times, and evaporating the suspension to determine the density (concentration). We then obtain a similar grain-size distribution curve.

When we analyse fine-grained sediments with a large clay fraction, or separate out clay fractions, it may be useful to use a centrifuge. We then increase the acceleration term g in Stokes' Law. The acceleration in the centrifuge can be calculated from the velocity of rotation and the length of the rotation arm.

There are also "sedimentation balances". Sediments suspended in a cylinder fall through the water column and deposit on a balance pan at the bottom of the cylinder. This balance records and writes out the increase in weight, and thereby precipitation from suspension, as a function of time. This gives a direct cumulative curve.

In recent times, other, new methods of analysing grain distribution have been developed. They are based on the refraction or dispersion of light which passes through suspensions (scattering). This method gives characteristic "scatter" which can be calibrated against samples of known grain size. Equipment has also been developed which uses X-rays instead of light to produce the characteristic scatter patterns.

It is important to note that *no* methods measure the nominal diameter. In methods which measure settling velocity, grain shape often plays a major part. A large, thin mica flake has a settling velocity which corresponds to that of a smaller spherical grain. We call the diameter which corresponds to the settling velocity the *effective diameter* (d_e). With the scatter method, flakey grains are attributed a different, probably greater diameter than that indicated by the settling velocity method.

Grain-Size Distribution in Solid Rocks

Lightly cemented sandstones can be disintegrated by means of ultrasound in the laboratory and then analysed in the normal manner. Carbonate cemented rocks may be disintegrated using acids. However, we must bear in mind that new clay minerals may have formed through diagenetic processes, and that some minerals may have been broken down mechanically or dissolved chemically. Consequently it is not certain that we are dealing with the original grain distribution. Post-depositional alteration (diagenesis) must be taken into account.

Well-cemented rocks must be analysed by means of a microscope in thin section. It is difficult to analyse the finer fractions in this manner (fine silt and clay), and we must always remember the "section effect", i.e. that in most cases we will not be seeing the greatest diameter. With spherical grains, the relation between the real diameter, d_r, and the observed diameter, d_o, can be expressed statistically: $d_r = 4/\pi d_o$.

Presentation of Grain-Size Distribution Data

Grain-size distribution is one of the many types of natural data which must be presented on a logarithmic scale for convenience. Wentworth's scale is based on logarithms to the base 2, and this is now the one most widely found in geological literature.

For the sake of convenience, these data are commonly plotted against a linear scale. The phi (φ) scale allows convenient interpolation of graphic data, where $\varphi = -\log_2 d$. The reason this negative logarithm is used is that normally most of the sediment grain diameters (d) are less than 1 mm, so these will have a positive phi value (Fig. 2.3a,b). It is convenient to plot grain-size distribution data as a function of phi values, especially on cumulative curves. In normal descriptions of grain size, however, it is most convenient to state grain size in mm, so the reader does not have to recalculate from phi values.

The simplest way of presenting grain-size distribution data is by means of histograms (Fig. 2.4). A histogram shows what percentage by weight of the grains falls within a particular size range. This type of presentation gives a good visual impression of the distribution of grains in the various size categories. In particular, it is easy to see how well sorted the sediments are, and whether the distribution of grain sizes is symmetrical, or perhaps bimodal, i.e. with two maxima.

A *cumulative* distribution curve shows what percentage of a sample is larger or smaller than a particular grain size. The steeper the curve, the better the sorting. The advantage of this curve is that it is easy to read off values which fall between the fixed

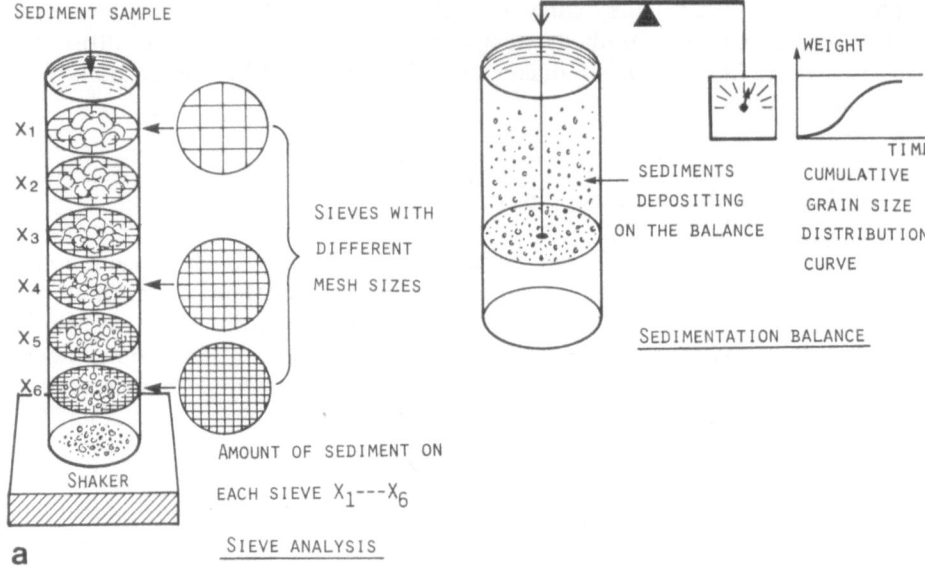

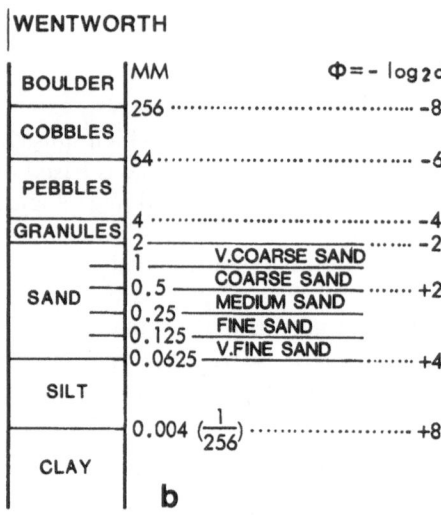

Fig. 2.3. **a** Sketch showing the principles involved in sieve analysis and use of a sedimentation balance. Sieve analysis is usually used for grain sizes down to 0.03–0.02 mm, but with wet-sieving even finer sediment grains can be sieved. The sedimentation balance gives us a direct expression of settling velocity, i.e. weight increase as a function of time. This is therefore *cumulative* grain-size distribution. **b** Grain-size classification of clastic sediments. The grain size (*d*) is often described in terms of φ values where $\varphi = -\log_2 d$

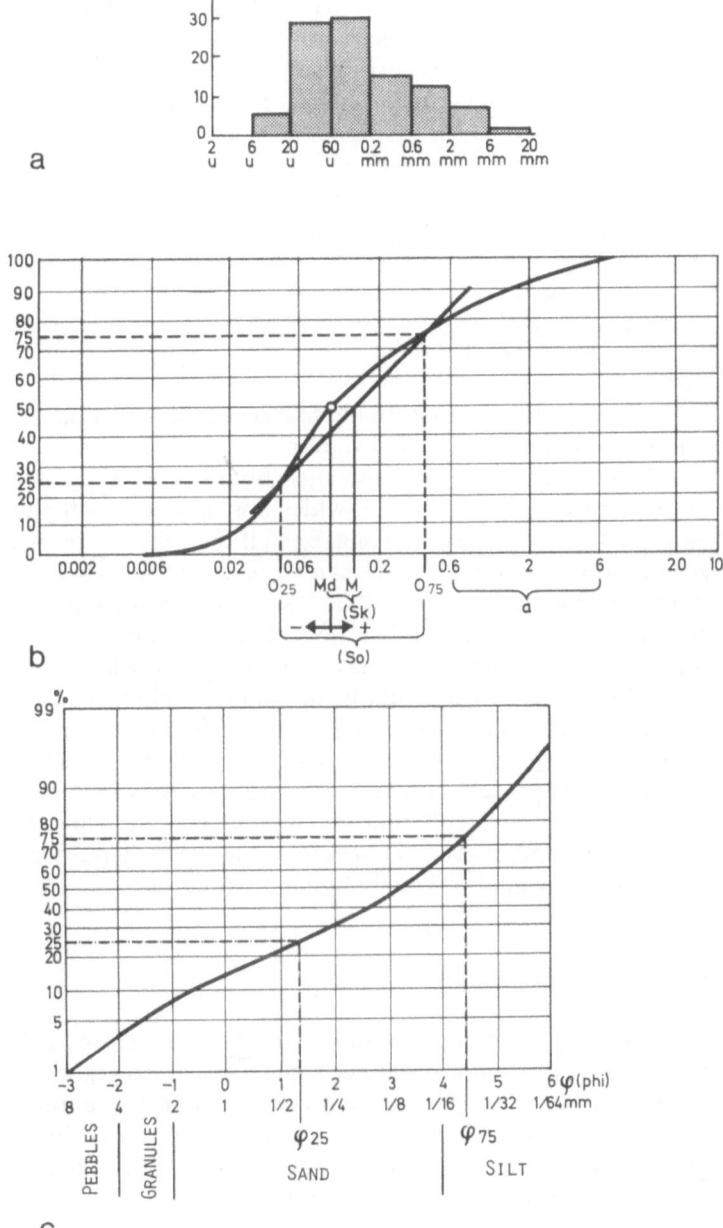

Fig. 2.4. Presentation of grain-size distribution data (**a**) histogram (**b**) cumulative curve (**c**) cumulative curve on probability paper. **a, b** and **c** represent the same sample. When plotted on probability paper, a logarithmic normal distribution, like a Gaussian curve, will plot as a straight line

points on which the curve is based (e.g. the sieve sizes), by extrapolation. However, it is not so easy to see symmetry and bimodality in this sort of curve. It has been claimed that naturally occurring sediments often have a lognormal distribution which approaches a Gaussian curve with the formula:

$$\gamma = \frac{1}{S\sqrt{2\pi}}\ e^{-1/2}\left(\frac{x-\bar{x}}{s}\right)^{2}.$$

Here x is the grain size and $\bar{x}$ is the mean value of all (n)x. S is the standard deviation, defined as:

$$S = \sqrt{\Sigma\ \frac{(x-\bar{x})^{2}}{n-1}}$$

68% of all x values fall within the first standard deviation; 95% of all x values fall within the second standard deviation.

   If, instead of plotting the distribution curves on normal paper, we use probability paper, distributions which are lognormal will plot as straight lines, the slope of which represents sorting. Even if the whole distribution is not lognormal, it often appears that the curve can be regarded as a composite of 2–3 lognormal grain-size populations. These populations often overlap, so that some sections of the curve represent a combination of two populations, each of which may be lognormal. Each population may represent grains which have been transported in different manners, for example by saltation, by rolling (bed load), in suspension etc.

   It is important that the samples we take for grain-size distribution analysis are representative, each representing only one bed which has been deposited through a sedimentary process. If we take a sample at the boundary between two beds, we will often get bimodal distributions and infiltration of fine-grained sediments into coarse sand and gravel. Sand which has been deposited in a coarse-grained conglomerate will also show bimodal distribution.

Grain-Size Distribution Parameters

Phi (φ) = $-\log_{2}d$ (after Folk and Ward 1957) where d is the grain diameter in millimetres (as previously defined). Definition φx: a grain size expressed in phi (φ) units such that x% of the sample is larger than this grain size (percentile). For example, φ30 means a grain size expressed as a φ value which 30% by weight of the grains exceed. The median, Md, = φ50. In other words, equal amounts of the sample are larger and smaller than this grain size.

Mean

$$M = \frac{\varphi16 + \varphi50 + \varphi84}{3}$$

Sorting or dispersion

$$So_{1} = \frac{\varphi84 - \varphi16}{4} + \frac{\varphi95 - \varphi5}{6.6}$$

Skewness

$$Sk = \frac{\varphi16 + \varphi84 - \varphi50}{2(\varphi84 - \varphi16)} + \frac{\varphi5 + \varphi95 - 2\varphi50}{2(\varphi95 - \varphi5)}$$

Kurtosis

$$K_G = \frac{\varphi95 - \varphi5}{1.44(\varphi75 - \varphi25)}$$

Grain-size distribution parameters after Trask and Krumbein. Px is the grain size (percentile) measured in mm which is such that x% of the sample is smaller than the grain size.

Median $Md = P50$

Mean

$$M = \frac{P25 + P75}{2}$$

Sorting

$$So_2 = \frac{P75}{P25} \quad \text{or} \quad \sqrt{\frac{P75}{P25}}$$

Skewness

$$Sk = \frac{P25 \times P75}{Md^2}$$

Kurtosis

$$K = \frac{P75 - P25}{2(P90 - P10)} .$$

The arithmetic mean M can also be defined as $\bar{X} = \Sigma x/n$. To calculate it we use a large number of points (n) on the curve.

Significance of Grain-Size Parameters

The *mean* diameter is an arithmetically calculated average grain size. The *median* diameter is the diameter of that fraction of grains with a size such that 50% by weight of the sample grains are smaller than it, and 50% are larger. This means that the area under the distribution curve must be the same on both sides of the line which marks the median diameter. In the case of completely symmetrical distribution curves, the mean diameter (M) and the median diameter (Md) will coincide (Fig. 2.5). If the sample has a wide spread (tail) towards the fine grain sizes (larger phi values) and a relatively sharp delimitation at the large grain-size end, we say that the sample has positive skewness. This will be typical of fluvial sediments. There will be a fairly definite upper limit to the grain sizes that rivers can transport, while there will be no sorting of the fine fractions which are transported in suspension. There are no processes in rivers which remove the finest grain sizes, but on the river bottom there is an upper limit to the grain size which can be carried as bottom load. Major variations in flow velocity, for instance during floods, will give poorer sorting.

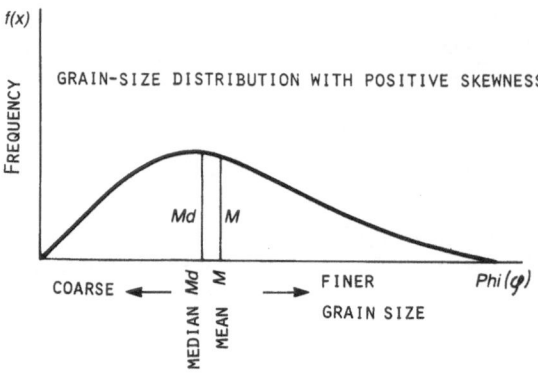

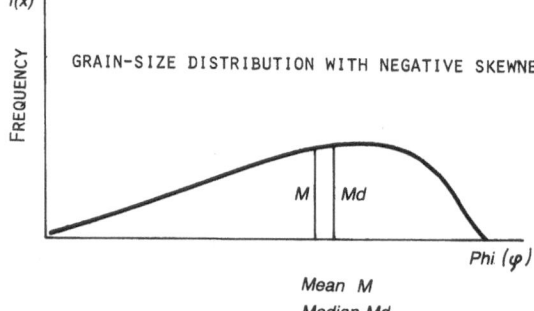

Fig. 2.5. Grain-size distribution curves with positive and negative skewness. Note that for sediments with positive skewness the mean grain size is finer (higher skewness value) than the median. With neutral skewness, the median diameter equals the mean

Eolian (wind) deposits also have positive skewness because there is an upper size limit to the grains which can be transported. Although the finest particles may be removed selectively, there will still be a "tail" of fine material. The fine material in dunes may also be protected by a cover of larger particles (*lag*) against further erosion and transport. Eolian deposits are very well sorted, however (Fig. 2.6).

Beach deposits, on the other hand, are negatively skewed, i.e. the distribution curve shows a definite lower limit, while there is often a "tail" of larger particles, i.e. granules and pebbles. The hydrodynamic conditions on a beach are such that each wave brings some sediment in suspension. Whereas sand grains, particularly medium to coarse sand, will rapidly settle from suspension and be deposited on the beach again, fine sand, silt and clay will remain in suspension longer and be transported further out from the beach. Beach deposits will be well sorted. Further out and at a depth of some metres (5–50 m), depending on how strong the waves are (and consequently the depth of the wave base), we will have poorer sorting because fine-grained sediments (clay and silt) will be deposited along with coarser-grained sediments which are carried out during storms.

Sediments which are deposited from suspension have poor sorting and positive skewness. This is very typical of turbidites. Clay suspensions which are deposited on land, or high-density suspensions (mud flows) have negative skewness because they often contain large clasts. Glacial transportation separates out particular grain sizes to only a very limited extent, and tills consequently have extremely poor sorting. *Kurtosis* is an expression for the spread of the extreme ends of a grain-size

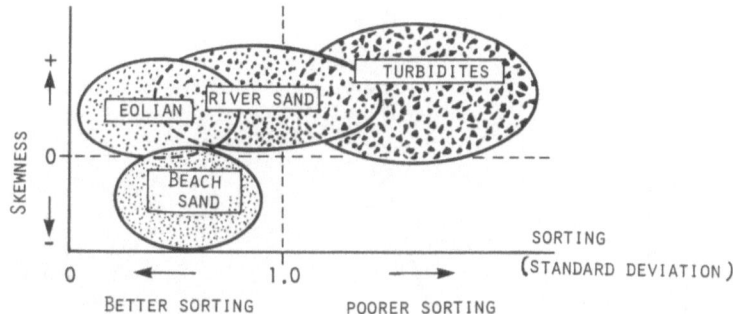

Fig. 2.6 Sorting and skewness of grain-size distribution parameters which are suitable for distinguishing between sediments deposited through various processes and in different environments of deposition

distribution curve in relation to the central part. This distribution parameter is used somewhat less than the others.

Grain-size distribution parameters can be calculated from a graphical representation of the distribution curve, or by means of relatively simple calculations from the primary grain-size distribution data.

We are often interested in the extreme values of a distribution curve. In conglomerates we can plot φ1 (1% percentile) against the median diameter on one diagram. This is called a Passega diagram, after Passega (1957, 1969), and thus relates the size of the largest 1% of the grain population to the median diameter. This tells us how large the largest clasts are compared to the average grain size, a feature which is an important transport mechanism indicator.

Well-sorted sandstones often produce a log-normal grain size distribution curve. Such a grain-size distribution curve will plot as a straight line on so-called "probability paper". A grain-size distribution curve may often consist of several populations of grains which produce straight lines and which have been transported by different processes (Fig. 2.7).

Use of Grain-Size Distribution Data

Grain-size distribution data ought not to be used uncritically or in isolation as the only indication of a depositional environment. The grain-size distribution curve we observe is a result of the interplay of many factors, not only of the depositional environment.

One of the most important points to bear in mind is the availability of grain sizes supplied to the area where the process is taking place. Strong currents or wave energy will not deposit coarse-grained sediments unless there are coarse-grained sediments in the area. A source area where sediments are formed by erosion and weathering will often supply very specific grain sizes. Chemical weathering of acid rocks (granites) will, for example, lead to the formation of sediments consisting of quartz grains corresponding in size to the quartz crystals in the granite, and clay

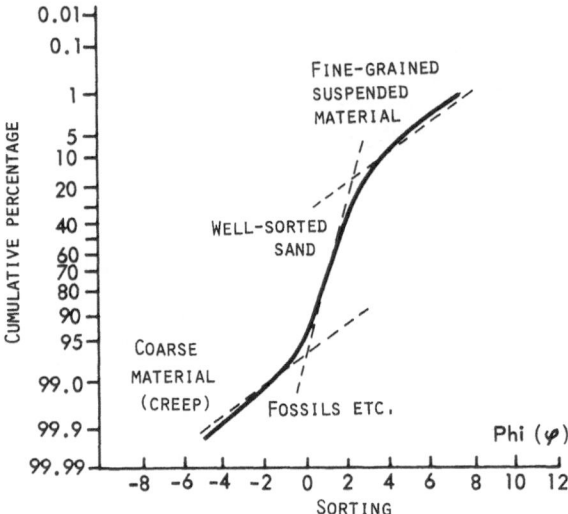

Fig. 2.7. A grain-size distribution curve may consist of several "populations" of sediment grains, each of which may have approximately log-normal distribution and form a straight line when plotted on probability paper. The different grain-size populations may have been transported by different mechanisms, but there will always be some overlap. (Visher and Howard 1974)

(kaolinite, smectite, illite) formed through weathering of felspars. Weathering of basic rocks (basalts and gabbros) will lead to the formation almost exclusively of clay minerals and practically no sand grains. The grain-size distribution in a depositional area also depends on the transport mechanisms to that area. The grain-size distribution we observe in a sediment therefore reflects both the hydrodynamic conditions in the area of deposition and the grain-size population of sediments available from the source area. A particular grain-size distribution does not point unambiguously to a particular environment of deposition because similar hydrodynamic conditions can exist in different environments.

To conclude: We should be cautious in our use of grain-size distribution curves as indicators of depositional environments. Whereas grain size is an important characteristic of sediments and should be noted, it is often pointless to do a large number of grain-size analyses when these are not needed to solve specific problems.

Grain Shape

We distinguish between three parameters:

1. *Roundness* is a property of surface shape — whether it is smooth or angular. Roundness can be defined as the sum of all (n) radii (r) of circles which can be inscribed by a section through the grain, divided by the radius (R) of the max. inscribed circle.

$$\text{Roundness} = \frac{\Sigma r/R}{n} \, .$$

However, this is difficult to measure in practice. Consequently the visual scale is most commonly used (Fig. 2.8).

Fig. 2.8. Visual scale for degree of rounding of sediment grains. **a** angular; **b** subangular; **c** sub-rounded; **d** rounded; **e** well rounded. (After Pettijohn 1957)

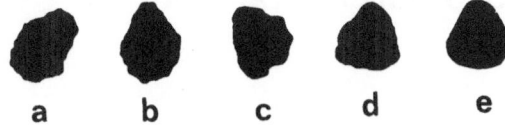

a   b   c   d   e

2. *Sphericity* is an expression for how much a particle deviates from a spherical form, and is defined as the ratio between the diameter of a circumscribed circle round the grain and the diameter of a sphere of the same volume (the *nominal diameter*). For grains which have approximately the shape of an ellipsoid with three axes with diameters d_L, d_I, d_S (longest, intermediate and shortest), the sphericity ψ (psi) can be defined as:

$$\psi = \sqrt[3]{\frac{d_s\,d_I}{d_L^2}}.$$

It may be practical to project particles against a background, and measure the projection sphericity, but this will vary with orientation.

3. We also use various expressions for grain shape such as (a) *discoid* or *bladed*, for grains which are flat, i.e. have one dimension which is very much smaller than the two others (b) *Prolate, roller,* for grains with one dimension considerably greater than the two others, and (c) *equant,* for grains with three relatively equal dimensions and (d) *oblate* for grains with one large, one medium and one small dimension.

4. *Surface textures* are concerned with the nature of the surface itself, whether it is rough, smooth, pitted, scratched etc. The surface structure of grains can best be studied under the scanning electron microscope. Glacial transport may give rise to very characteristic surface structures such as striations and faceting. Eolian sand grains may develop fine pitting on their surfaces due to the collisions of grains during transport.

In studying roundness we have to take into account the mineralogy, hardness, cleavage etc. of the grains, and distinguish between mechanical rounding and chemical solution. Large grains become rounded far more rapidly than smaller ones because the impact energy released in collisions with other grains declines in proportion to the cube root of the radius. Grains less than 0.1 mm in diameter undergo little rounding even when carried long distances in water. Blocks may be rounded after only a few hundred metres or kilometres of transport. In comparing textures, we must always compare grains of the same size.

3 Sediment Transport

A Little About Hydrodynamics

Sedimentary grains can be transported by water or by air. In order to understand the transportation processes we must know a little about the hydrodynamic (or aerodynamic) principles involved. When a liquid or gas flows in a channel or pipe it exerts a force (shear stress) against the walls or bottom. This force is counter acted by friction from the walls.

Pure water without suspended sediment is a Newtonian fluid which obeys Newton's law:

$$\tau = \mu \, \frac{dv}{dh}.$$

A Newtonian fluid has no shear strength, so it will be deformed by an infinitely small shear stress (dv/dh).

$\tau = $ *shear stress,* which is an expression of force per unit area (N/m^2). μ is the dynamic viscosity expressed in poise (g/cm/s or 0.1 Newton s/m^2), dv/dh is the change in velocity (dv) or velocity gradient (deformation velocity) as a function of distance from the boundary (dh). The viscosity of pure water decreases with increasing temperature. Suspended material may, however, also affect viscosity. The concentration of suspended material must be quite high (15–25%) before the viscosity increases significantly. If the water contains a large percentage of swelling clay minerals (smectite), however, the viscosity will increase at lower concentrations. The kinematic viscosity ν is the dynamic viscosity (μ) divided by density ρ, i.e. $\nu = \mu/\rho$ and units are cm^2/s.

We distinguish between laminar flow, where each point in the liquid moves along a straight line parallel to the bed, and turbulent flow, where each point follows an irregular path so that eddies form. Reynold's number (Re) is a dimensionless number which describes flow in channels and pipes. It is defined as:

$$\mathrm{Re} = \frac{v \cdot D \cdot \rho}{\mu}$$

Here v is the mean velocity, D is the depth of a channel or the diameter of a pipe in which fluid is flowing, ρ is the fluid density and μ its viscosity. If Reynold's number exceeds a certain value, about 2000, the flow changes from laminar to turbulent. The density of water is 1 g/cm^3 and viscosity 0.01 g/cm/s. We see that the boundary between laminar and turbulent flow corresponds to:

$$\frac{v \cdot D \cdot 1 \text{ g/cm}^3}{0.01 \text{ g/cm} \cdot \text{s}} = 2000.$$

$v \cdot D = 20$ cm²/s. This means that for the flow of water to be laminar the product of velocity (cm/s) and depth (cm) must not exceed 20. If the velocity is 1 cm/s, there will be turbulence if the depth (D) is greater than 20 cm. In practice, then, flow in rivers and channels is always turbulent. For turbulent flow the expression for shear stresses

$$\left(\tau = \mu \, \frac{dv}{dh} \right),$$

which applies to laminar flow, is no longer adequate. The shear stress in turbulent flow will then increase as a function of velocity because of the eddies which produce an *eddy viscosity* (Fig. 3.1).

The total shear stresses will then be:

$$\tau = (\mu + \eta) \, \frac{dv}{dh} \, .$$

Here η is the coefficient for turbulent viscosity.

Flow in Rivers and Channels

For all types of water flow the forces acting on the water must be in equilibrium. In most cases it is the force of gravity which balances bed frictional forces. In order to understand geological processes in connection with erosion, transport and deposition of sediments, it is important for us to be aware of the relationships which govern the flow of water in channels.

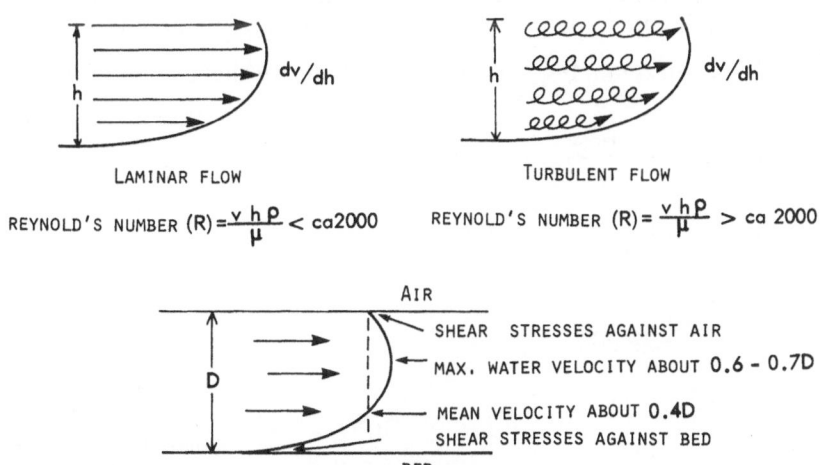

Fig. 3.1. Diagram showing principles of turbulent and laminar flow

If the channel has a cross-section A and we look at a stretch L of the channel, the force of gravity will be:

$$F_1 = \rho \cdot g \cdot L \cdot A \cdot \sin\alpha,$$

where ρ is the density of water, g is the force of gravity (constant) and α is the angle of slope of the channel. The resistance to flow consists of frictional forces against the bed and against the air. If we disregard friction against the air, the frictional forces are:

$$F_2 = \tau \cdot L \cdot P,$$

where τ = shear stress (force per unit area) and L·P is the area of the bed on which the forces are acting. P is the wet perimeter and is the length of a line along the bed in a section across the channel. If the water flow has a steady velocity, the force of gravity F_1 will just equal the frictional force (F_2) (Fig. 3.2). Consequently

$$\tau L \cdot P = \rho g \cdot L \cdot A \cdot \sin\alpha$$

or

$$\tau = \rho \cdot g \cdot \frac{A}{P} \cdot \sin\alpha.$$

A/P is the cross-section of the channel divided by the wet perimeter, and we call this the hydraulic radius — R. For flow in a pipe,

$$R = \frac{\pi \left(\dfrac{D}{2}\right)^2}{\pi D} = \frac{D}{4}$$

The shear stress will then be $\tau = \rho \cdot g \cdot R \cdot \sin\alpha$ or $\tau = \rho \cdot g \cdot D/4 \cdot \sin\alpha$. In broad channels the wet perimeter, p, approaches the breadth (b) of the river and the hydraulic radius R then approaches the depth of the channel (h). Darcy-Weisbach's equation (Fig. 3.2) gives us the relationship between the slope of a pipe with diameter (D) and the velocity (v) of flow through it. f represents the coefficient of friction along the pipe wall, and will depend on how smooth it is.

If we put $\sin\alpha = f \cdot \dfrac{v^2}{D2g}$ into the expression for shear stress, $\tau = p \cdot g \cdot D/4 \sin\alpha$, we

obtain: $\tau = \dfrac{f \cdot \rho \cdot v^2}{8}$.

FLOW IN A CHANNEL

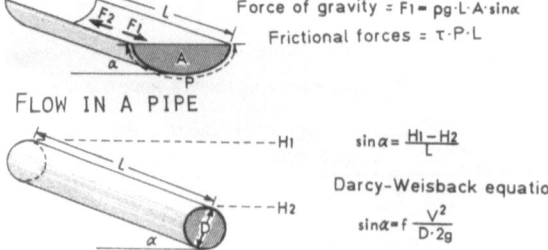

Force of gravity = $F_1 = pg \cdot L \cdot A \cdot \sin\alpha$
Frictional forces = $\tau \cdot P \cdot L$

FLOW IN A PIPE

$\sin\alpha = \dfrac{H_1 - H_2}{L}$

Darcy-Weisback equation :

$\sin\alpha = f\dfrac{v^2}{D \cdot 2g}$

Fig. 3.2. Flow of water in channels. Darcy-Weisback's equation $\sin\alpha = f\dfrac{V^2}{D.2g}$ gives a relationship between the frictional forces (F_2) and the forces of gravity (F_1)

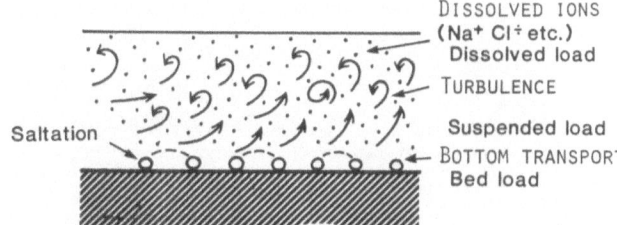

Dissolved ions
(Na⁺ Cl⁺ etc.)
Dissolved load
Turbulence
Saltation
Suspended load
Bottom transport
Bed load

Fig. 3.3. Different forms of transport in water

We see that the shear stresses increase in proportion to the square of the velocity.

This relation between shear stress and flow velocity can also be used for flow in channels where we have bed load transport (Fig. 3.3). Solving the two equations above with respect to the velocity (v) we obtain:

$$\bar{v} = \sqrt{\frac{8 \cdot g}{f} \cdot R \cdot \sin\alpha} \, .$$

If we replace 8g/f with *Chezy's number*, C, which is a function of the friction, we obtain

$$v = C \cdot \sqrt{R \cdot \sin\alpha} \, ,$$

which is Chezy's equation. The number (C) depends on the roughness of the bed and on the *shape* of the channel, particularly its *sinuosity*.

Often used in engineering for calculating the velocity of water in channels, is Manning's formula:

$$v = \frac{1 \cdot 49 \cdot R^{2/3} \sin\alpha^{1/2}}{n} \, ,$$

where n is the coefficient of roughness of the bed, n = 0.01 corresponds to a smooth metal plate and n = 0.06 to a shifting bed of gravel or coarse gneiss. It is of great practical importance to be able to calculate water velocity and thereby the erosion potential of artificial channels.

The Froude number is a parameter which is often used to describe water flow.

$$F = \frac{v}{\sqrt{g \cdot H}},$$

where v is the average velocity, H is depth of water and g the force of gravity. The Froude number is the ratio between the kinetic energy of the water masses (which is proportional to the square of the velocity) and the forces of gravity, which are proportional to the depth, H. For low Froude numbers the water flows out of phase with the bed forms and current ripples or cross-bedding develop. This is called a lower flow regime. When the velocity, v, becomes high in relation to the depth of water, H, rapid or shooting flow develops where the waves come into phase with the boundary irregularities (Fig. 3.4), which represent an upper flow regime.

The transition between lower and upper flow regimes corresponds to a Froude number of 0.6–0.8.

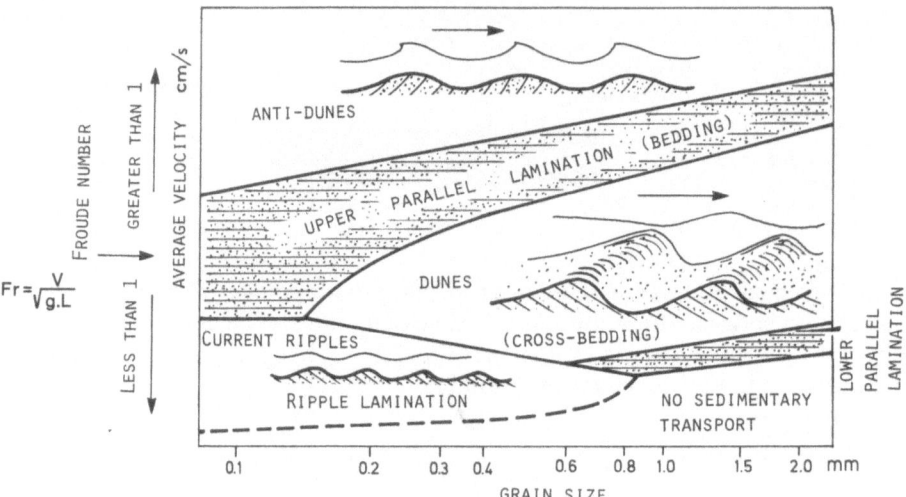

BEDFORMS AS A FUNCTION OF FLOW VELOCITY AND GRAIN SIZE

Fig. 3.4 Relationship between flow velocity and depth (Froude number), grain size and sedimentary structures

Sediment Transport Along the Bed Due to Water Flow

What actually gives flowing water the capacity to carry sediment, and how are sediment particles transported?

We have seen that flowing water exerts shear forces against the stream bed. Frictional forces are converted into turbulence in the overlying water, and have the effect of transporting sediment particles along the stream bed. Under moderate flow conditions the largest particles will be transported along the bed or just above the bed (bed load) (Fig. 3.3). This takes place partly through rolling or slow creep, partly through saltation, i.e. the grains jump along the bed.

Saltation can be partly explained through Bernoulli's equation:

$$P + g \cdot h + \frac{v^2}{2} = C \text{ (constant).}$$

Here P = pressure, h = height above the stream bed, and v = velocity. We see that water which flows over a sediment grain on the bed will have a greater velocity than water which flows under the grain. Bernoulli's equation predicts that the pressure above the grain must be less than the pressure adjacent to the grain (P), and when this difference becomes sufficiently great in relation to the grain, it will be possible to lift it from the stream bed. This "airplane wing effect" does not, however, work when the grain is in the water above the stream bed, and the grain will then drop to the bed again.

The condition for sediment grains being transported in suspension is that their settling velocity must be less than the upward vertical turbulence component. This means that the grain must be transported upwards through the water at least as fast as it falls downwards. The magnitude of the vertical turbulence upwards will be a function of the horizontal velocity (about 1:8). Under normal flow conditions (< about 1 m/s) only clay and silt will be transported in suspension. Under high flow energy conditions, e.g. during floods, sand and gravel may also be transported − at least partly − in suspension.

It has been clear for a long time that there is a connection between velocity and the size of the sediment grains which can be transported. Hjulstrøm was among the first to try and quantify this relationship in his laboratory experiments with water flow in channels in Uppsala, Sweden. Erosion and transport are a function of shear stress against the stream bed. This in turn is a function not only of water velocity but also of depth. Hjulstrøm's diagram (Fig. 3.5) applies to channels about 1 m deep. Other factors which complicate these relationships are the viscosity (and hence temperature) of the water and the density and shape of the sediment grains.

With small grain sizes (silt and clay) there is a great difference in the flow velocity required to sustain transport and to erode a particular grain size. This is because cohesion between sediment particles is such that once they are deposited, it is more difficult to erode sediment, particularly clay.

Note that fine-grained sand (about 0.1 mm) is the easiest to erode. On the other hand, fine-grained particles remain in suspension for a long time at low velocities. Flocculation of small particles to form larger ones, or "pelletisation", through clay being eaten by organisms, is important for the formation of many fine-grained clay sediments, and complicates their hydrodynamic interpretation.

Bed Forms

Sediment transport by flow in water or air causes the formation of bed forms, which are morphological features resulting from the interaction between particular types of flow and the sediment grains. A particular type of bed form is formed only within certain ranges of flow rate and also requires certain grain sizes.

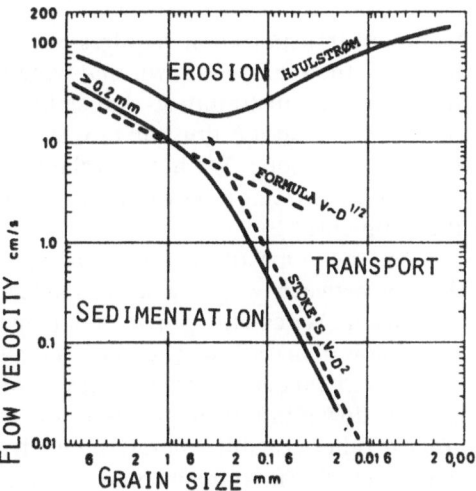

Fig. 3.5. Erosion, transport and sedimentation in water as a function of grain size and flow velocity (modified Hjulstrøm's diagram). Note that fine-grained particles can be transported at very low flow velocities, but once settled, relatively high velocities are required to erode them, because of the cohesiveness of clay

Current ripples form in fine-grained sand at relatively low rates when the velocity exceeds the lower limit for sediment movement. Ripples may also form in coarser sand but there is an upper limit of 0.6–0.7 mm for the grain diameter. If only coarse sand is available, plane bed lamination (lower stage) may form under flow conditions that would otherwise have produced ripples.

Ripples and dunes have a stoss side upstream where erosion takes place and a lee side on which deposition takes place (Fig. 4.1a). They therefore move laterally as a result of the combined effect of erosion and deposition. Ripples are less than 3–5 cm high, and may have a wavelength of up to 40 cm. The ripple index is an expression of the ratio of the wavelength divided by the wave height, and varies between 10 and 40. Sections through ripples show inclined foreset laminae, a structure called "small-scale cross-lamination". Waves may produce oscillatory flow resulting in symmetrical ripples (foreset laminae pointing in both directions).

Dunes are similar to ripples in shape, and form in coarse-, medium- and fine-grained sand, but require significantly higher flow velocities for their formation. Dunes range in height from 5 cm up to several metres, and wavelengths may exceed 10 m.

Cross-sections through dunes show "large-scale cross-stratification", which is often referred to as "cross-bedding".

"Plane beds" (upper stage) may form when the shear stress against the bed exceeds the values which produce dunes. In cross-section we only see planar lamination, which is an internal structure, but on the bedding surface we may see very small ridges, which define a lineation called primary current lineation, parallel to flow.

At even higher velocities in relatively shallow water, standing waves may produce antidunes when the Froude number exceeds 0.8. The antidunes which are produced when standing waves are in phase with the bedforms develop with a low-angle cross-lamination which dips up-current.

Different Types of Sediment Transport

Sediment transportation requires energy to overcome the friction in sediments, liquid or air. We have shown that water or air flowing over a surface exerts shear forces on the substratum so that sediment can be transported. With *traction currents* of this kind it is thus the movements of the water which cause the sediments to be carried along. We have a relatively low concentration of sediment in water (air), and the water (air) will flow in approximately the same manner as without the sediments.

Another type of sediment transport is due primarily to the difference in density between water with suspended sediments and clear water outside the suspension. We call this *gravity flow* (Fig. 3.6) and it includes turbidity currents and debris flow or mass flow. The force of gravity acts on mixtures of sediment and water and movement occurs because they have a higher density than their surroundings. Gravity flow is thus distinguished from traction currents by the fact that it can take place in otherwise still water. It is the release of potential energy from the sediments which sets the mixture in motion and sustains the transport, while in traction

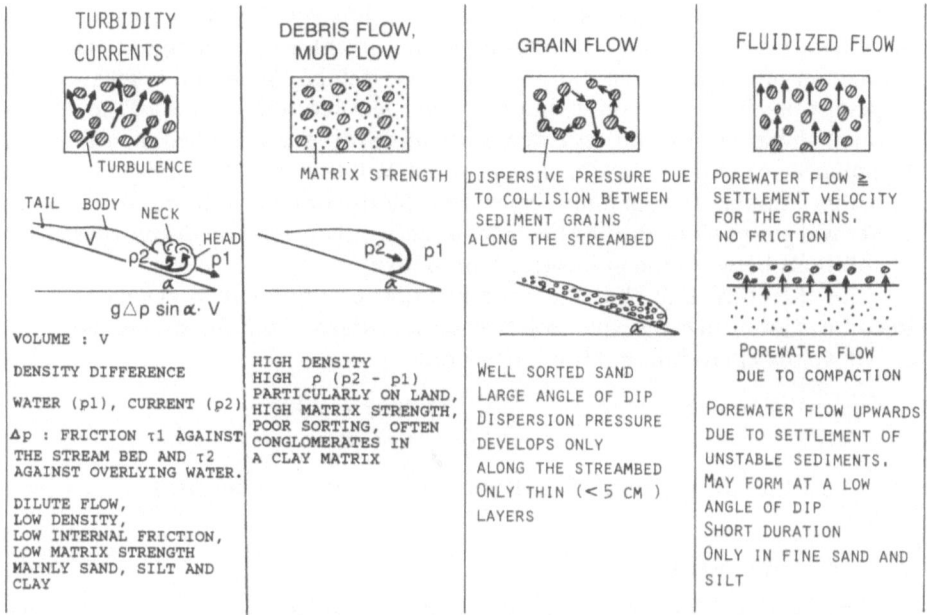

Fig. 3.6. Sketch of principles behind different types of sediment transport on submarine slopes

currents it is the water which transports the sediments. However, there are transitions between these two fundamentally different processes and we often have combined effects.

Turbidity Currents

If sediment which is carried in suspension by turbulent currents due to wave or tidal energy is borne out onto a slope, it will also gain a gravitational component because the suspension is heavier than the surrounding clear water. It may then turn into a density current (*turbidity current*).

Turbidity currents consist of suspensions of sediment in water. The currents are propelled by the difference in density between the suspension and the water around it which contains less suspended material. The forces working downslope are:

$$F = g \cdot \Delta\rho \cdot A \cdot l \cdot \sin\alpha,$$

where g is the gravity constant, $\Delta\rho$ is the difference between the density of the current and that of the surrounding water, A is the cross-section of a turbidity current with a length l, and α the angle of the slope. Acting against the movement are frictional forces (F_2), which, as long as the current is not accelerating, must be equal to the gravitational forces. These are shear forces against the bed, τ_1, and against the overlying water, τ_2, plus internal friction and turbulence within the current, which keep the sediments in suspension. In order for the sediment grains to remain in suspension, the turbulence must be sufficiently strong to have an upward component which corresponds at least to the settling velocity of the coarsest grains. Turbulence is greatest near the bottom of the current, where change in velocity as a function of height above the streambed, i.e. velocity gradient, is greatest. The largest grains in suspension will be concentrated near the bottom of the current, because the level of turbulence decreases upwards. Near the bed, in addition to turbulence, we also have shear stresses which will transport the grains in virtually "pseudo-laminar" flow in a thin layer over the streambed. If the concentration of large sand grains along the bed becomes large, we also get *dispersive energy* because of collisions between the grains (see "grain flow").

We therefore find that the concentration of sediment in suspension, and maximum grain size in suspension, decrease upwards from the streambed.

If we disregard internal friction, we obtain

$$g \cdot \Delta\rho \cdot V \cdot \sin\alpha = (\tau_1 + \tau_2) \cdot A, \qquad V(\text{Volume}) = L.B.H.$$
$$ F_1 F_2 \qquad\qquad A(\text{area}) = L.B.$$

where L is the current length, B is the breadth and H is the height. The shear stress is then:

$$\tau_1 + \tau_2 = g \cdot \Delta\rho \cdot H \cdot \sin\alpha.$$

The shear stresses are proportional to the square of the velocity ($\tau = C\,v^2$). The velocity (v) of a turbidity current is then:

$$v = C^1 \sqrt{g \cdot \Delta\rho \cdot H \cdot \sin\alpha}.$$

Here the coefficient C^1 includes the coefficient of resistance for friction against the sea floor and against the overlying water. This corresponds to Chezy's number for fluvial flow, so in many ways we can regard a turbidity current as an underwater river.

We see from the above equation that thick turbidity flows will have a higher velocity than thin ones and that thick flows can flow on gentler slopes than thinner ones. This is because the shear stress against the bottom and the overlying water is nearly independent of the thickness of the flow. The flow velocity also increases with increasing density of the sediment-water mixture in the flow, but high density flows will have higher internal friction and require higher velocities to keep the material in suspension.

A turbidity current can be divided into head, neck, body and tail. The sediment particles in the head area move somewhat faster than the front of the current itself. This leads to sediment being swept upwards and then backwards towards the neck, where it mixes with water from the overlying water. From there they are carried backwards to the body and tail, where we find a finer-grained, thinner suspension. When the turbidity current loses velocity, the largest particles in the head will settle out of suspension first because of reduced turbulence. Gradually smaller and smaller grains will settle out of suspension, and we get deposition of a bed which is fairly massive, without internal structure, but which becomes finer upwards. In most cases, apart from in proximal turbidites, we also find deposition of some fine material in this layer, so there is poor sorting. An example of this is Unit A of the Bouma Sequence (Fig. 3.7).

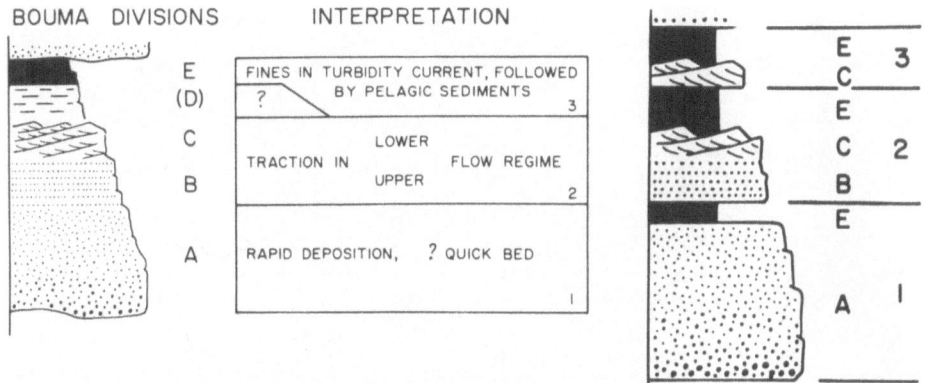

Fig. 3.7. The Bouma model for turbidites. Unit *A* is graded or massive sandstone. Unit *B* is a planar laminated sandstone. Unit *C* has small scale cross-lamination, often climbing ripples. Unit *D* is laminated siltstone. Unit *E* is a hemipelagic deposit between turbidites or the mud and silt deposited from the tail of the turbidity current

As settlement from suspension slows down, the water will have time to sort the grains further, and we find lamination and bedding structures. The B Unit exhibits parallel lamination which may be due to flow near the *upper flow regime.* The C Unit exhibits current ripples and convoluted laminae and represents the lower flow regime. The D Unit has parallel lamination and was probably deposited from the tail, which consists of very fine-grained sediment. The E Unit consists of pelagic material, fine-grained clay and fossils deposited during the long periods (often thousands of years) between turbidity flows, or of very fine turbiditic material (Fig. 3.8). The E Unit is therefore not necessarily a part of the turbidite sequence. We often find Mesozoic and Cenozoic deposits to have a high carbonate content due to their coccolith and foraminifera content. The Bouma sequence, first described by Arnold Bouma in 1962, is an ideal sequence in the sense that in most cases we do not find all the units developed. In some sequences, particularly those thought to have been deposited close to the base of submarine slopes, where the gradient is still fairly steep, we will find only the coarsest parts of a turbidity current deposited, only the A Unit or a sequence of A + B. We call these *proximal turbidites.* Beyond the foot of the slope or out on the ocean abyssal plain the finest-grained fractions of a turbidity current tend to be deposited. In these areas we often only find alternation between C-D-E, or only D-E. These are called *distal* turbidites. In many cases it may be difficult to distinguish between distal turbidites and alternations between silt and clay formed by traction currents at great depths. Proximal turbidites which are well sorted may also resemble coarser sediments deposited by powerful traction currents in submarine channels.

At the base of turbidity sequences, particularly at the base of the A Unit, we often find well-developed erosion structures, particularly *flute casts* and *groove casts.* Flute casts are formed by the turbulence of turbidity currents when they pass over a substratum which consists of finer-grained sediments. The structures, which point upcurrent, are produced by fixed vortices in the turbulent flow eroding into the sediment surface. Groove casts are formed by larger grains being dragged along

Fig. 3.8. Turbidite sandstone beds separated by thin shale layers. The beds young towards the *right*. The Brøttum Formation at Maihaugen, near Lillehammer in Norway

the bottom. Even though flute casts and groove casts are typical of turbidites, they cannot be used as proof that we are dealing with turbidites, because similar structures can also be formed by various types of traction currents where there is turbulence and transport along the bottom, e.g. in fluvial environments. Sequences resembling Bouma sequences may also be produced by processes other than turbidity flows, for example rapidly accelerating fluvial flows. To assist our interpretation we should therefore look at the entire sequence and also try to obtain palaeoenvironmental information from fossils etc.

Formation of Turbidity Currents

For a turbidity current to form, there must be sediment in suspension. Therefore we need a mechanism which brings sediments into suspension. The simplest mechanism involves starting with fluvial sediments which contain suspended material as they enter a sedimentary basin. In marine basins, however, the difference in density between river (fresh) water and salt sea-water is so great that even if river-water carries a heavy load of sediment in suspension, it will in most cases not be heavier than salt water. In consequence there will not be a positive density contrast, which is the prerequisite for the formation of turbidity currents. In lakes, on the other hand, river-water is often heavier, due both to suspended material and to its being colder than the water on the surface of lake. When it flows into the lake, therefore, the river-water will be able to follow the bed down the slope, and turn directly into a turbidite. This means that instead of a normal constructive delta being formed, a turbidity fan is formed on the lake floor. As a result, some of our best examples of turbidity currents are from lakes, particularly in Switzerland, where cold river-water, partly melt-water containing a lot of sediment, flows into the great lakes on the molassic plains.

In marine basins river sediment will mix with marine water, and sediments will be deposited on the delta slopes. If the slope becomes too steep (see p.83), we get slides, which may result in more turbulence and a thinner suspension, so that a turbidity current may be generated. When fine-grained sediments are deposited on slopes they have a very high water content. Compaction, sometimes caused by earthquakes, will cause upward flow of porewater and may result in liquefaction. This causes the sediments to begin to flow even on gentle slopes because friction is reduced, and they may then turn into turbidity currents. Strong currents can also bring sediments into suspension, and these may later turn into turbidity currents which flow down the slope from the shelf. We believe that large amounts of sediment may have been liberated through subsea *slumping* or liquefaction. The slide on the Grand Banks in 1929, which snapped a number of telephone cables, probably started as a slumping and liquefaction then turned it into a turbidity current. The different times when the telephone cables snapped have been used to calculate the velocity of the turbidity current, which seems to have been very high (> 100 km/h). The cables on the upper parts of the slope, at least, were probably broken by slumping or settling movements in the sediments. What was actually measured may have been the velocity of the wave which was transmitted through the sediments, leading to collapse and slumping of the sediments. Velocities calculated on the basis of snapped cables consequently do not necessarily correspond to the velocity of the sediment particles.

The turbidite theory was first launched at the end of the 1930's on the basis of studies of Tertiary sediments in California. Studies of microfauna, particularly foraminifera in sediments and slates, indicate that the sandstones were deposited in deep water. The macrofauna, however, consisted of molluscs, which indicates shallow water (Fig. 3.9). Previously it had been supposed that deposition of sand was largely limited to shallow water facies, and that deep ocean sediments were almost exclusively fine-grained.

The only way these observations could be explained was by sand with molluscs being transported from a shallow water facies down to a deep water facies, and there acquiring deep water foraminifera. Kuenen and Migliorini (1950) developed the theory further. They were able to relate turbidity currents to graded beds which were common in many ancient rocks, i.e. in the Alpine flysch.

Fig. 3.9. An erosion channel with a turbidite infill. Note that the bed fines upwards. Here we can recognise the A, B, C and D Units of the Bouma sequence. From Santa Paula Creek, the Pliocene Ventura Basin, California

At the time when the turbidite theory was developed, it was believed that as soon as one penetrated below the wave base or tidal level, there were only very slow ocean currents. We now know (Shepard et al. 1979) that there are very swift traction currents, particularly in submarine valleys (*canyons* — Fig. 3.10). These traction currents are capable of transporting coarse sand and at times even coarser material. Turbidity currents are thus no longer the only transport mechanism on the continental slope and in the deep ocean. We must therefore try to decide in each individual case whether anything in ancient deposits indicates traction currents or suspension currents. This is often very difficult, and we are probably often dealing with deposits which have been transported by flows which have both a traction component and a turbidity or mass flow component. *Imbrication* (shingle structure) is common in deepwater conglomerates, and in a single exposure they may be very difficult to distinguish from fluvial conglomerates. In deepwater conglomerates, however, we often find inverse grading at the bottom and normal grading towards the top of a layer, so that the coarsest material concentrates in the middle. This is less common, but not unknown in fluvial conglomerates. So it is mainly on the basis of palaeobathymetric indications in the sediments with which conglomerates are associated that we can determine whether we are dealing with deepwater conglomerates.

High Density Mass Flows — "Debris Flows" and "Mud Flows"

Debris flows occur both on land and under water, and represent a type of mass transport where the ratio sediment/water is very much greater than in turbidity currents. The water content is only about 40–80%, resulting in high viscosity and high internal friction during flow. This also means that the density of the mass is from 1.5 to 2.0 g/cm³, while most turbidity currents have a density of 1.1–1.2 g/cm³ or less. The high density of a debris flow means that all clasts have increased buoyancy because of the dense matrix. The high viscosity of the matrix also means that large stones do not sink rapidly towards the bottom of the flow. In flows with high density and viscosity, blocks may remain near the surface of the flows until the flow solidifies through loss of water or reduced gradients, and becomes quite rigid as a result of increased density and cohesion. We call this "matrix strength". Debris flows may be rich in stones and other coarse material. Mud flows typically have a more clay-rich matrix. But there is no clear distinction between the two.

Because of the high matrix strength, little sorting takes place in debris flows (Fig. 3.11). Large blocks are often concentrated at the front or on the sides of the flows, and there may be less coarse material near the base of the flow because of the deformation here.

The velocity of these debris flows declines rapidly with increased density due to their high viscosity. Flows with a particularly low water content will only flow slowly down a slope, and we get transitions into what we call *solifluction* (creep). Sediment flows with a higher water content can move faster, however. In debris flow with high internal friction most of the shear forces will be released along the bottom of the flow, so that the overlying mass moves more or less as a mass with little internal deformation. This helps to reduce the total frictional resistance to movement. Shear strength and viscosity tend to decrease with increasing rate of shear, and this means that when a movement first gets going it will tend to accelerate. Debris flows and mud flows normally have *thixotropic* properties: the shear forces must reach a critical threshold before deformation (shear) takes place, and the material loses much of its shear strength following deformation. In clay containing smectite (montmorrillonite) this property is particularly well developed. Under shear stress water will be released. The house-of-cards packing of clay minerals is destroyed, the clay particles tending to develop parallel alignment, with the release of water, which will lubricate the movement. Some of the water in the bottom layer in which deformation is taking place will be lost to the other sediments, however, and friction will mount again. If large clasts enter the basal shear zone, friction will also increase. In consequence, debris flow often proceeds jerkily.

The stability of mud on slopes and the flow properties of mud flows depend on the clay mineral composition of the mud and on the geochemistry of porewater and adsorbed ions.

Where Do We Find Debris Flows?

Debris flows are often described in continental deposits, but are also found on submarine slopes. In water the density difference between sediment and sur-

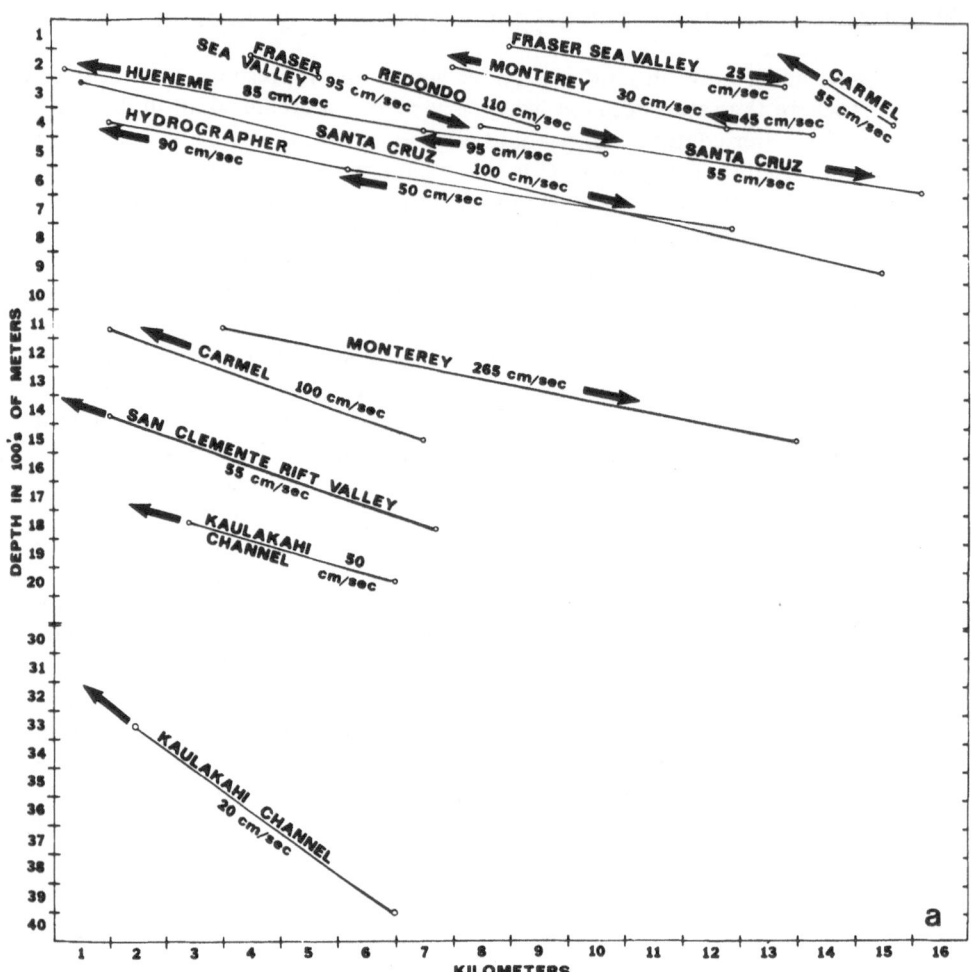

Fig. 3.10a. Reversion traction currents in submarine channels controlled by tidal cycles. The flow up channel is often as strong as the flow down. (Shepard et al. 1979).

roundings is far less than on land, so that the angle of the slope must be greater for flows with the same internal friction. Submarine mud flows, on the other hand, will not dry up, thereby increasing their viscosity, and they can easily take up more water as they move.

Debris flows are particularly common in desert deposits. This is because of the often powerful rains which mobilise sediments which in a wetter climate would have been transported by fluvial processes. In addition, there is little vegetation in deserts to bind the sediments, so they are more easily set in motion. Also of great significance is the fact that the clay mineral smectite is formed particularly through weathering in desert environments (see p. 134). Clay containing smectite will expand when it begins to rain, and prevent the water from filtering rapidly through the soil

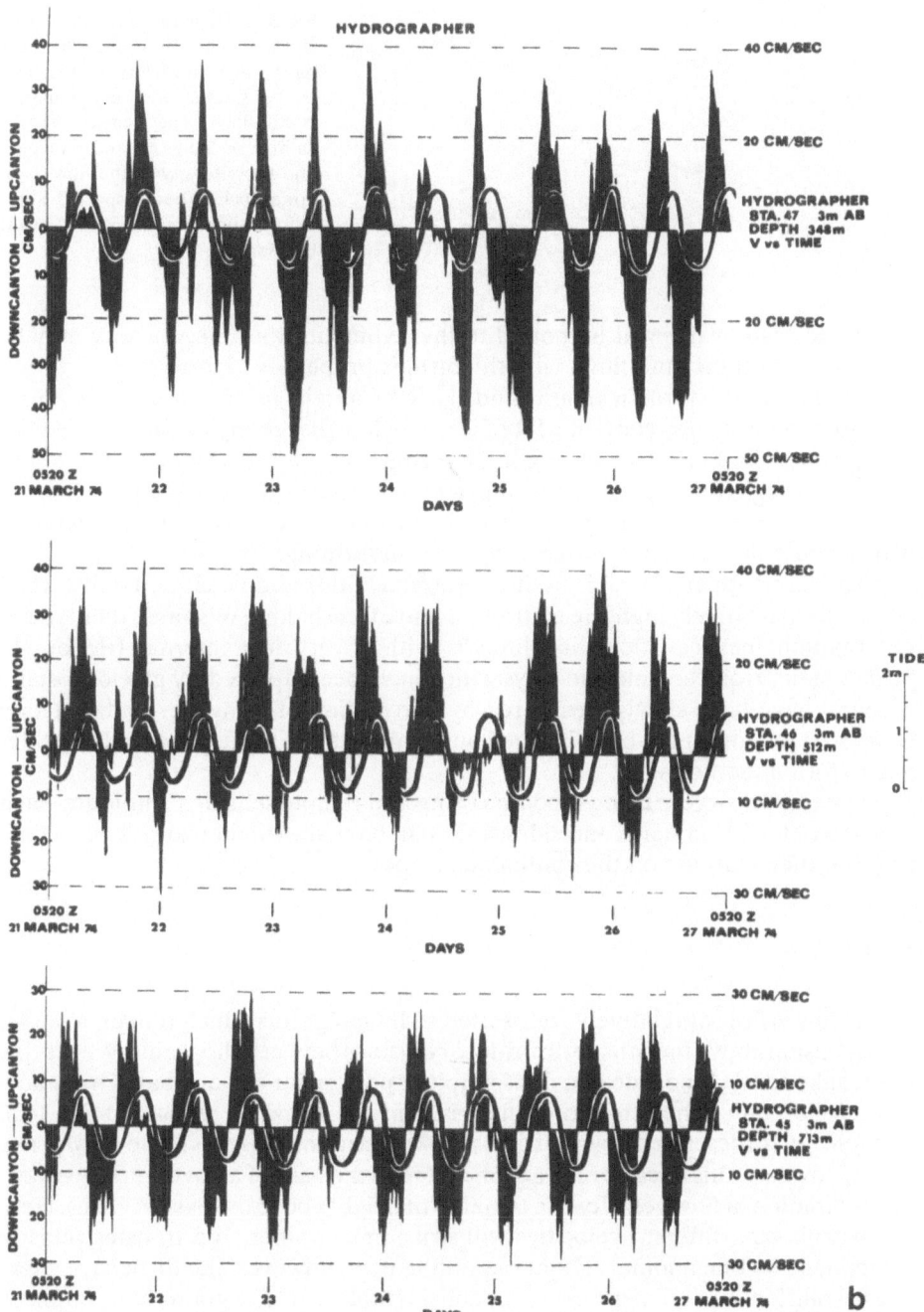

Fig. 3.10b. Current velocity recordings in one canyon (The Hydrographer) showing that the currents correspond to tidal cycles. (Shepard et al. 1979)

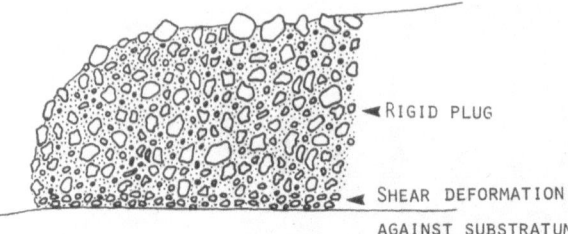

Fig. 3.11. Diagram showing the principles of debris flow. Such flows can have a high clay content (mud flow) or be sandier and conglomeritic (sand flow). The density may be as high as 2.0 g/m³ and buoyancy and the viscosity of the flow allow large blocks to be transported. Shear deformation is concentrated at the bottom

profile. Instead, water will be bound to the sediments, and the viscosity may be reduced enough for mud flows with thixotropic properties to form.

Mud flows also form in areas round glaciers (periglacial environments). This is because moraines may contain a lot of ice, which melts, giving the moraine a high water content and low shear strength. The sediments may also increase their water content through freezing (frost heaving) and this leads to great instability when the ice melts in spring or summer. Periglacial and arctic areas further lack vegetation which would prevent mud or debris flows from forming.

The term "quick clays" is used for extremely thixotropic clays. Undisturbed clays have a relatively high shear strength, but after shaking or some other type of deformation they can flow like liquids, with a very low internal friction. In Scandinavia, Holocene marine clays which have been uplifted by glacio-isostatic rebound have been slowly weathered by percolation of rainwater and meteoric water so that sodium has been leached out, making them more prone to landslides and to form mud flows.

Systematic surveys using modern coring and remote sensing techniques, such as underwater TV cameras and sidescan sonar, have shown that large-scale debris flow is rather common on the continental slopes.

Grain Flow

Grain flow is flow of relatively well-sorted sediment grains which remain in a sort of suspension above the substratum due to collisions between the grains. We see this if we make a little landslide in a dry sandpit or pour sugar out of a bag. Grain flow can develop only when the initial flow is near the angle of repose (about 34°). Bagnold (1956) described how collisions between sediment grains led to a *dispersive stress*. However, this stress is only significant near the base of a flow, where we have rapid variation in flow velocity as a function of height above the base (dv/dh). Here grains with very different velocities will strike one another, and transfer velocity components to one another. Higher up in the flow, however, the dispersive stress due to collisions between grains will be considerably less. The grains have far more similar velocities, despite turbulence. The dispersive pressure developed near the base cannot support a thick layer of overlying sediment. There is therefore an upper limit to how thick grain flows can be, and Lowe (1976) has shown that grain flows cannot exceed about 5 cm.

Thick horizontal sand layers which have previously been described as grain flows must therefore have been deposited by another mechanism. Ordinary traction of well-sorted sand also involves a sort of dispersive stress for grains which are transported along the bottom. Sand grains which avalanche down the lee side of sand dunes form small grain flows and are probably one of the few significant examples of natural pure grain flow.

Liquefied Flow

Liquefaction is the name given to a process whereby sediments lose most of their internal friction, and consequently act almost like fluids. This is the case when the pore pressure is equal to the weight of the overburden. When sediments are deposited, they have a high water content and the sediment grains are packed in an unstable manner. As the overburden increases, the stress on the grain contacts increases, and the framework of the sediment grains may collapse suddenly. Earthquakes produce tremors which may cause this structure to collapse, but it can also take place purely as a result of stress. When the packed framework of grains which was formed during deposition is destroyed, the grains can pack more closely together. For this to be able to happen, however, water must flow out of the bed as the porosity decreases. This leads to an upward flow of porewater and fine sediment particles which may be as great as or greater than the settling velocity of the grains. This process is called liquefaction. The force of gravity, acting on the contact between the grains, is therefore neutralised, and friction between the grains tends towards zero, resulting in liquefaction. If we measure the pressure in the porewater, we find that it increases during settlement (compaction) when the unstable grain framework is destroyed. At one stage the pore pressure will be approximately as great as the weight of the overlying sediments. We can then use Coulomb's Law:

$$\tau = C + (S\text{-}P) \tan \varphi,$$

where τ = shear strength, C = cohesion, S is the weight of the overlying sediment, P = pore pressure and φ is the angle of friction (about 34°). S-P is the effective stress. When the pore pressure, P, approaches the weight of the overlying sediments (S), the friction component (S-P) tan φ approaches zero. Fine-grained sediments like clay have considerable cohesion (C), and this will often prevent clay sediments from sliding even if there is little friction (Fig. 3.12). Once a deformation plane forms, there will often be movement mainly along it due to cohesion in the rest of the clay. If we have coarse-grained sediments, coarse sand and gravel, compaction will lead to excess water flowing out so rapidly that the overpressure will drop very rapidly or not succeed in building up properly because of the high permeability (Fig. 3.13). It is therefore silt and fine sand, which are most susceptible to liquefaction, which cause high-velocity subsea flows (Andersen and Bjerrum 1968) (Fig. 3.14). Liquefaction can, as already mentioned, be triggered by tremors, e.g. earthquakes, and stress. Stresses on sediments (e.g. soils) due to buildings, fills etc. can lead to collapse of the grain frameworks and cause liquefaction. Lowering of the groundwater table on a slope, e.g. down towards the coast, has a similar effect because of reduced buoyancy in part of the sediment column. Extremely low tides or a combination of a strong ebb and a land wind can trigger a slide in otherwise stable coastal sediments. The stability of slopes can be estimated by calculating the gravitational forces acting on a particular volume of sediments in relation to

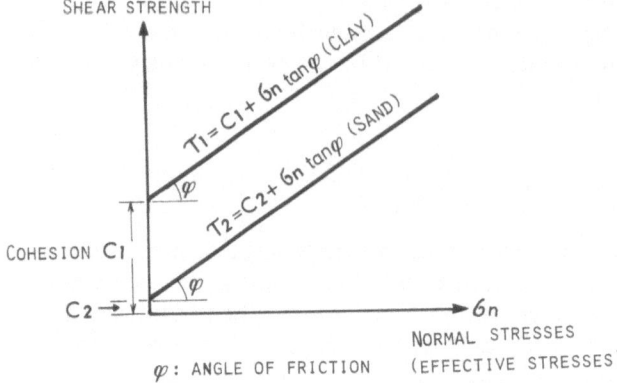

Fig. 3.12. Relationship between shear strength and effective stresses in sediments according to Coulomb's law. The shear strength is the sum of a cohesion component and a frictional component, $\delta n \tan\varphi$, δn = the weight of the overlying sediments (S) minus the pore pressure P. The shear force is $s \cdot \sin\alpha$ where α is the angle of dip. Failure occurs when $s \cdot \sin\alpha > \tau$

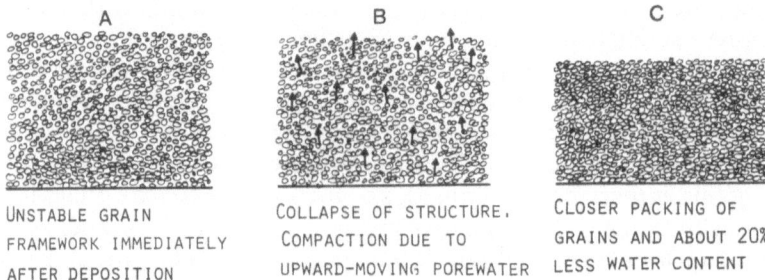

A	B	C
UNSTABLE GRAIN FRAMEWORK IMMEDIATELY AFTER DEPOSITION	COLLAPSE OF STRUCTURE. COMPACTION DUE TO UPWARD-MOVING POREWATER	CLOSER PACKING OF GRAINS AND ABOUT 20% LESS WATER CONTENT

Fig. 3.13. Schematic diagram of how sediment becomes "liquefied". Grains larger than medium-grained sand will not be liquefied due to high permeability, so that the excess water escapes and overpressure rapidly evens out

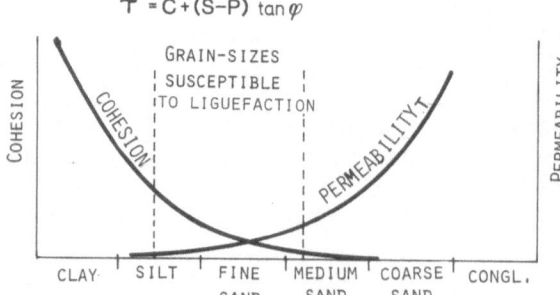

Fig. 3.14. Clay-rich sediments will to a large extent resist liquefaction even though friction is reduced due to overpressure. Silt and fine sand will therefore be most easily liquefied

frictional forces. The pore pressure (P) in the sediment can be measured by drilling, but it may suddenly increase due to increased compaction or deformation. Slides may sometimes be prevented by drilling wells which release the excess pore pressure, so that the effective stresses, and therefore the frictional forces, increase.

4 Description of Sedimentary Rocks and Facies

What should we measure and observe when we study sedimentary rocks?

It is difficult to observe or take measurements of rocks entirely objectively and consistently. Most types of measurements and observations have a considerable degree of inherent uncertainty, and the validity and usefulness of results often depend on the experience and skill of those carrying out the field work. It has turned out to be very difficult to observe structures which one does not recognise and understand the significance of. Sedimentary structures which are easily distinguished today by relatively inexperienced students were as a rule overlooked in descriptions, even by experienced geologists, not more than 20–30 years ago. The structures were simply not known from textbooks or articles.

A good description of a stratigraphic profile depends on good theoretical knowledge of sedimentary processes, and of experience from studies of similar rocks.

It is easy to forget to record or measure some of the properties of a rock. In order to obtain a more homogeneous description and avoid forgetting anything, it may be a good idea to have a well-established routine or even a checklist. However, it is not always necessary or desirable to include all the points on this list. If our investigation has a definite and limited objective, we measure only the properties we think will be of importance to it. A list of features which can be observed (measured or registered) in sedimentary rocks:

1. Textures — grain size, sorting, grain shape etc.
2. Grain orientation — fabric.
3. Sedimentary structures and their orientation.
4. Fossils
   A Preservation or impressions, casts or the fossils themselves, and their mode of occurrence.
   B Trace fossils.
5. Colour.
6. Resistance to weathering and erosion.
7. Composition (a) Mineral (b) Chemical.
8. Thickness of strata and geometry (variations in thickness).
9. Variations in textures and composition within a bed, e.g. increase in grain size upwards or downwards in the bed (grading or inverse grading).
10. Type of contact between beds (e.g. erosion contact, conformable contact, gradational contact etc.).
11. Association or any tendency to statistical periodicity in the features of the strata in a profile — bed types, structures etc.

These observations form the basis for defining facies, which are a synthesis of all the data listed above which can be used to group certain types of rocks. They may be genetic facies, i.e. strata which one assumes have formed in the same manner. All strata which contain criteria which indicate that they were deposited in shallow water can be described (in reality interpreted) as shallow water facies. In the same way you have deep water facies, evaporite facies, stagnant water facies etc. The facies concept can also be used to distinguish between different rock compositions (lithologies), e.g. carbonate facies, sandstone facies etc.

Sedimentary Structures

By sedimentary structures we mean structures in sedimentary rocks which have formed during or just after deposition. We distinguish between *primary* structures which are formed at the time of deposition of the sediments, and *secondary* structures which are formed after deposition (Table 4.1).

Layering and Lamination

Most sedimentary rocks exhibit lamination or bedding, but we also have massive (unlaminated) rocks. Lamination records variations in the sediment composition which follow the contours of the sedimentary surface during deposition. The variations may be due to different grain sizes, sorting or mineral composition. Laminae are less than 1 cm thick. Variations greater than 1 cm are called *beds*. A bed will contain sediments which have been deposited by the same sedimentary processes. Some sedimentary processes, for example deposition of a turbidite bed, may be fairly rapid. Migration of a sand dune to give cross-bedding takes a somewhat longer time, while deposition of a clay bed from suspended material may take a very long time. For all that, it is the same process which deposits the beds. However, this does not mean that the sediment composition does not vary within a bed.

Graded beds have a grain size which tends to decrease upwards within the bed. The opposite is called *inverse grading. Normal grading* may be due to deposition from suspension, when the largest particles tend to fall to the bottom first, as with turbidites, or to flow velocities dropping off during deposition in a river.

Inverse grading may be due to increasing flow velocity but if the increase in velocity is too high, the result will be erosion. The supply of coarse material during transport and deposition may also produce inverse grading. In high density sediment currents (debris flows) we may get inverse grading. Smaller particles sink more easily to the bottom between the large particles because of the shear stress. Massive beds, or at any rate reasonably massive beds without lamination or bedding, may be formed during very rapid deposition of sediments from suspension. X-ray photography of apparently massive sediments nevertheless usually reveals the presence of weak lamination. Massive beds may also arise through intense bioturbation which has destroyed the primary lamination.

Table 4.1. Classification of sedimentary structures

Primary Bedforms (Formed during deposition)

	Internal structures
1A Plane beds	Planar laminations
1B Ripples	Ripple cross-lamination and small-scale cross-lamination
1C Dunes	Large-scale cross-stratifications (cross-bedding)
1D	Convolute lamination
1E	Graded bedding

Erosion structures on the underside of beds (sole markings)

2A Flute casts
2B Tool marks
   Groove casts
   Prod marks, bounce marks
   Chevron marks

Erosion structures on the upper side of beds

3A Rill marks
3B Wind erosion
3C Raindrop imprints

Secondary (formed after deposition)

Water escape
4A Dish structures (immediately after deposition)
4B Sandstone dykes
4C Sand volcanoes

Load structures (inverse density gradient)

5A Load casts
5B Ball and pillow structures
5C Clay diapirs

Cracks

6A Dessication mudcracks
6B Shrinkage cracks, synaeresis
6C Frost cracks (polygons)

Deformation structures (due to gravity)
7A Slumping. Growth faults

Current Ripples and Cross-Bedding

Current ripples are regular undulations on the surface of a sediment, with a relief of less than 3–5 cm, and spacing of less than 50 cm between the ripples. Larger structures of a similar type are called dunes. Both ripples and dunes are formed through sand being transported along the bottom and deposited in sloping strata on the lee side of the structure (Fig. 4.1a,b). In consequence they always have slanting laminations (*foreset beds*) which may lie at an angle (angle of repose) of up to 35° to the surface of the bed, but such high angles are rather rare. Current ripples may be straight or form arched or concave sinal patterns (*sinuous crests*). Ripples with a symmetrical cross-section (symmetrical ripples) are formed by waves as a rule. Asymmetrical ripples are formed by a predominantly unidirectional current and their steeper side faces downstream. Ripples with a high sinuosity are also asymmetrical in most cases.

Tongue-shaped (*linguoid*) ripples have a very high sinuosity and asymmetry and are usually formed in shallower water or under higher velocities than straight-crested types. Wave ripples in particular may be split up into two ripples. This is called bifurcation. In intertidal zones ripples formed at high tide may be eroded at low tide, and so become flattened. At high tide the tidal flat is submerged in water and may also be partly below the normal wave base, resulting in deposition of clay, which has a tendency to collect in the troughs between the current ripples. Current ripples with thin lenses of clay between them constitute *flaser bedding* (see Fig. 4.1c). Wind may cause waves which move in different directions, particularly at very low water, so that we find two or more sets of ripples at an angle to each other (*interference pattern*). In most cases each bed with current ripples represents a sort of equilibrium with deposits reflecting current patterns, and the ripples are horizontally delimited by periods of erosion and/or clay deposits. Isolated sand lenses in clay are called *lenticular bedding* (Fig. 4.1c).

Normally current ripples form completely horizontal beds. In some cases, however, we find examples of current ripples appearing to climb upstream in relation to the horizontal plane. They form several sets of cross-laminated beds delimited by erosion boundaries, but with small internal erosion planes. These are called *climbing ripples*, and are due to sedimentation taking place so rapidly that there is no equilibrium between erosion and sedimentation, as there is with normal ripples. Climbing ripples are therefore typical of environments with rapidly declining flow velocity and consequently a high rate of sedimentation.

Cross-bedding (large-scale cross-stratification) is seen in cross-sections through dunes. Each lamination is called a *foreset bed* and represents the lee side surface of a migrating dune. If the dunes have straight crests the foreset beds on the lee side will form a straight transverse plane (*tabular cross-bedding*) (Fig. 4.2). Curved dunes have rounded foreset beds, and in *trough cross-bedding* the laminae have a rounded surface which is concave in the downstream direction. The foreset lamination may form a relatively sharp angle with the underlying bed, or may have a more tangential contact. The latter is typical of trough-shaped sets.

Eolian dunes may be several metres high, and their cross-bedding will then be of a corresponding size. (Fig. 4.1b).

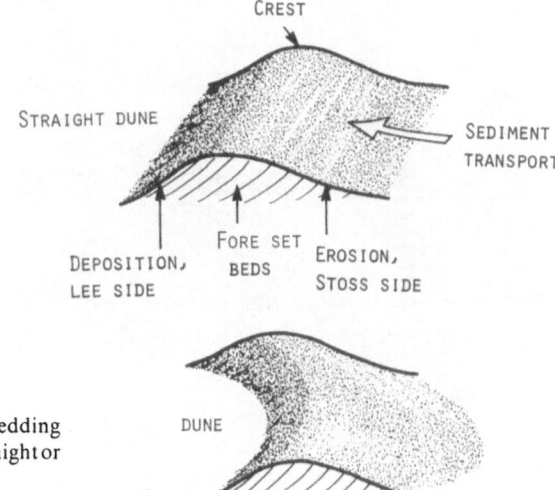

Fig. 4.1a. Sand dune with cross-bedding (foreset beds). The dunes may have a straight or curved crest

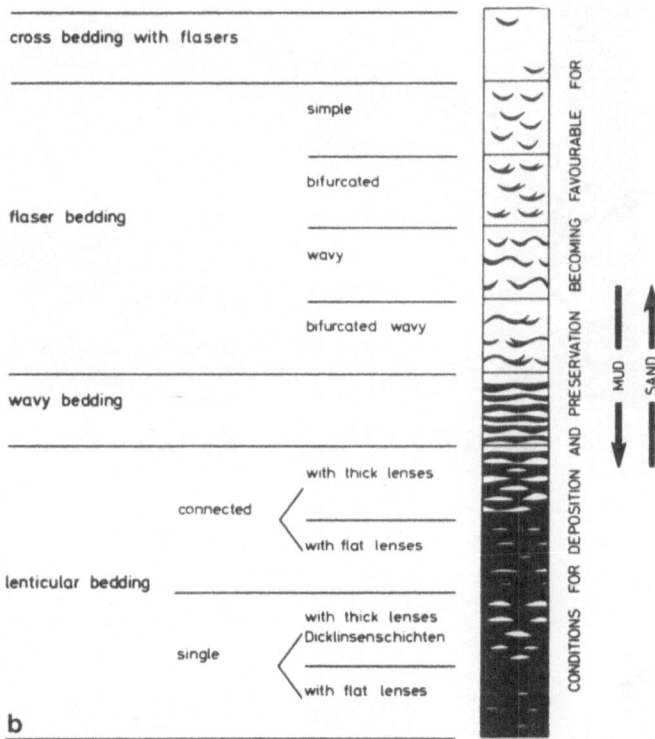

Fig. 4.1b. Classification of bedding types in sediments with different sand/clay ratios. (Reineck 1970)

Fig. 4.1c. Eolian dunes several metres high in the Navajo Sandstone (Jurassic). From Zion National Park, Utah

Fig. 4.2. Cross-bedded sandstone which represents the filling of a fluvial channel above a coal bed. The grain size fines upwards somewhat. It is largely trough cross-bedding represented here. From the Tertiary sequence of Spitzbergen. (Photo A. Dalland)

Sandwaves are dune-like structures which are a greater distance apart than dunes and have a less sharp relief (ratio height:distance apart). Sandwaves may be formed at flow velocities (of 30 cm/s to 150 cm/s). Sandwaves often have current ripples on their surface. This may also be the case with dunes. Sandwaves are typical of subtidal environments, e.g. the North Sea, where they may be 3–5 m high with a wave length of up to 600 m. Sand ribbons are also typical of shallow subtidal environments (20–200 m), and they may be up to 20 km long, 200 m wide and less than a metre thick.

Bedforms are a function of both flow velocity and grain size (Fig. 4.2). Current ripples are formed only in silt and medium-grained sand, while dunes require medium to coarse sand. Upper-slope plane beds which have an internal structure of planar lamination are formed when the Froude number is about 0.6–0.8. Plane beds typically develop in beach sand. A well-developed lineation on the bedding surface parallel to the direction of sediment transport is typical of plane beds (Fig. 4.3). Antidunes are formed under an upper flow regime. The absolute flow velocities required are a function of water depth:

$$F = \frac{v}{\sqrt{gh}}.$$

Below the ordinary, fair-weather wave base we find different types of sedimentary structures from those formed through constant wave action. During storms in near-shore areas traction current may develop and carry fine material (fine sand) from the beach out to greater depths. This material is deposited as *hummocks* consisting of parallel laminae which form a dome-formed structure (Fig. 4.4). Because deposition of beds of this type takes place during short periods

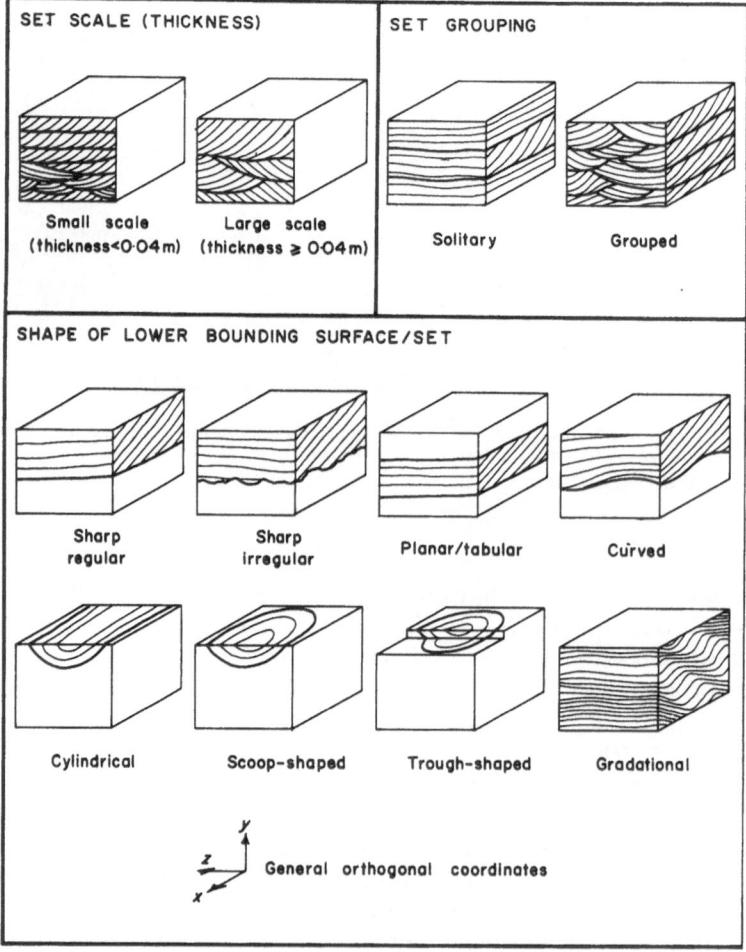

Fig. 4.3. Classification of cross-bedded deposits (After Allen 1982)

during storms, they are often reworked by bioturbation near the top. Hummocky stratification is common in sections through some shallow marine deposits and wave-dominated deltas.

Erosion Structures on the Underside of Sand Beds (*Sole Structures*)

When sand is deposited rapidly over finer sediments (silt and clay) there will also in very many cases be an initial phase of erosion so that erosion hollows are formed. These hollows fill with sand, and we get a sort of *cast*. We see these erosion structures on the underside (sole) of sandstone and conglomeritic strata. We seldom see these surfaces unless the bed is steeply inclined or inverted. Where there are overhanging rocks, it is important to study the underside of the bed.

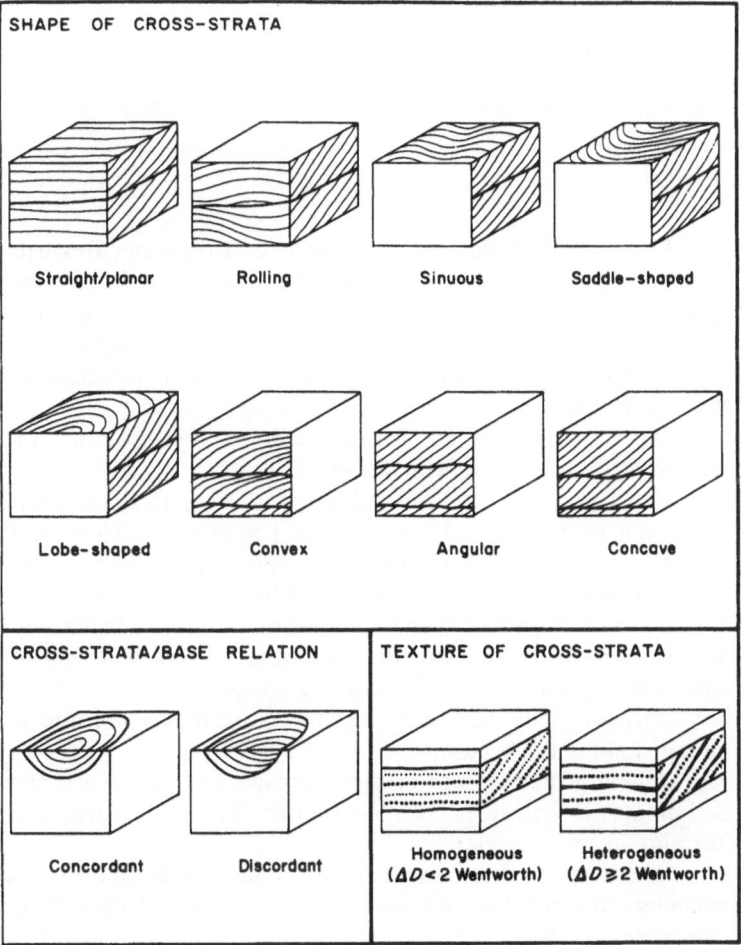

Fig. 4.3. Part II

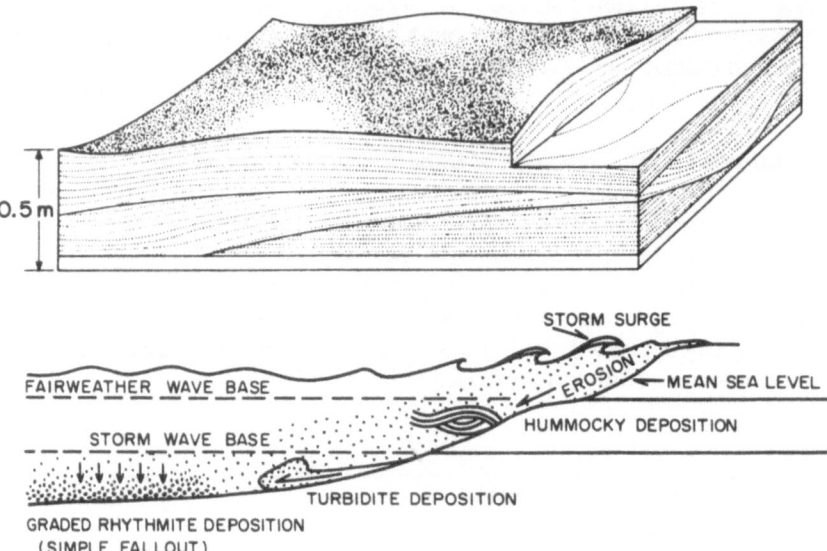

Fig. 4.4. Hummocky stratification. (After Harms et al. 1975; Walker 1982)

Flute casts are formed by static vortices in the water above the sediment surface. Their sharp end points upstream, and they broaden in the downstream direction. Flute casts (Figs. 4.5 and 4.6) are good indicators of current direction, which is measured along an axis of symmetry through the structure. Flute casts are typical on the underside of turbidite beds, but also occur in fluvial sandstones as a result of fluid turbulence.

Similar structures are formed around objects which project up from the bottom, e.g. stones and large fossils.

"Gutter marks" are longitudinal grooves up to 20 cm deep and rather narrow (less than 20–30 cm) with a spacing of 1 m or more. They are the results of channelized flow, and flute casts or groove casts may be found along their margin.

Ridges and furrows may also be formed through erosion of the substratum. Some are U-shaped or V-shaped channels. Erosion structures formed by objects which are transported along the bottom are called "tool marks". The objects may be fossil fragments or larger sediment particles.

"Groove casts" are long, narrow erosion furrows due to something being dragged along the bottom.

"Chevron marks" are erosion furrows with V-shaped structures in the clay sediments on both sides. The V-structure closes downstream, and is due to a cast forming in cohesive clay.

"Prod marks" show where an object has dug down into the clay and then been plucked out again by the current. As a result the steep side of prod marks is the downstream side.

"Bounce marks" are more symmetrical marks due to objects being swept along the bottom.

Fig. 4.5. Flute casts on the lower surface (sole) of a coarse-grained sandstone bed in the Ring Formation, Rena, South Norway. Note the small flute casts on the large flute cast structure. Flute casts are casts of the erosion structures formed in the finer-grained underlying bed by vortices

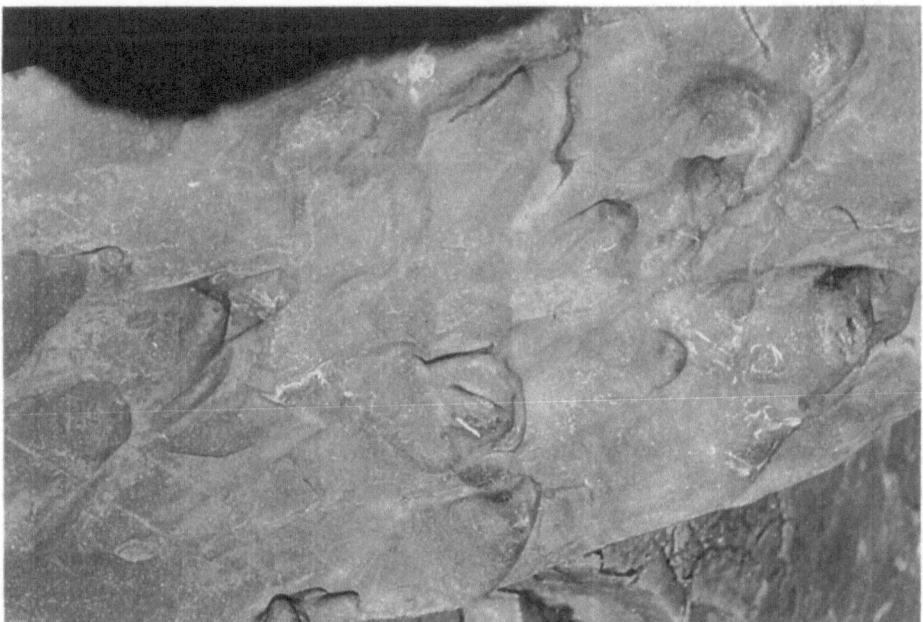

Fig. 4.6. Flute cast structure from the lower surface of a late Precambrian sandstone bed in Nyborg, Varanger, in North Norway

Structures on the Upper Surface of Beds

We also find erosion structures on the upper surface of sand beds. Water running over a plane surface, e.g. beach sand, will produce small-scale, branched (dendritic) erosion marks called *rill marks*. The flow of water may be due to runoff from big waves or from groundwater seapage at low tide. These structures are good indicators of inter- or supratidal environments, but they are seldom preserved because they are often destroyed when the water level rises again.

Larger-scale erosion produces erosion channels which may have small-scale erosion structures like flute casts and groove casts superimposed on the base.

Raindrop imprints are often preserved on bedding surfaces, and are good indicators of subaerial exposure.

Deformation Structures

Sediments are often unstable immediately after deposition, and later movements will deform the primary structures.

Deformation may be caused by four main factors:

1. Shear stress due to water or sediment movement, e.g. convolute lamination.
2. Expulsion of porewater (liquefaction, dewatering), e.g. dish structures.
3. Heavier beds above lighter beds (inverse density), e.g. load casts and ball and pillow structures.
4. Gravitational deformation. Gravitation due to sliding, folding and faulting on a slope, e.g. slumping.
5. Shrinkage, e.g. due to dessication.

It is important to distinguish between these types of deformation structures, because they have completely different implications for environments of deposition. They are not mutually exclusive, however, and may be found in close association in the same sequence.

Convolute lamination forms in fine sand or silt, and is due to laminae, e.g. current ripples, being deformed almost as they develop. Folded and inverted (overturned) lamination is typical, and the structure is deformed in the downstream direction due to the stress induced by the water movement, and to the unstability (almost a state of liquefaction) of the sediments (Fig. 4.7).

Dish structures (Fig. 4.7) are thin, clay- and silt-enriched dish-shaped laminae, in sandy sediments. Their structure is due to the porewater which flows upwards immediately after deposition, containing clay and silt which become trapped in these thin laminae. *Sand dykes* are intrusions of sand into cracks in a sediment, due to overpressure in the porewater in the sediment. The overpressure reduces the friction between the grains and injects water with sand into the cracks. Overpressurised porewater with sand may also rise right to the surface and form *sand volcanoes*.

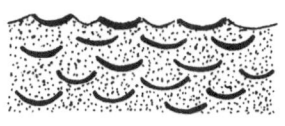

CONVOLUTE LAMINATION
STRUCTURES WITHIN A BED
FORMED DURING DEPOSITION.
EROSION OF THESE STRUCTURES
BEFORE THE NEXT BED IS DEPOSITED

SLUMPING
FOLDING OF A NUMBER OF BEDS
SIMULTANEOUSLY DUE TO GRAVITY-
INDUCED SLIDING FOLLOWING
FURTHER DEPOSITION.

DISH STRUCTURES
CLAY- AND SILT-ENRICHED, ROUNDED
DISH-SHAPED LAMINAE FORMED BY
DEPOSITION FROM UPWARD-FLOWING
WATER IMMEDIATELY FOLLOWING
DEPOSITION.

Fig. 4.7. Sedimentary structures due to liquefaction and soft sediment deformation during the deposition process. See text

Well-sorted sand has a porosity of 35–45%, whereas recently deposited mud contains 50–80% water. When sedimentation is rapid the mud has no time to lose its water, and we then find heavier layers over lighter ones (Fig. 4.8a). This is an unstable situation, and the sand layer may then sink down into the underlying silt and clay and form *load structures.* On the lower surface of a sand layer we often see pillow-formed depressions with clay which has oozed up around them. This process may continue and form isolated sand pockets in the underlying clay: *ball and pillow structures* (Fig. 4.8b). Primary structures such as flute casts often sink in and are deformed by *loading*.

On a larger scale, too, sand, e.g. channel sand, will sink down into surrounding clay (see "deltas"). Poorly compacted clay and silt will be lighter than surrounding sediments, and flow upwards to form clay diapirs.

Gravitational deformation occurs in sediments which are deposited on slopes. These forces can be resolved into a vector normal to the bedding, and a shear stress parallel to it. The vector which acts along the bedding is proportional to the sine of the angle of dip, and acts as a compaction force which can slide and fold the beds. In the upper part of the slide, tensional deformation is prominent, producing faulting, while compression occurs at the top of the slide producing folding. The result is called *slumping* (Fig. 4.9). Slumping occurs most easily where we have rapid sedimentation and therefore relatively thick beds with a high water content and low shear strength. Deformation takes place when the shear stress exceeds the shear strength. The shear stress increases with the thickness of the unconsolidated sediments, but there is not usually a corresponding increase in shear strength with increasing thickness. Slumping may resemble convolute lamination, but normally affects more than one bed. Gravitational deformation also leads to faults on a greater or lesser scale. Sliding of large volumes of sediment down slopes produces slope scars. *Growth faults* and other types of "listric" faults are a result of large-scale

a FORMATION OF "LOAD CASTS
AND BALL AND PILLOW' STRUCTURES

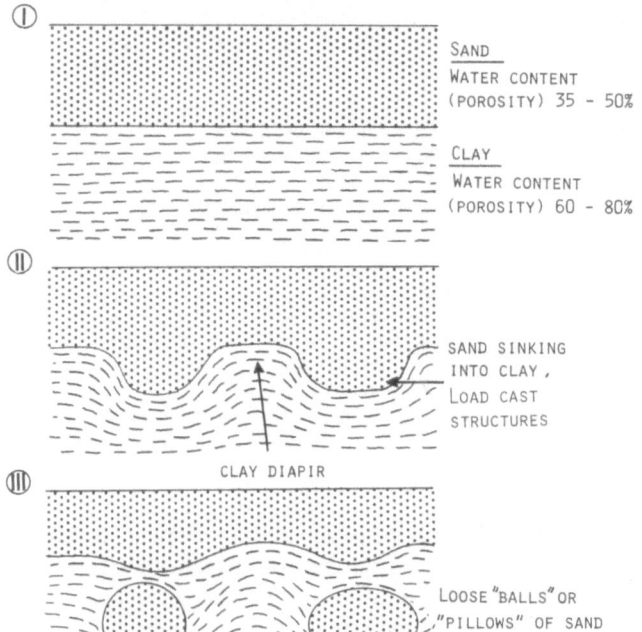

(I)

SAND
WATER CONTENT
(POROSITY) 35 - 50%

CLAY
WATER CONTENT
(POROSITY) 60 - 80%

(II)

SAND SINKING
INTO CLAY,
LOAD CAST
STRUCTURES

CLAY DIAPIR

(III)

LOOSE "BALLS" OR
"PILLOWS" OF SAND
(BALL AND PILLOW)

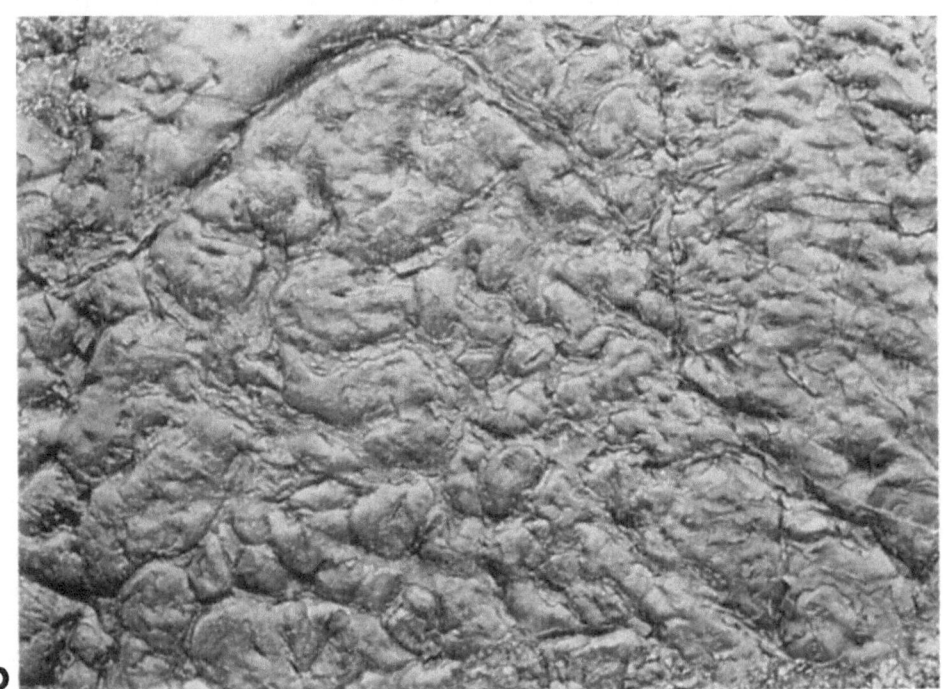

b

Fig. 4.9. Sediment beds which have been folded immediately after deposition (through slumping) due to sliding on submarine slopes. Note that the overlying beds are undeformed. From the Ridge Basin (Miocene-Pliocene), California. Scale John Crowell

gravitational deformation in the upper part of the area which is under tension (p. 84).

Dessication cracks or mud cracks are examples of contraction or shrinkage of sedimentary beds due to dehydration. Cracks frequently form regular polygons, often hexagons or orthogonal sets. Dessication cracks form only in clay and silt, and the cracks are often refilled with sand, resulting in good contrast. The formation of dessication cracks requires that the beds be exposed to the air. In certain cases shrinkage structures may also form underwater through dehydration of clay minerals (smectite) as a result of variation in the salinity of the porewater (*synaeresis*). Shrinkage structures are less regular than dessication cracks and are usually not interconnected.

When the porewater in sediments freezes to ice and remelts, we also find expansion and contraction which results in polygonal marks and ice wedges.

Fig. 4.8. a. Load cast and ball and pillow structures form because well-sorted sand and gravel are heavier than poorly consolidated clay and silt, so that sand sinks into the clay. **b** Load cast structures at the base of a sandstone bed in the Late Precambrian Ring Formation at Rena, Southern Norway. The picture covers an area of 3 × 4 m. (Bjørlykke et al. 1976)

Concretions

Concretions are round, flat or elongated structures which consist of cement which has been chemically precipitated in the pores of the sediment. The most common types of concretion are carbonate (calcite and siderite) and silica (chert). Sulphides, particularly pyrite, also form concretions.

A characteristic feature of concretions is that any laminations in the sediment pass through the concretions. This shows that the concretion has been formed through passive filling of its pores. As the overburden increases, the sediments around the concretion will be subject to compaction, while the concretion cannot be compressed because the pores are full of cement. A concretion therefore has a cement content which corresponds to its porosity at the time of formation (Raiswell 1971). Carbonate concretions in clay have a carbonate content reflecting 60–70% porosity if the matrix does not contain carbonate. Concretions in calcareous rocks, i.e. marls, contain clastic or biogenic carbonate in addition to carbonate cement, and therefore have more carbonate than the matrix. In consequence the carbonate content cannot be taken as an indication of the porosity at the time of formation. Concretions often contain fossils, and these show no sign of compaction. The fossils may have been dissolved away, leaving only impressions around the concretions, or they may have been deformed through compaction. In carbonate sediments, particularly chalk, there are silica concretions. These are formed through precipitation of finely divided amorphous silica to form a type of chert called *flint*. The source of the silica is usually amorphous biogenic silica, frequently sponge spicules.

Trace Fossils

Trace fossils are structures in sedimentary rocks which have been left by organisms which have lived on and burrowed in the sediment. Such organisms are extremely sensitive to changes in the composition of the nutrient content, the sedimentation rate, the currents etc. of bottom sediments (Fig. 4.10).

Trace fossils are therefore good indicators of the depositional environment, and ought to be described along with other secondary structures. Trace fossils can be classified taxonomically, i.e. according to the animal which left the traces. But very often it is not possible to decide which animal has made the various traces, and for this reason a descriptive, morphological classification of trace fossils has been developed.

The morphological classification defines genera and species (ichnogenus and ichnospecies), solely on the basis of the form of the trace fossils, and these may be difficult to link up with ordinary biological genera and species. The same species may form several different types of trace depending on the sediment composition and on its mode of life. Different animals may leave traces which are so similar that they are classified as one trace fossil. We can define and classify trace fossils (toponomy) on the basis of biological nomenclature, but it is important to remember that names of trace fossil types refer only to the morphology of the actual traces. Thus a trace fossil species (ichnospecies) does not correspond to a particular biological species.

Trace fossils can also be classified according to where they occur in relation to the bed:

1. On top of beds (e.g. a thin sand or carbonate layer Epichnia).
2. Within the bed (Endichnia).
3. On the lower surface of the bed (Hypichnia).
4. Outside the bed (Exichnia).

The traces in the sediment may consist of an impression made by an animal, which has been preserved through passive filling with sediment. We then get a *mould* of the animal or a *cast* in the overlying bed. Organisms which burrow into sediments often secrete a cement which ensures that the walls of their burrows do not collapse. These secretions also contribute to preservation.

When worms eat sediment, for example, it passes through their digestive organs and their burrow refills with a sediment which has a somewhat different composition from the surrounding sediments. This is particularly noticeable in burrows at the interface between two strata with different compositions. The amount of bioturbation reflects nutritive conditions and the sedimentation rate. With very rapid sedimentation there will be less time for organisms to burrow through the sediments. Where we have very slow sedimentation or a hiatus, the sediments will often be thoroughly churned up by bioturbation, and thereby homogenized. Bioturbation is most widespread in marine environments, but can also be found to a lesser extent in freshwater sediments.

Trace fossils tell us a lot about the environment of deposition, but they must be viewed in the context of the physical sedimentary structures. In high energy environments, e.g. above the fair-weather wave base, we only have vertical trace fossils, e.g. *Skolithos* or *Diplocraterion*. The high current velocity prevents these organisms from crawling around on the bottom, and they have to burrow down into the sediment and live by filtering nutrients out of the water. To remain at the same level beneath the sediment surface they must move upwards or downwards in their holes, depending on whether erosion or sedimentation is proceeding in the area. Structures or fillings which reflect such adjustments are called "spreiten".

In modern marine environments we can study a number of organisms which create bioturbation structures. The most common are worms like *Arenicola* which make U-shaped traces in fine-grained sand and silt. Such traces (arenicolites) are also to be found in older rocks. Burrowing bivalves also create various types of trace. Arthropods like crabs and prawns also make burrowing structures in beach sand (*ophiomorpha*). These vertical trace fossils are grouped together in an ichnofacies called the *Skolithos* facies. *Thallassinoids* are more horizontally aligned networks of arthropod burrows.

Below the wave base and in other protected environments, for example the intertidal zone, we find traces in the horizontal plane from organisms which live off blue-green algae and other organic material on the surface of the sediment. In this *neritic* zone (Fig. 4.10) we find a number of different types of traces from organisms which eat their way through sediments, and which form different patterns. This is

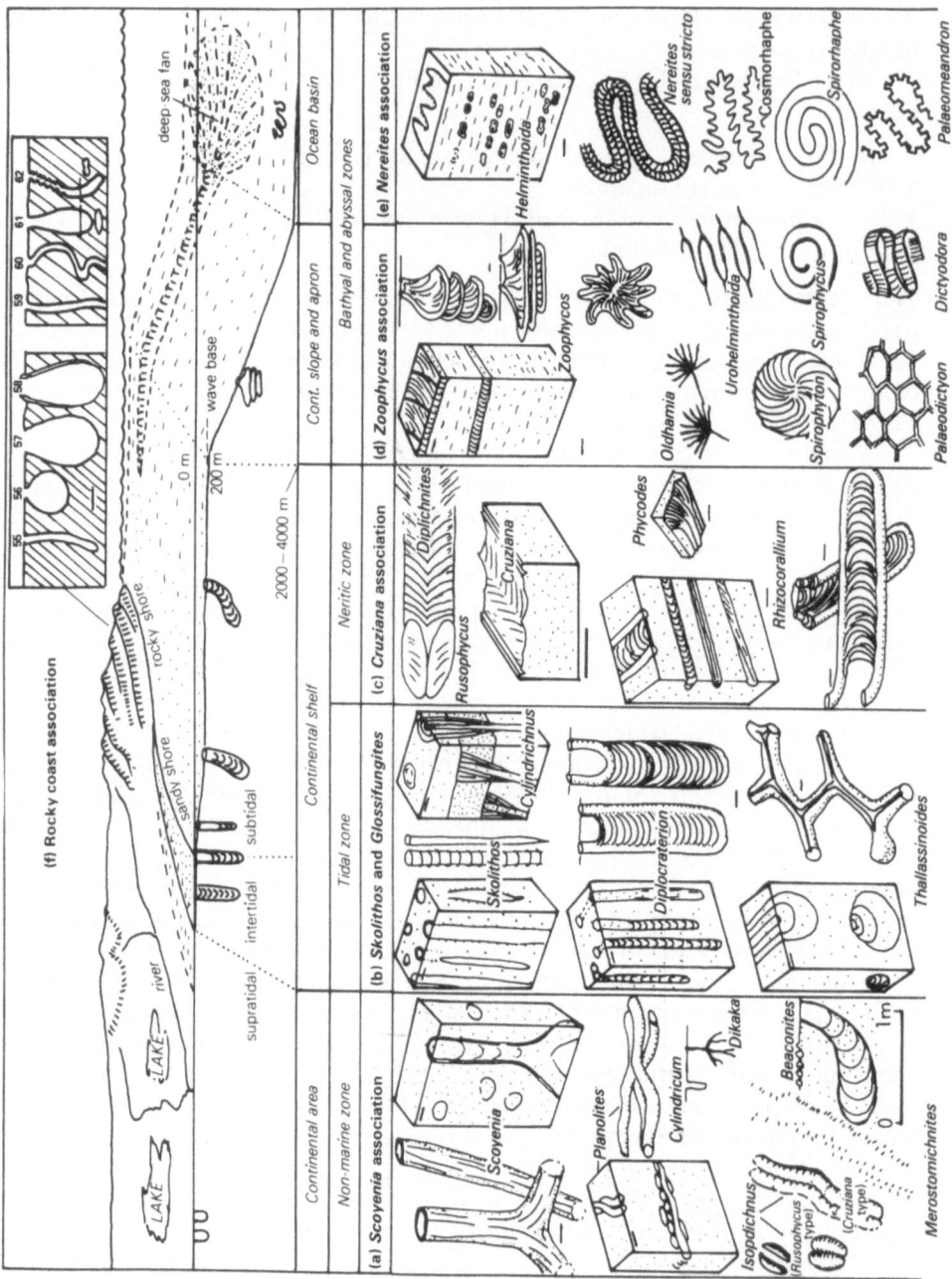

Fig. 4.10 – Part I

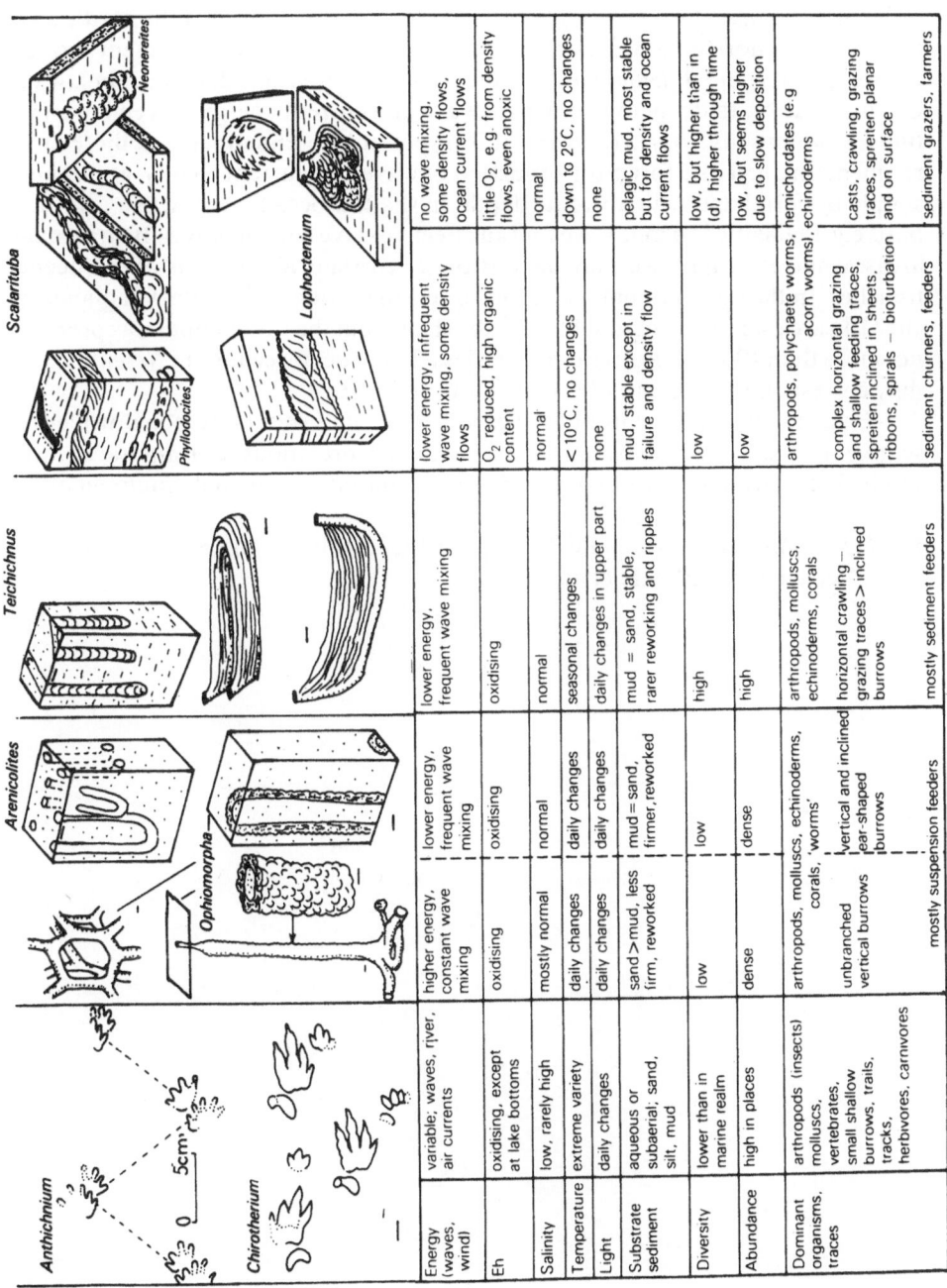

	Anthichnium / Chirotherium	Ophiomorpha	Arenicolites	Teichichnus	Phyllodocites / Lophoctenium	Scalarituba / Neonereites
Energy (waves, wind)	variable; waves, river, air currents	higher energy, constant wave mixing	lower energy, frequent wave mixing	lower energy, frequent wave mixing	lower energy, infrequent wave mixing, some density flows	no wave mixing. some density flows, ocean current flows
Eh	oxidising, except at lake bottoms	oxidising	oxidising	oxidising	O_2 reduced, high organic content	little O_2, e.g. from density flows, even anoxic
Salinity	low, rarely high	mostly normal	normal	normal	normal	normal
Temperature	extreme variety	daily changes	daily changes	seasonal changes	< 10°C, no changes	down to 2°C, no changes
Light	daily changes	daily changes	daily changes	daily changes in upper part	none	none
Substrate sediment	aqueous or subaerial; sand, silt, mud	sand > mud, less firm, reworked	mud = sand, firmer, reworked	mud = sand, stable, rarer reworking and ripples	mud, stable except in failure and density flow	pelagic mud, most stable but for density and ocean current flows
Diversity	lower than in marine realm	low	low	high	low	low, but higher than in (d), higher through time
Abundance	high in places	dense	dense	high	low	low, but seems higher due to slow deposition
Dominant organisms, traces	arthropods (insects) molluscs, vertebrates, small shallow burrows, trails, tracks, herbivores, carnivores	arthropods, molluscs, echinoderms, corals, 'worms' unbranched vertical burrows	arthropods, molluscs, echinoderms, corals vertical and inclined ear-shaped burrows	arthropods, molluscs, echinoderms, corals horizontal crawling—grazing traces > inclined burrows	arthropods, polychaete worms, echinoderms complex horizontal grazing and shallow feeding traces, spreiten inclined in sheets, ribbons, spirals – bioturbation	hemichordates (e.g acorn worms), echinoderms casts, crawling, grazing traces, spreiten planar and on surface
		mostly suspension feeders	mostly suspension feeders	mostly sediment feeders	sediment churners, feeders	sediment grazers, farmers

Fig. 4.10 – Part II

Fig. 4.10. Common trace fossils in different sedimentary facies. (Collinson and Thompson 1982)

called the *Cruziana* facies. *Rusophycus* and *Cruziana* are typical of the neritic zone and represent horizontal traces of arthropods which feed on the sediment surface. *Rhizocorallium* and *Teichichnus* are other types of organisms that occur below the *Cruziana* facies. In deeper water where wave and current energy is even lower, we find *Zoophycus* and *Nereites* facies. It is important to remember that these environments are first and foremost a function of wave and current energy, and cannot simply be correlated with absolute depth. In shallow ocean areas with a low wave base, e.g. the Baltic Sea today, we find an effective wave base of only 5–10 m in many areas, while elsewhere we may have stronger currents along parts of the deeper trenches. In the epicontinental Cambro-Silurian marine sedimentary sequence of the Oslo area we find *Nereites* facies in the shales, but the water depth was probably not more than 100–200 m, perhaps even less. In the deep oceans the *Nereites* facies may correspond to a depth of several thousand metres.

Trace fossils are very useful facies indicators and should be noted whenever sedimentary sections are examined for facies interpretations. Certain trace fossils can be linked with animals that have fairly specific environmental requirements.

Further reading: Collinson and Thompson 1982; Reineck and Singh 1975; Allen 1982; Leeder 1982; Ekdale et al. 1984.

5 Sedimentary Facies

The word "facies" is used in a number of geological disciplines. A term such as "metamorphic facies" is thoroughly entrenched. Sedimentary facies have also long been identified in sedimentology to distinguish between sedimentary rocks which differ in appearance and have formed in different ways. The term facies can be used both descriptively and genetically. We use terms like "sandy facies", "shaley facies", "carbonate facies" and we are referring here to properties of the rock that can be observed or analysed objectively. We use terms such as "shallow water facies", "deep water facies", "turbidite facies", "deltaic facies" "intertidal facies", "eolian facies", reef facies etc., depending on which environment we believe the rocks represent. In these examples, the word "facies" represents an interpretation, and is therefore not very suitable for describing sedimentary rocks objectively. For this reason it is important that we define the objective criteria (observations) on which we are basing our interpretations.

What we can do, then, is first describe and take measurements on a series of beds in the field, and on the basis of certain criteria divide the series into facies. The criteria will generally be texture, sedimentary structures, mineral composition, and, if present, also fossils. Interpreting a facies in terms of environment of deposition is often very difficult, especially because few criteria are completely unambiguous diagnostics of a particular environment. In some cases it may be useful to use statistical methods for distinguishing between different facies and for describing facies sequences.

What we observe and measure is a selection of the properties of the rock. When we describe a sedimentary rock, we observe the results of processes which have acted in the environment in which the rock was deposited. The sedimentary structures we observe tell us something of the hydrodynamic conditions during deposition. Organic structures and fossil content help us reconstruct the ecological conditions at the time of deposition in the basin. Evidence of chemical processes such as weathering, diagenesis and precipitation of authigenic minerals also supply important information about the environment of deposition. Figure 5.1 is a diagrammatic representation of the major environments of deposition on a continent and the various transition stages to deep water.

Facies Analysis

Following an interpretation of the facies of each bed or sequence of beds, we may try to group these into what we call *facies associations*. These associations of facies can be related to large-scale processes in particular environments. An association of facies coarsening upward from marine shales into delta front sand or into coal

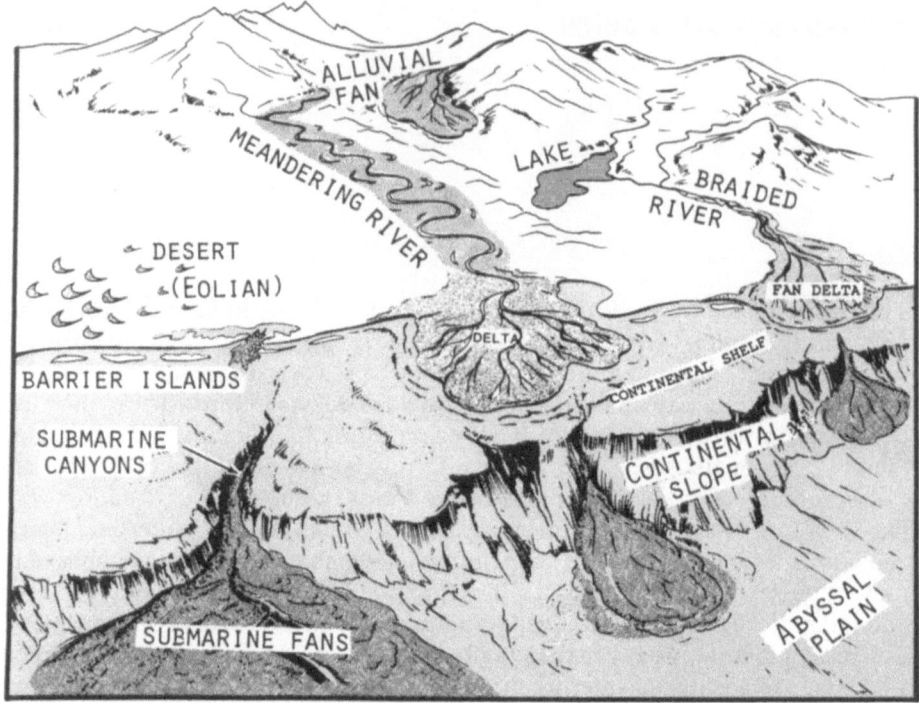

Fig. 5.1. Schematic representation of sedimentary environments on a passive margin

beds of the delta plain facies may represent a facies association typical of delta progradation.

Other facies associations may represent transgressions or regressions in a shallow marine environment.

If we have a continuous sequence without breaks (unconformities) the vertical succession represents the lateral succession of environments. This, in essence, is Walther's Law, called after the German geologist Johannes Walther.

Alluvial Fans

Alluvial fans are deposits of sediments which have been transported by fluvial processes or different forms of mass transport (e.g. mud flows and debris flows) after which the sediments have been deposited in valleys or on slopes fairly near the erosion area. The fans are formed in areas with a considerable relief, and are usually associated with faults which keep raising the erosion area in relation to the area of deposition. This tectonic sinking of the alluvial fans is a necessary condition for their being preserved in the geological series. In glaciated areas we may find alluvial fans on steep slopes produced by glacial erosion, and these may consequently be unrelated to active faults. Alluvial fan deposits in older rocks may therefore supply important information about tectonic movements during deposition. Alluvial fans which have formed in areas with a high relief but little tectonic activity, will merely represent a stage in the transport of the sediments. They will be eroded again if they are not rapidly covered by a transgression.

In the erosion area on the uplifted block, V-shaped valleys (or canyons) form, and these collect erosion products from the area that drains into the valley. The mouth defines the top of the fan (apex), where we start to find deposition and not only erosion. The fan apex is usually near the main fault plane when we have tectonism. Sediment transport from here will tend to follow the steepest slope downwards, and the sediments will therefore be spread out in a fan (Fig. 5.2). If there is a short distance between the adjacent valleys, and consequently between fan apices, the fans will coalesce. If major drainage systems develop, larger fans will form further apart from one another. The transport of sediment down and across a fan depends largely on the climate. In areas with a relatively humid climate fluvial processes will account for sediment transportation even high up on the fan. In arid climates the water table under the fan will be deep down, and water will rapidly filter down into the upper part of the fan during rains. The slope of the fan may then increase due to deposition on the upper part of the fan. As a result the sediments may be transported with a high sediment/water ratio as debris flows or mud flows. The upper part of the fan may consist of large blocks or pebbles which form an open network system through which finer-grained sediments can pass. In this way the sediment is sieved, and *sieve deposits* are formed.

Down slope the fan channels usually split up into a number of smaller channels. This reduces the hydraulic radius of the channels, and in consequence their velocity and capacity for carrying sediment is lowered. Sediment will therefore become finer-grained down-slope, even if there is no reduction in the gradient.

The water table will be deepest at the top of the fan, and shallowest at the foot. Alluvial fans are good groundwater reservoirs, easy to tap because of their porosity and permeability. The circulation of groundwater through an alluvial fan leads to strong oxidation of the sediments, giving them a red colour due to iron oxides, and in most cases any organic material will be completely oxidised.

In areas with arid climates there will be a great deal of evaporation from the groundwater which flows out of the fan, and the ions in solution will be precipitated as carbonate (*caliche*) and iron oxide. Water flowing beyond the alluvial fans may collect in playa lakes and form evaporites that are often associated with arid fans.

Humid fans will be dominated by fluvial channels on the lower part of the fan. They may drain a major portion of the fan, and have a large hydraulic radius (see p. 18) and thereby a greater capacity for transporting sediment, even if the slope of the fan is not very great. In the lower part of the fan cross-bedding will be fairly pervasive, while the upper part will tend more to consist of massive conglomeratic beds.

The foot (distal end) of the fan will merge into the other sediments which cover the valley floor. These may be lacustrine or fluvial deposits. In arid areas there will be *playa* deposits (from lakes which dry up each year) or rivers which are dry for much of the year (*ephemeral lakes and streams*) (Fig. 5.3).

A characteristic feature of transport and sedimentation across and around arid zone alluvial fans is that sedimentation takes place during short periods in connection with the rains. However, the water rapidly disappears into the ground so

Fig. 5.2. a Alluvial fan. Death Valley, California. **b** Dessicated playa lake with a block which is being moved by the wind over the smooth clayey surface. Death Valley, California

that channels which have plenty of water during maximum runoff gradually increase their sediment/water ratio, and eventually dry up to form poorly sorted conglomeratic beds. It is important to remember that flow conditions during actual deposition are not representative of transport conditions in general here. The sedimentary structures and sorting we observe are in fact only representative of the final deposition phase.

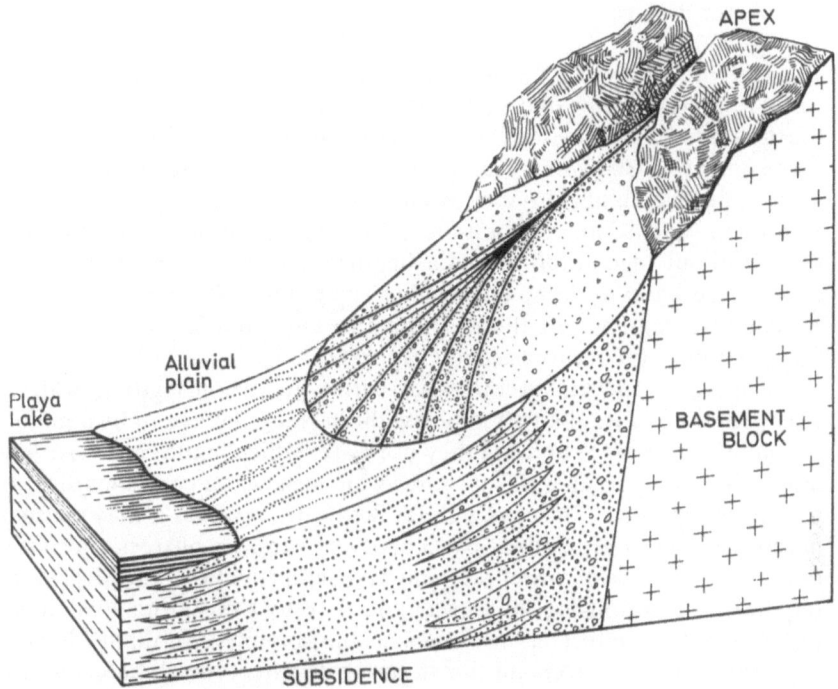

Fig. 5.3. Schematic representation of an alluvial fan developed along a basement fault. The degree of progradation of the fan into the basin depends on the relief along the fault, but also on the rainfall and the catchment area on the upthrown block

What Factors Determine the Composition and Geometry of Alluvial Fans?

We have discussed how alluvial fans are formed in areas with a relatively high relief, and how we need continuous elevation of the source area and subsidence of the depositional area in order for thick series of alluvial fans to form. Grain size in the sediments on the fan is a function of weathering and erosion in the source area and of transport capacity outwards across the fan. With fluvial transport, the velocity (v) and thereby transport capacity

$$v = C \cdot \sqrt{R \cdot \sin\alpha}.$$

Here R is the hydraulic radius (see p. 18), and we see that velocity is a function of the slope α and hydraulic radius of the channel. The hydraulic radius is a function of water depth in the channel during flooding, and this in turn is a function of precipitation and the drainage area. We often find that alluvial fans along a fault or some other steep slope vary greatly in size, even when the relief is approximately the same. In addition to relief, then, the drainage area of the various canyons on the raised block is a vital factor for determining the volume of coarse material deposited on the fan.

Further reading: Koster and Steel (1984).

The Water Budget

The precipitation which falls in a particular area places an upper limit on flow into water reservoirs. Some of the precipitation evaporates immediately upon reaching the ground. In hot, desert areas where the air is very dry, evaporation may correspond to 2–3000 mm/year, i.e. more than the normal amount of precipitation in most places. Plants also contribute to water loss through transpiration, and in areas with vegetation, evaporation and transpiration may reach up to 2000 mm/year. Without vegetation we find a significant amount of evaporation only when there is free water or damp ground right up to the surface. Even if the water table is shallow (less than 1 m), there is little direct evaporation, but some water will be drawn up by capillary forces and evaporate.

Plants are important in connection with evaporation of groundwater; tree roots may penetrate more than 10 m below the surface. Trees can be a major drain on groundwater reserves because they use water which could otherwise have seeped into wells, rivers or lakes. Dense forests may use water corresponding to 200–500 mm of precipitation. In areas with limited water reserves it may consequently be necessary to limit vegetation, particularly of varieties of trees and bushes with a relatively high transpiration rate and no useful function. However, vegetation is important for stabilising the topsoil and thereby preventing erosion. Vegetation may also have an effect on the *albedo*, i.e. the amount of sunlight which is reflected from the earth's surface, and this in turn may affect the climate and increase the rainfall.

Runoff is the fraction of precipitation which reaches the rivers. The percentage of runoff which enters the rivers depends on how much precipitation evaporates and how much filters down to the water table. The percentage of precipitation which filters down to and recharges the groundwater is of particular importance. Rivers also have groundwater added to them if their surface is below the water table. If their surface is above the water table they will lose water. In consequence, rivers cannot be considered in isolation from groundwater reserves.

Infiltration of surface water into the ground depends on the permeability of the soil cover and on the vegetation. Clayey sediments allow far slower infiltration of surface water than sand or gravel. Where there are soil types with swelling clay (smectite group), which are very common in desert areas, the first rains will cause these minerals to swell, further reducing their permeability. Consequently, infiltration is reduced, and the runoff which creates floods increases. If the earth is dry above groundwater level, it is often somewhat cracked and therefore permeable. However, water has to overcome the capillary forces which act against water percolating down through dry, fine-grained sediments. We may find a layer of air in the pores between the groundwater table and the water filtering down from above (trapped air). This has the effect of reducing infiltration until the air is able to escape upwards as bubbles.

Sand and gravel are good groundwater reservoirs, and in connection with fluvial sand deposits there will also be considerable groundwater resources. It is often advisable to produce water from wells in sand adjacent to rivers rather than taking the water directly from the river. The groundwater there has been filtered free of suspended material, and ion exchange with clay minerals may remove chemical pollution, giving it higher quality both as drinking water and for industrial use. Alluvial fans are porous and have a relatively good potential as water reservoirs, and we often find springs at the foot of fans. In Scandinavia glaciofluvial deltas are important water reservoirs, while areas with clay-rich moraines supply little water to wells which do not reach right down to bedrock. In dry areas the depth of the water table can be most simply charted through seismic recordings.

Because fresh water is lighter than salt water, it will flow over salt porewater and form pockets of fresh water under islands (Fig. 5.4). Fresh water may also flow from the continents, following permeable beds underneath continental shelves.

Hydrogeology is an important subject which deals with groundwater in sediments and rocks. Fresh water of high quality (little pollution) is scarce in many regions, and in recent years in particular a large number of geologists have been involved in mapping groundwater reserves. The goal of this research is:

1. To estimate the size and rate of annual renewal of groundwater reservoirs.
2. To plan appropriate and prudent use of groundwater.
3. To find places where waste such as household, industrial and radioactive waste can be deposited without the groundwater reservoirs being harmed.

Further reading: Davies and de Wiest 1966.

Glacial Sediments

Glacial sediments consist in principle of two types:

1. Sediments deposited directly by ice.
2. Sediments deposited near ice, or in an area characterised by a glacial environment.

Glacial erosion, transport and deposition is a result of the flow of ice. Ice is a material which has considerable shear strength, but which exhibits plastic flow when loaded.

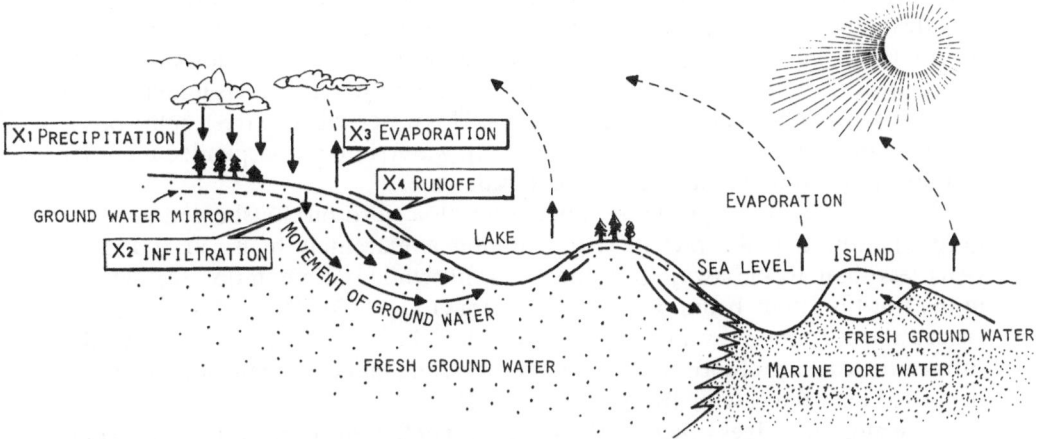

Fig. 5.4. Diagram showing the circulation of groundwater on land and under marine basins

If we apply the formula for Reynold's Number: $R = v \cdot L \cdot \rho / \mu$ (see p. 16) we see that the viscosity (μ) is high and the velocity (v) is low, so that the Reynold's Number will be low and flow laminar. The velocity of ice, v, varies from a few metres to several hundred metres a year. Just as with mud flow, the high shear strength and viscosity mean that there is no sorting and large rock fragments (clasts) undergo little rounding. Tills are therefore generally poorly sorted, and often contain angular clasts. Stones lying in the base of the ice are ground against the substratum and acquire ice-faceted surfaces with glacial scour striations. The best-preserved striations are found in carbonate rocks and siltstones. Elongated pebbles usually become oriented with their long axis parallel with the direction of ice-flow.

In glacial environments, sediments are also transported by the water from the melting ice. Under the glacier the ice will be subject to pressure solution (wet-based ice). Near the edge of the ice, where it is thinner, surface meltwater runs down under the glacier. In recent years it has been realised that a considerable amount of sediment transport in glaciers takes place in the water between the ice and the substratum. Near the front of the glacier, tunnels form in the stagnant ice, and here there is large-scale fluvial transport and the tunnels fill with sediment. Crevasses in the ice also fill with sand and gravel, which form mounds or ridges of bedded material when the ice melts (*kames*). At the same time sediments which have been supported by or deposited on top of the ice collapse or slide outwards. Sediments which were distributed through the ice will become concentrated and remain as uneven deposits (*hummocks*) of poorly sorted material (ablation deposits). Remnants of ice which become covered by sediments leave ice hollows (*kettles*) when they melt.

Large quantities of meltwater flow from the front of glaciers, particularly during mild periods, and deposit glaciofluvial or glaciomarine sediments. The material near the ice is poorly sorted and rounded and may also be deformed (folded) by the advance of the ice. Water under pressure beneath the ice may flow up, transporting coarse sand and gravel upslope. The sediments deposited may then show bedding dipping towards the ice. The sediments are rich in water, and mud flows may form. Till material may also be transported and redeposited as mud flows (*flow till*). Glaciers which flow into lakes (possibly ice-dammed) or fjords may form well-developed glaciofluvial deltas or ice-front terraces. Beyond the marine ice-front terraces and deltas, marine clays are deposited. Meltwater flows out from under the ice, and the coarsest material may be deposited in tunnels under the ice or at the front where the ice calves into the sea. This water, which is fresh or brackish, will flow upwards because it is lighter than salt water, and the suspended material will precipitate out, due to flocculation in salt water.

Detailed studies of glacial sedimentation in a fjord on Spitsbergen (Kongsfjorden) show that the rate of sedimentation is extremely high (50–100 mm/year) less than 10 km away from the ice front, while sedimentation further out in the fjord is about 1 mm per year (Elverhøi et al. 1983). The 1000-km² glacier erodes about 1 mm per year. It is clear that glacial erosion is very rapid when the substratum consists of softish sedimentary rocks, and that most of the sediments are deposited near the calving front (Figs. 5.5, 5.6).

The moraine ridges which lie along the coast of Southern Norway, Sweden and Finland are examples of concentrated sedimentation ahead of a calving front

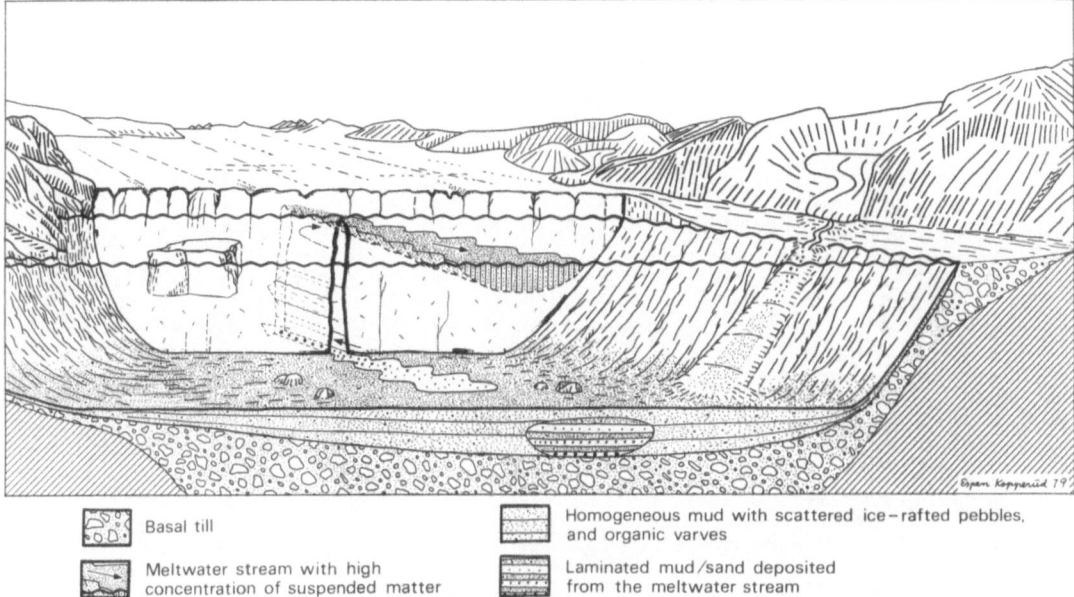

Basal till	Homogeneous mud with scattered ice–rafted pebbles, and organic varves
Meltwater stream with high concentration of suspended matter	Laminated mud/sand deposited from the meltwater stream

Fig. 5.5. Section through a fjord (Kongsfjorden, Spitsbergen) with calving ice front. (Elverhøi et al. 1980)

in the Scandinavian de-glaciation area. These moraines have been reworked at the top, and have a layer of larger pebbles and boulders.

In the North Sea we find Quarternary deposits which are several hundred metres thick. Glacial erosion has incised down into the underlying Tertiary and Mesozoic sediments. Moraines formed through the erosion of clay-rich sediments contain little coarse-grained material and may be difficult to distinguish from marine or glacio-marine deposits. However, they have a higher shear strength and lower water content due to compaction caused by ice-loading. Moraines and other glacial deposits on land tend to be eroded and we mainly find deposits from the last glaciation. The glacial deposits in the North Sea tend to be better preserved because this basin is constantly subsiding and as a result we find a more complete stratigraphic record there. Continental glaciations have therefore tended to leave poor stratigraphic records in the centre of the glaciated area where erosion takes place for the most part. The most important record is to be found in surrounding basins, where deposition takes place.

On the continental shelf around Antarctica, sedimentation is slow (1–5 cm/ 1000 years), particularly outside the area with an ice shelf. The slow accumulation of sediment is due partly to the low rate of erosion by the Antarctic ice and partly to the fact that some of the sediments which are carried out from land on the sole of the glacier are released through melting of icebergs outside the shelf. The shelf around Antarctica is considerably deeper (about 500 m) than other continental shelves. The reason for this is not clear yet, but extensive, long-term glacial erosion right out to the edge is probably the decisive factor.

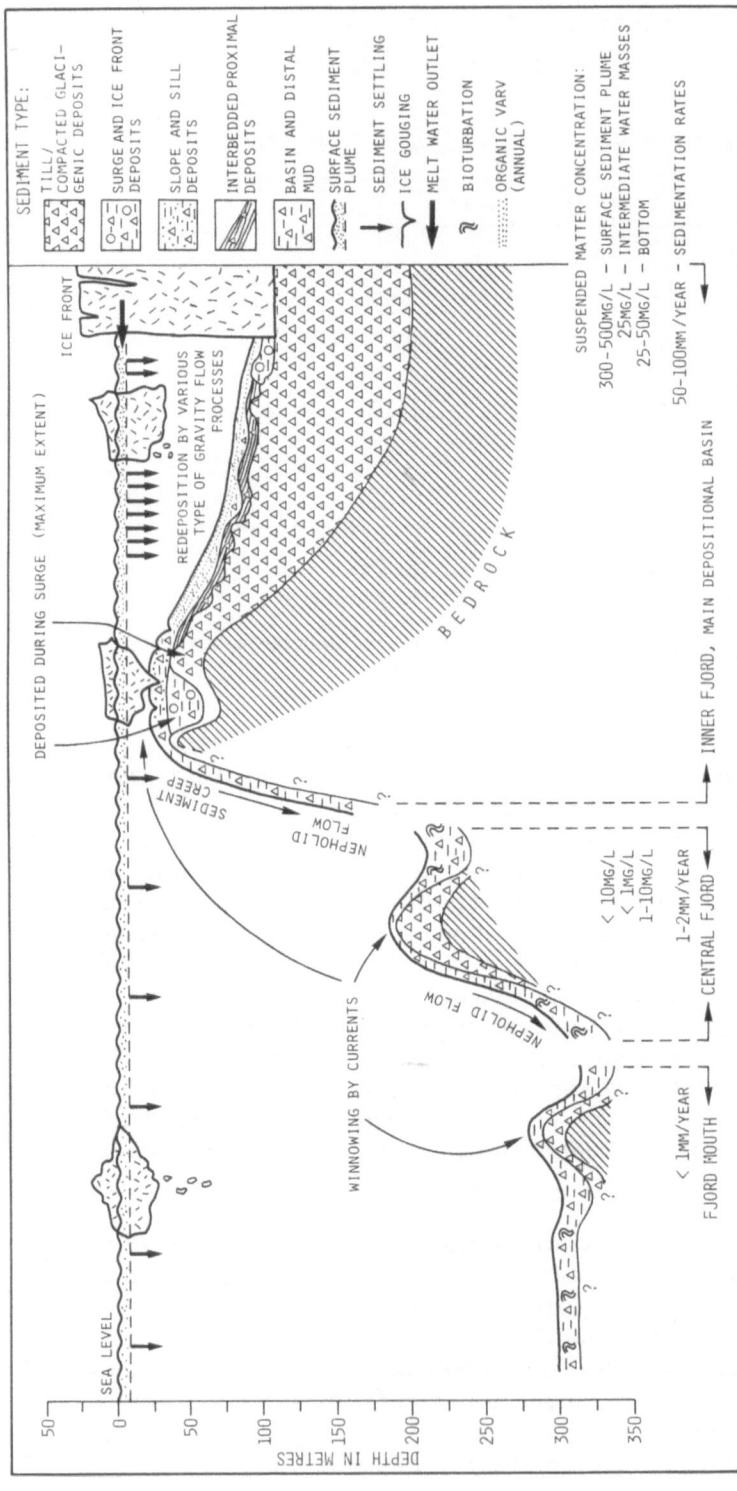

Fig. 5.6. Section through Kongsfjorden, Spitsbergen (Elverhøi et al. 1983). Note how the sedimentation declines away from the ice calving front

Glacial sedimentation is not limited to the Quaternary period. In Antarctica there have been glaciers for at least 15 million years, i.e. back to Miocene times, and we now know of glacial sediments from several long periods in the geological time record. In the Precambrian there were several glaciation periods and tillites and glacio-marine deposits from the late Precambrian (Varangian) are found widely on all the continents. It is difficult to know exactly which were contemporaneous and to reconstruct the palaeogeography.

During the Ordovician the South Pole was in West Africa, and in the Sahara there are extremely well-preserved glacial deposits with ice-scoured rock, moraines, glacio-fluvial sediments and eskers. In the Amazon basin there are glacial deposits which were laid down near the South Pole in Devonian times. During the Carboniferous-Permian eras large areas of Gondwanaland were glaciated, and glacial sediments of this age are found in South Africa, Australia, South America and Asia. In Oman, Permian glacial sediments form petroleum reservoir rocks.

Glaciation involves considerable variation in sea level, which very strongly affects sedimentation in deltaic and shelf areas. During geological periods with glaciations in polar regions, the ocean basins were also better ventilated with oxygenated water from the polar regions.

Further reading: Crowell 1982; Hambrey and Harland 1981.

Desert

Deserts are areas with little or no vegetation. There may be a number of reasons for this, and we can distinguish between different types of desert:

1. Deserts due to low precipitation and high evaporation in hot climates.
2. Deserts due to low precipitation in cold climates.
3. Deserts due to soil erosion.

The first two types of desert are largely governed by meteorological factors. Areas which lie on the lee-side of major mountains will be dry. At about 30°N and 30°S high pressure areas predominate, resulting in low precipitation. Since the prevailing wind at these latitudes will come from the east, we find most of the deserts on the western side of mountain chains. This is true of Asia Minor, the Sahara, California and Nevada in the northern hemisphere, and Australia, South Africa and Chile in the southern hemisphere. Inland areas surrounded by mountains tend to have a desert climate.

Iceland has good examples of desert regions which are due to a cold climate, and has problems with soil erosion due to lack of vegetation in many areas.

Shortly after the withdrawal of the ice across Scandinavia there was a desert with considerable eolian deposits, which formed before the vegetation cover developed. Loess deposits are fine-grained sediments which form largely through eolian erosion and transport from deserts and also from glacial sediments exposed after glacial retreat. Large areas of China, for example west of Beijing, are covered by up to several hundred metres of Quaternary loess.

The lack of vegetation naturally means that eolian transport and deposition are important in deserts, but only limited desert areas are covered by wind-blown sand.

Large areas consist of bare rocks and mountains with little sediment. In other regions there is only wind erosion (deflation), which leaves the ground covered by a layer of stones which protect it from further erosion. Even in desert areas fluvial transport may dominate. When we fly over the Sahara, we see many areas which are dominated by fluvial structures. Even though several years may pass between rains in this area, a heavy fall of rain may transport so much sediment that the fluvial drainage pattern remains for many years.

Large expanses of wind-blown sand are called *ergs*. They may be formed by the coalescence of different eolian bedforms. *Barchans* are crescent-shaped isolated eolian dunes with a convex erosion side and a concave lee side (Fig. 5.7). Barchans are found mainly on the edges of the erg area, where there is not enough sand to form a continuous thick cover. Transverse dunes are common within ergs. *Seif dunes* are long, straight sand dunes which may occur within an erg but are also found in areas with incomplete sand cover. In the central zones very large dunes form which may be over 100 m high and 1 km long.

How Can We Recognise Eolian Deposits in Older Rocks?

Although modern (recent) eolian deposits have a very distinctive appearance, it is not always easy to identify wind-deposited sediments in older rocks. The surest indicators are:

1. Lack of fossils or organic material.
2. Large-scale cross-bedding, up to 20–30 m.
3. Wind ripple marks on top of slanting cross-bedded surfaces (foreset beds).
4. High degree of oxidation which gives a red colour.
5. Good sorting – largely medium- to fine-grained sand.

Some of these features may also be found in sandstones deposited in shallow marine environments, but there are also important differences.

Cross-bedding exceeding 5–10 m is very rare in marine environments. A high degree of oxidation plus lack of marine microfossils are fairly reliable indicators of desert deposits.

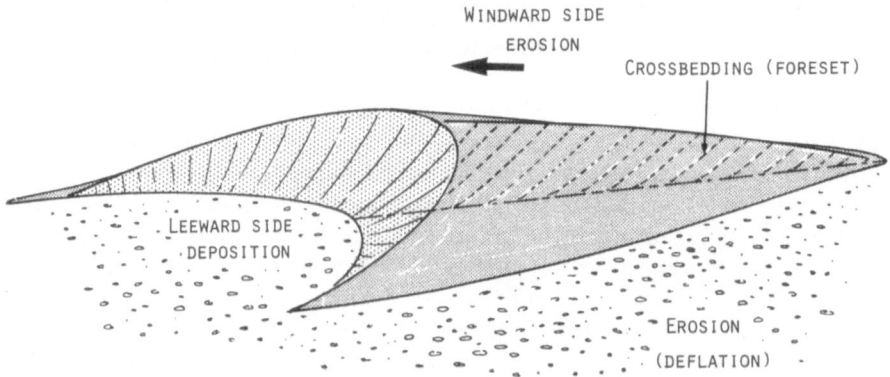

Fig. 5.7. Eolian sand dune (barchan)

The lack of thin silt or clay beds between the cross-bedded layers, and the general association with overlying and underlying sediments are also important criteria. One of the best-known examples of an eolian sandstone is the Navajo Sandstone (Jurassic) in the Rocky Mountain area (Fig. 4.1b). In Northern Europe there are eolian sandstones in the Upper Permian (Rotliegendes) and the continental Triassic (Bunter) (Fig. 5.8).

In pre-Devonian times all continental environments were deserts in the sense that they lacked vegetation. Also in these rocks, however, we can distinguish between dry and wet climates, largely from the nature of the fluvial or eolian deposits.

Lacustrine Deposits

How Are Lakes Formed?

In principle we have three types of lake:

1. Lakes of glacial origin:
   a) lakes formed through glacial erosion,
   b) lakes formed through damming by moraines or by the glaciers themselves.

2. Lakes of tectonic origin:
   a) lakes formed in areas of rapid tectonic subsidence (rifting) or more uniform subsidence,
   b) lakes formed as a result of damming by horsts which have been elevated through faulting, or damming by lava etc.

3. Lakes formed by sedimentary processes, e.g. *oxbow lakes* in fluvial environments, and delta top lakes.

Lakes of glacial origin will have a relatively short lifetime by geological standards. In 10,000–100,000 years most of the lakes in Scandinavia and North America will have filled up with sediment, if we do not have another glaciation.

Lakes of tectonic origin, however, will continue to subside. If the rate of subsidence keeps pace with the rate of sedimentation, a lake will continue to exist (see Fig. 5.9).

Extensive carbonate beds (freshwater carbonates) and diatom deposits are also common in lacustrine basins. Some of the largest lakes formed by rifting are found in East Africa, where a number of lakes occur in the actual rift valley system. Lake Victoria, which lies between two rift valley systems, was formed by tectonic movements about 100,000 years ago.

In humid climates all lakes will have outlets in the form of rivers or via the groundwater, but in arid climates lakes develop into evaporite basins without outlets. Normal, non-saline lakes differ from ocean basins in several ways:

1. Low salinity leads to slower flocculation of clay particles than in marine basins, producing more distinct lamination.
2. Low wave and tidal energy and weaker currente mean less erosion and resedimentation. Seasonal variation in the influx of sediments may produce annual cycles in lacustrine sediments.

a

b

Fig. 5.8. a Eolian cross-bedding 4–5 m high in the Permian "Yellow Sand" of North-East England (Old Quarrington Quary, Durham). This sand is equivalent to the "Rothliegendes" sandstone of the North Sea. **b** Barchan from the desert of Mauritania

Two main types of lakes:

A) BASINS FORMED THROUGH GLACIAL EROSION.

Glacial lakes: They will fill with sediment
within a relatively short period, geologically
speaking (10^4 - 10^5 years). Low preservation
potential in a geological series if they lie
above sea level, e.g. Fennoscandia and N. America.

B) BASINS OF TECTONIC ORIGIN.

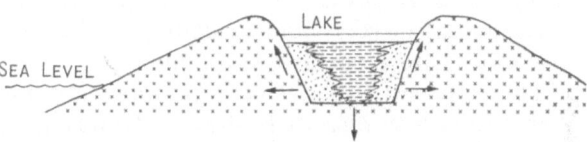

Fig. 5.9. Two main types of lacustrine basin: A) formed by glacial erosion, B) formed by faulting (rifting) or other types of tectonic subsidence

Subsidence simultaneous with sedimentation.
High preservation potential, e.g. East Africa,
California.

3. Limited water circulation makes it easier for the water to develop layering based on density, i.e. there is a thermocline between dense layers at the bottom and less dense water near the surface. This may result in limited oxidation of organic material and the development of organic-rich sediments in the deeper parts of the lakes.
4. River water will generally have approximately the same density as lake water, and will therefore mix well and deposit sediments rapidly. Cold (glacial) river water or river water with a lot of suspended material will be heavier than lake water, however, and will form turbidity currents along the bottom.
5. Lake sediments have a distinctly different fauna from marine basins. The geochemistry of lake sediments and the composition of carbonates and evaporites are also different from those of marine sediments.

By comparison with marine deltas, lacustrine deltas are among the most constructive of all (e.g. the Gilbert Delta) because erosion in the basin is so limited. Nevertheless, wind conditions, the size of the lakes and the composition of the sediments will be crucial factors. In large lakes we might have relatively high wave energy which could erode sediments from the river mouth and deposit them laterally on beaches.

Wind will also cause water to well up from the bottom, increasing the circulation of oxygen along the bottom. Density inversion, at about 4°C, also helps to circulate water in cold areas when the temperature falls to 4°C. In warmer regions this density inversion does not occur, and this helps to stabilise the stratification within the body of water and maintain reducing conditions near the bottom.

In dry regions we may get evaporitic lakes and lakes which dry up after each flood (*ephemeral lakes*).

How Can We Recognise Ancient Lacustrine Sediments?

This is often difficult, particularly in Precambrian rocks which do not contain fossils. But the most important indicators to look for are:

1. Lack of marine fossils. It may also be useful to find out whether the sediments contain microfossils. Diatoms and algae may be important components and they have a high tolerance for variations in salinity.
2. Very fine laminations in clay and silt sediments.
3. Bioturbation structures are found in lacustrine sediments, but they are less common than in marine sediments.
4. Chemical analyses of freshwater sediments will reveal a low concentration of trace elements which are enriched in sea water (e.g. Cl, Br and B). Minerals formed in freshwater lakes may have characteristic isotopic compositions. Evaporites in lakes consist mainly of carbonate minerals and sulphates and chlorides are normally less common.
5. The content of sulphur-bearing authigenic minerals must be relatively low since there is little sulphate in freshwater. However, organic material in lakes contains a fair amount of sulphur which can be precipitated as sulphides by sulphate-reducing bacteria.
6. Lack of tidal structures and similar marine structures are also an important indicator.

Amongst the most important ancient lacustrine deposits are the Karoo deposits of South Africa and East Africa, from the Carboniferous to the Jurassic periods. These sediments contain important coal deposits. In eastern China there are large areas of Cretaceous and Tertiary lacustrine sediments, which are also important petroleum-bearing sediments.

The Green River Formation (Eocene) of Colorado, Wyoming and Utah, is a lacustrine sediment of great extension which represents one of the most important petroleum source rocks in the world.

Rifting associated with the opening of the Atlantic Ocean in Mesozoic times resulted in a large number of lacustrine basins, and sediments deposited in such basins now underlie the continental shelves. In dry regions these turned into evaporite basins.

Further reading: Matter and Tucker 1978.

River Deposits

Transport of sediments in rivers depends on the gradient and cross-section of the river, which determine the flow rate, and on the composition and concentration of sediment. The processes involved in sediment transport determine which sedimentary structures are formed, and also the geometry of the sandstone deposit.

In rivers which flow only during and directly after the rains, but are otherwise dry (*ephemeral streams*), equilibrium between flow and sediment load is not achieved, so that even fine-grained sediment is deposited in the channel when it dries up, leaving a mud-cracked surface. Intensive oxidation of sediments, including silt and clay beds, is typical of such fluvial deposits. There is little organic production in areas with a dry climate, and clay and silt particles will therefore contain little organic material which could act as a reducing agent. A low water table during dry periods also contributes to oxidation of organic matter.

Most rivers have marked seasonal and annual flow variations. The flow velocity increases appreciably with flow volume because the friction per unit volume of water is inversely proportional to the water depth. When water flow is greater than the capacity of the channel, the water flows out over the banks. The water outside the channel will normally be very shallow and have a low flow velocity. Fine-grained sediments, which have been transported in suspension, are then deposited. We call sediments of this type *overbank sediments*. The amount of overbank sediment is a function of the amount of fine material in suspension and the duration of the flood period. Overbank sediment may build up into elevated banks called *levées*. The resulting soil along modern rivers is often very rich and ideal for cultivation. In post-Devonian sedimentary rocks we often find traces of plants, commonly roots, in levées. Levée deposits have parallel lamination and in some cases also current ripples, particularly *climbing ripples*, which are typical of rapid sedimentation. The primary sedimentary structures, however, will often be partly or wholly destroyed by traces left by plant roots.

Major floods may cover the areas beyond the levées as well, and clay and silt will be deposited on these *flood plains*. Many rivers which carry a large amount of suspended sediment gradually build up their beds through deposition in their channels and on the levées, so that the surface of the water may be considerably higher than the surrounding plain. During floods the water may then break through the levée and flow down from the channel to the plain, forming temporary lakes several metres deep. The lower Mississippi River is considerably higher than the surrounding area, including New Orleans. The major Chinese rivers have also built themselves up above the surrounding countryside, so floods can cause very great damage.

However, the stability of the levées will determine how much a channel can build itself up, and clay-rich sediments form stronger levées than sand and silt because of their cohesiveness. When the levées give way, sand will flow out of the channel and deposit sandy sediments in fans or *crevasse splays* on the flood plain. They are characterised by thin sand beds, often fining upwards, with an erosion base in the proximal part (nearest the channel). However, we may also find small sequences which coarsen upwards as a result of progradation of this fan, because grain size decreases towards the edge of the fan.

Channel Shapes

We distinguish between relatively straight fluvial channels and curved channels. Curved channels are shaped rather like a sine curve. To describe the degree of curvature we speak of high and low sinuosity. Sinuosity is defined as the ratio between the length along the channel and the length of a straight line through the meander belt, i.e. the length of the river valley. Meandering streams have a single channel and by definition a sinuosity greater than 1.5.

Braided streams have a branched course, but in most cases the river channels are fairly straight. These rivers are branched because the river channel is not very stable, and because sediment is deposited in the middle of the channel, forming small islands or bars. Formation of braided streams therefore favours the deposition of coarse sediments containing little clay or silt. A higher stream gradient gives greater energy and greater erosion of the sides of the channel. An abundant supply of sediment leads to sediment being deposited more rapidly than it can be eroded and transported onwards, resulting in the formation of deposits in the channel. Braided streams are therefore also typical of areas where the velocity declines somewhat, e.g. when a river widens onto a plain after passing through a narrow valley. The velocity and flow will in most cases vary considerably with time. When the river is low it will flow round bars of sand and gravel, while at high water and with strong currents sand bars will migrate as they are eroded on the upstream side and accumulate deposits on the downstream side. Gravel bars, on the other hand, have rather complex patterns of erosion and deposition.

Meandering rivers have a wavelength which is a function of the breadth of the river. The wavelength is approximately 11 times the breadth or 5 times the radius of the river. There is also a relatively regular ratio (about 7:1) between the breadth and depth of the river.

Meandering rivers move in loops, with the greatest velocity at the outer bank so that erosion takes place there. The velocity at the inner bank is much lower, so sediment is deposited there. The flow velocity and shear forces against the bottom also decrease upwards towards the top of the bank on the inner side of the swing, and sediments are deposited in a *point bar* which reflects the hydrodynamic conditions. At the lowest point there is a lag conglomerate followed by large-scale cross-bedding due to the flow in the upper part of the lower flow regime. Then come current ripples, and finally fine-grained sand, silt and clay sediments corresponding to the lower part of the lower flow regime at the top of the profile (*overbank sediments*). This produces a fining-up sequence from sand to silt and clay (Fig. 5.10).

The secondary flow, in a vertical section at right angles to the downstream flow direction, moves from the outer bank where erosion takes place along the bottom, and up the inner (point bar) bank where deposition takes place. This is a result of a difference in hydrostatic pressure because the surface of the water slopes inwards towards the inner bank due to centrifugal forces. If we combine this movement with the main flow of water down-river, we find a corkscrew or *helical* movement (Fig. 5.11). At each bend in the river the helical flow reverses direction. The point bar becomes asymmetrical, with coarser material on the upstream side, so that a perfect fining-upward profile is not developed. The upper fine-grained part and the overbank deposit will also be lacking. Variations in the depth of water in the channel will also lead to departures from the ideal fining-upward sequence.

During floods the water may flow over the point bar and form a little channel, or *chute*. This may be widened by further erosion to become the main channel. This process is called *chute cut-off*. The sinuosity may also become so high that erosion cuts a channel through a narrow neck (neck cut-off), making a straight course to a lower swing of the river. The whole meander will then be abandoned by the river, and a small, curved *oxbow lake* is left, which will fill with clay, silt and organic matter.

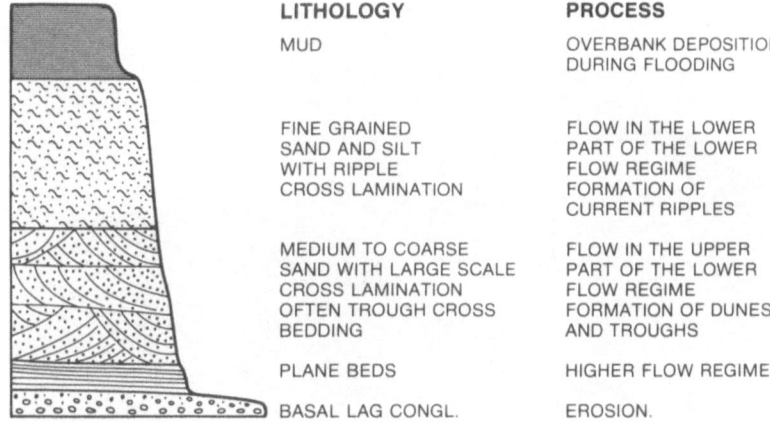

LITHOLOGY

MUD

FINE GRAINED
SAND AND SILT
WITH RIPPLE
CROSS LAMINATION

MEDIUM TO COARSE
SAND WITH LARGE SCALE
CROSS LAMINATION
OFTEN TROUGH CROSS
BEDDING

PLANE BEDS

BASAL LAG CONGL.

PROCESS

OVERBANK DEPOSITION
DURING FLOODING

FLOW IN THE LOWER
PART OF THE LOWER
FLOW REGIME
FORMATION OF
CURRENT RIPPLES

FLOW IN THE UPPER
PART OF THE LOWER
FLOW REGIME
FORMATION OF DUNES
AND TROUGHS

HIGHER FLOW REGIME

EROSION.

Fig. 5.10. Schematic representation of a fining-upwards sequence deposited in a meandering river

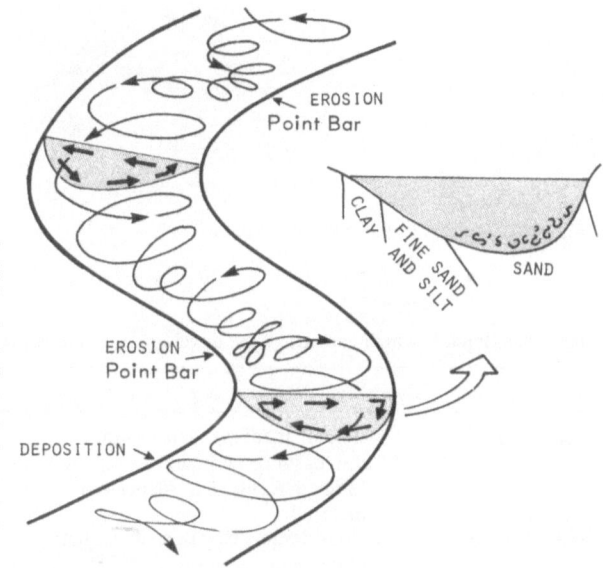

Fig. 5.11. Diagram showing principles for water flow in meandering rivers. Note how the spiral flow component in the channel seen in vertical section reverses direction at each bend (helical flow). This leads to point bar deposits being asymmetrical, with most erosion at the upstream side and deposition on the downstream side. The vertical profiles are therefore different and deviate from the ideal coarsening up point bar sequence

Rapid subsidence and low sand/mud ratios will increase the stability of the channels and we may find an anastomising channel distribution where there is little or no lateral acretion of the channel (Fig. 5.12).

When the whole river channel shifts course (avulsion) the abandoned channel will fill up with mud and form a clay plug.

Summary of Fluvial Sedimentation

Fluvial processes are fairly simple in principle. We know the physical laws which govern the flow of water in channels and transport of sediment in water. The great

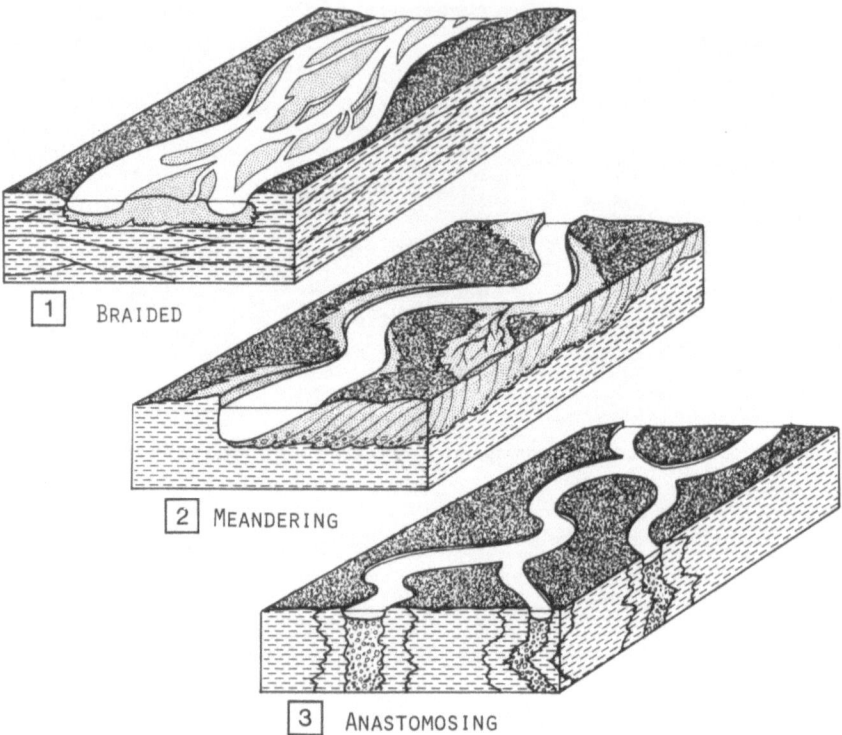

Fig. 5.12. Different types of fluvial facies

Table 5.1. Factors which encourage the development of meandering and braided rivers

	Meandering	Braided
Grain size	Largely medium to fine sand, thin basal conglomerate	Coarse sand and gravel
	Large percentage of suspended material	Small percentage of suspended material compared to bed load
Sediment type	Considerable clay and silt content. Great channel stability	Low clay and silt content, less channel stability due to the lower cohesion of sandy sediments
River gradient	Very small	In most cases greater than meandering rivers
Water flow	Relatively moderate variations	Great variations in water flow
Vegetation	Stabilises channels	Lack of vegetation increases bank erosion

variation in composition, structure and geometry demonstrated by fluvial deposits is due to all the variables which influence transport and sedimentation. As we have seen, the most important factors are:

1. Climate, particularly precipitation and seasonal distribution of precipitation
    a)   in catchment areas
    b)   along river valleys — vegetation stabilizes river banks.
2. The drainage area
    a)   Size
    b)   Type of rock being supplied
    c)   Topography — tectonic uplift.
3. Subsidence of the alluvial plain.

The topography along the fluvial system is a function of erosion, deposition and tectonic movements. In order for thick fluvial series to be deposited and preserved, the fluvial plain must represent a tectonic subsidence area.

   The drainage area and precipitation determine the rate of flow in the fluvial system, and we have seen that variation in the rate of flow (e.g. floods) helps to determine the shape of the river channel and the amount of overbank sediment deposited outside the channel.

   Drainage areas with a lot of shale and other fine-grained sedimentary rocks will produce sediments with a high clay and silt content. The products from weathering of eruptive rocks will contain a considerable amount of clay, mainly kaolinite, illite and smectite, which increases the cohesiveness of the sediments and stabilises the fluvial channels. The climate along the river plain may be very different from that in the drainage area. It is the former which determines how much vegetation there will be in the area. In humid climates the vegetation will help to stabilise the river channel and reduce the velocity of the floodwater outside the channel.

   The flow of rivers on the river plain will also depend on the water table. The river will contribute water to the groundwater if it is lower than the surface of the river, and groundwater will flow into the river if the reverse is the case.

Further reading: Miall 1984; Collinson and Levin 1983.

Delta Sedimentation

Delta sedimentation is a large and complex area of sedimentology. It may be useful to frame a number of questions on which we can focus when describing processes and deposits.

1. How are deltas formed?
2. Which factors determine the quantity and compositions of sediments which are deposited?
3. What determines the distribution of different facies and the geometry of the sedimentary units which are deposited in the delta?

The largest deltas are formed where there are large drainage areas with sufficient precipitation to produce high runoff.

The drainage systems are strongly influenced by mountain chains folding and rifting, and the largest deltas are formed along passive plate margins where extensive drainage systems can unite into large rivers. The greater part of the drainage from North and South America and Africa flows into the Atlantic Ocean. Only minor rivers enter the Pacific Ocean from the American continent, and relatively little of the drainage from Africa enters the Indian Ocean. The major rivers follow old drainage systems, the main features of which have existed since the Mesozoic, and which often seem to have been governed by Mesozoic rifting along the Atlantic continental margin.

Other drainage systems, such as that of the Mississippi River, are controlled by mountain chains.

As mentioned in connection with fluvial sediments, the erosion area will determine the composition and grain size of the sediments which are transported by rivers and which are deposited on the delta. The Mississippi, for example, drains large areas of Palaeozoic and younger sediments which contain a lot of shale, and the river therefore contains a load with a high percentage of clay and silt which is transported in suspension. Areas of metamorphic and acid eruptive rocks will give sandier sediments, particularly where there is rapid erosion in comparison to weathering.

Deltas develop outwards into a sedimentary basin, and form a surface just near the water surface called a delta top. The waves break against the *delta front*, beyond which is the *delta slope*.

The formation of a delta can be depicted as a battle between the fluvial development of the delta and the erosion of the delta by marine forces. We therefore distinguish first between river-dominated deltas, which prograde far out into the basin and which consist largely of fluvial sediments, and deltas which are eroded more rapidly by marine forces (tide and waves) and consist largely of marine sediments (Fig. 5.13).

River-dominated deltas are often referred to as constructive deltas, since they tend to prograde more rapidly into the basin without much marine reworking. Tide- and wave-dominated deltas are often called destructive deltas.

We may therefore distinguish between three main types of deltas: (1) river-dominated (2) tide-dominated and (3) wave-dominated. Most deltas fall somewhere between these three extremes. The Mississippi, however, comes near the river-dominated end in a marine environment. The Rhine is fairly wave-dominated, and the Ganges typically tidal.

Table 5.2. Factors which favour river-dominated and basin-dominated deltas

Favours river-dominated deltas (constructive deltas)	Favours wave- and tide-dominated (basin-dominated deltas)
Copious supply of fluvial sediment	Large tidal difference
High clay and silt content in the fluvial sediments	High wave energy
Vegetation on the delta	Slow subsidence of basin
Little difference in density between river water and water in the basin	Strong coastal currents outside the basin
Broad shelf outside the delta which protects against wave energy and strong coastal currents.	

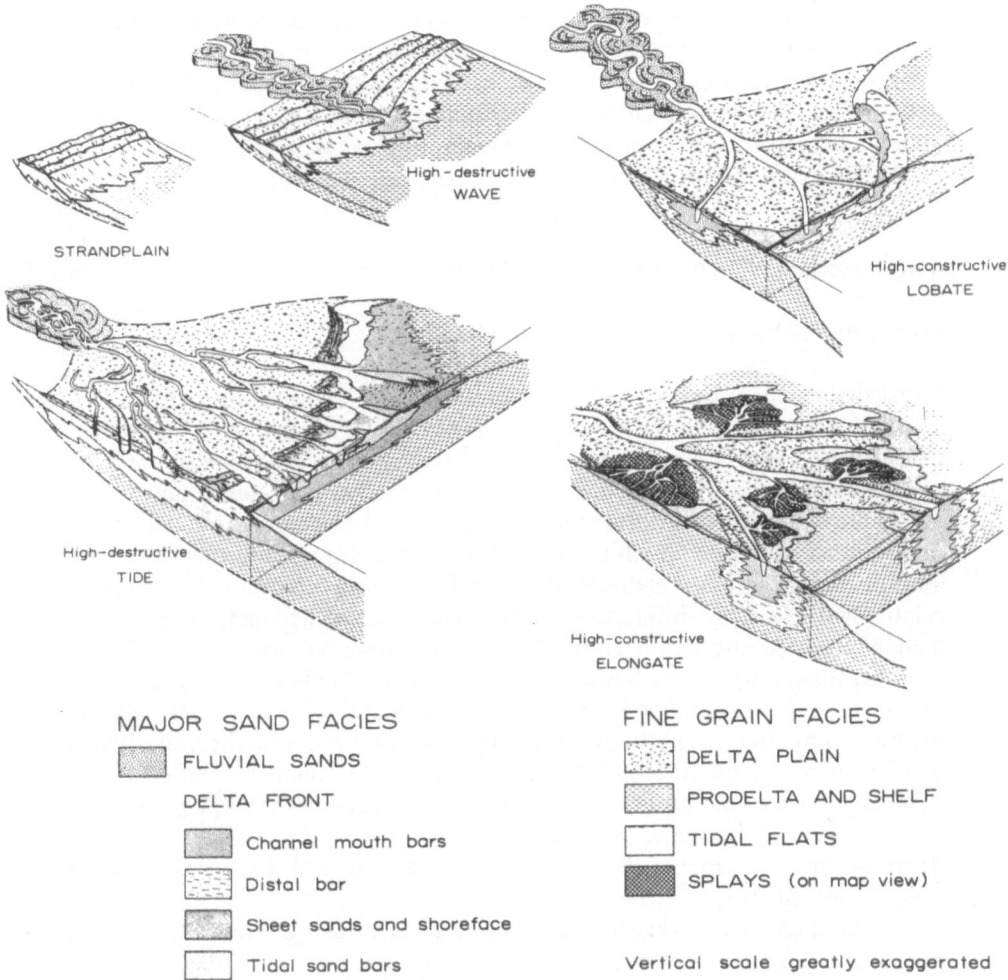

STRANDPLAIN

High-destructive
WAVE

High-constructive
LOBATE

High-destructive
TIDE

High-constructive
ELONGATE

MAJOR SAND FACIES

▦ FLUVIAL SANDS

DELTA FRONT

▦ Channel mouth bars

▦ Distal bar

▦ Sheet sands and shoreface

☐ Tidal sand bars

FINE GRAIN FACIES

▦ DELTA PLAIN

▦ PRODELTA AND SHELF

☐ TIDAL FLATS

▦ SPLAYS (on map view)

Vertical scale greatly exaggerated

Fig. 5.13. Classification of delta types based on relative amount of dominance by river processes (constructive) and marine processes (destructive). (Fisher et al. 1974)

This is a very useful division, but it is important to remember that there are other factors than merely sediment supply, wave and tidal energy which play a part (Fig. 5.13).

The difference in density between the river water and the water in the marine basin plays a major role. In most cases river water will be lighter than salt water, even if it contains a good deal of suspended material. In deltas river water will therefore flow far out over the salt water (*hypopycnal* flow) before salt and fresh water mingle. If the river water and the water in the basin have the same density (*homopycnal* flow) there will be more rapid mixing of the water masses in *axial* flow, and the sediments will settle out of suspension more rapidly. *Hyperpycnal* flow, where the river water is denser than the water in the basin, takes place only in lakes as a rule, and leads to flow of river water along the bottom of the delta slope with erosion of the delta front and formation of turbidites on the basin floor.

In studies of modern deltas we must also take into account the fact that most deltas are rather out of balance as regards progradation in relation to sea level. This is because the Holocene transgression raised the sea level about 120–130 m, so that deltas retreated in relation to periods of Quaternary glaciation. Beyond the present deltas, where the rate of deposition is not too rapid, we find delta deposits which are 12–25,000 years old, which corresponds to the last advance of the last glaciation.

River-Dominated Deltas (Mississippi Type)

Why is the Mississippi a river-dominated delta?

The Mississippi drains a huge area of the North American continent (about 3.2×10^6 km^2) with an average precipitation of 685 mm/year. It carries vast quantities of sediment (about 5×10^8 tonnes/year) with a high clay and silt content, and the gradient of the lower part of the river is very low (about 5 cm/km).

In the 200–300 years during which bathymetric measurements have been taken in the area, it has been possible to record considerable progradation. The high clay and silt content gives the sediments great cohesion and the channel is consequently relatively stable. Clay and silt sediments which are deposited on the delta plain have a high porosity and water content (60–70%), however, and they will undergo considerable compaction. Sand will be deposited chiefly in the channels and as mouth bars, and since well-sorted sandy sediments will have only about 40% porosity immediately after deposition, they will then sink into the underlying clay. This leads to the channel not building itself up so much in relation to its surroundings as it would otherwise have done. Consequently it will be longer before the channel has to change its course. This means that the channel sand which is deposited may be thicker than the depth of the channel, because of contemporaneous subsidence.

The channel has a very low sinuosity despite its low gradient. In the event of floods large quantities of silt and clay are deposited in overbank areas between channels, and these sediments are stabilised by vegetation, despite the relatively rapid tectonic subsidence and sediment compaction. In the Mississippi delta, sedimentation is very rapid so that the channel and the levées build up above their surroundings. This means that the average gradient of the channel decreases. Sooner or later the channel will have to find a new route (through avulsion) to the ocean which is shorter and which therefore has a somewhat greater slope. We can see that the Mississippi has constantly shifted course, and has consequently been a focus of sedimentary activity in historic as well as present times (Fig. 5.14a). The present course has extended the modern delta far out, and should have been abandoned for

Fig. 5.14. a Sedimentation in the modern Mississippi delta. When the river water breaks through the levées, crevasse channels and splays are formed, which help to fill the areas between the channels. (Coleman and Prior 1980). **b** Delta lobes which show how the sedimentation has changed during the past 7000 years. Each lobe of the delta appears to be active for 1000–1500 years (Coleman and Prior 1980)

Fig. 5.14a,b

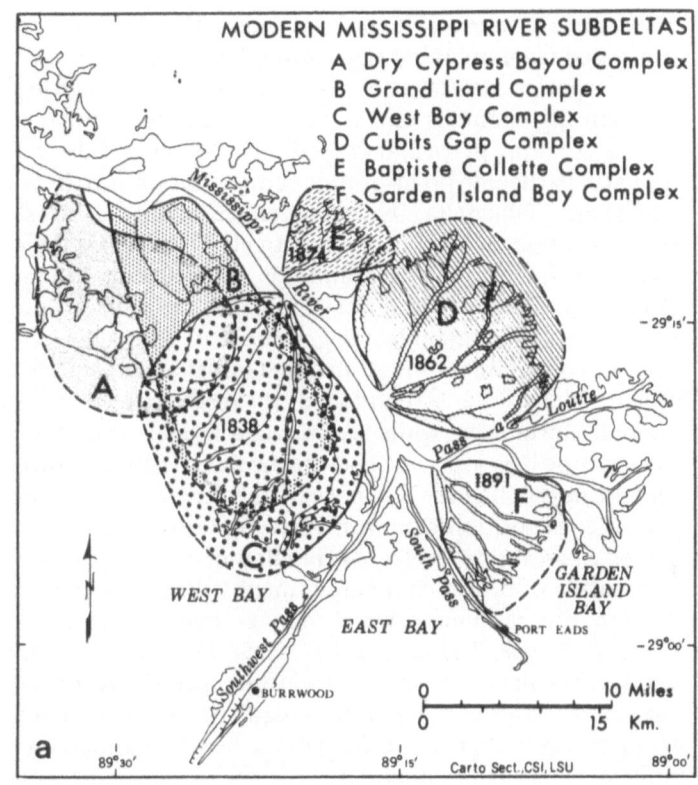

MODERN MISSISSIPPI RIVER SUBDELTAS
A Dry Cypress Bayou Complex
B Grand Liard Complex
C West Bay Complex
D Cubits Gap Complex
E Baptiste Collette Complex
F Garden Island Bay Complex

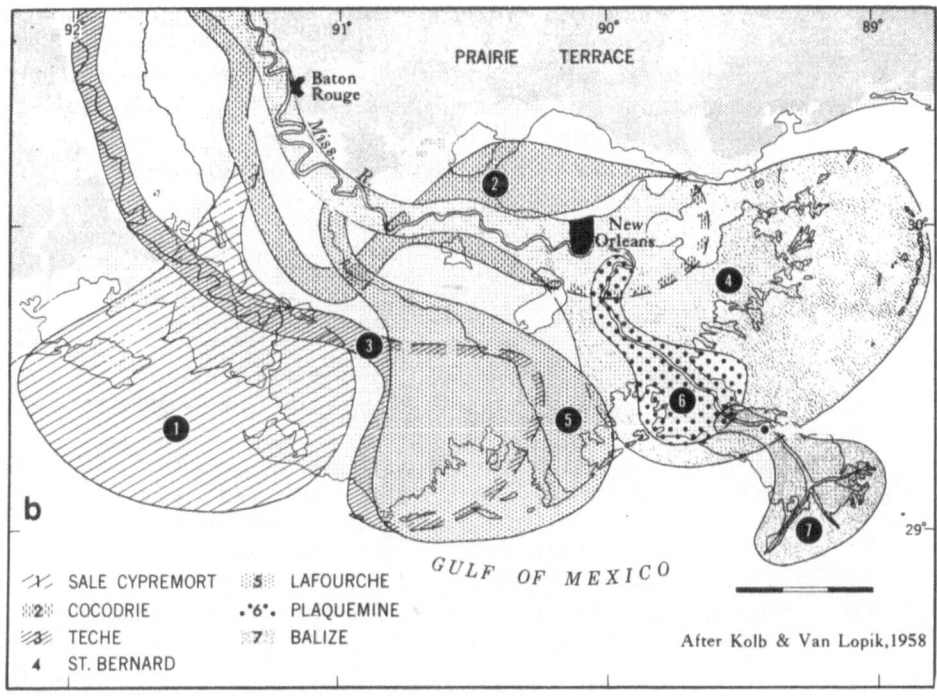

SALE CYPREMORT   5 LAFOURCHE
2 COCODRIE   6 PLAQUEMINE
3 TECHE   7 BALIZE
4 ST. BERNARD

After Kolb & Van Lopik, 1958

a shorter course towards the south west. However, the river has been artificially prevented from breaking right through in this direction because of its importance for transport to and from towns like New Orleans, which lies on the present channel. The shifting channels on the delta account for a great deal of the sedimentary processes taking place and the sediments deposited there. It is important to realise that at a particular point in time sedimentation is only taking place on relatively small parts of the delta. Over a longer period the channel shifts distribute sediment over the whole delta (Fig. 5.14b). In clay-rich deltas the channel and mouth-bar sand will sink into the clay, and this reduces the frequency of channel switching (Fig. 5.15). The long strings of sand which may then be preserved in the mud-rich environment are called *bar-finger sand*.

We call sediments deposited in intervals between channel switches *delta lobes*. When a delta lobe is abandoned, it subsides slowly because of compaction and tectonic subsidence, while sedimentation takes place elsewhere. The abandoned lobe may be below sea level before the fluvial supply returns to this part of the delta and the next delta lobe is deposited in the same area. At the abandoned surface we often get deposition of thin marine deposits – carbonate or shale – during this period of local transgression (abandonment facies). In vertical profile we observe alternation of fluvial channel sediments: levée deposits, crevasse splays and possibly marine sediments, which may be marine clay deposited between channels (*interdistributary bay facies*) (Fig. 5.15). Delta lobe shifting might be expected to produce regular cyclic vertical facies changes. However, it has proved to be difficult to find any statistically significant regularity. Because so many factors are involved, it appears as though the shifting of the fluvial channels on the delta plain results in quite irregular facies fluctuations.

Even small variations in sea level have a strong influence on delta sedimentation. In a section where a fluvial facies gives way to a marine bed, it is often difficult to know whether this represents a transgression due to elevation of the sea level or whether this part of the delta was abandoned by the fluvial system and is subsiding. Only if we can correlate a transgressive bed over the whole area can we assume that it is due to changes in sea level. Transgressive carbonate or thin sandstone beds are most useful for regional correlations.

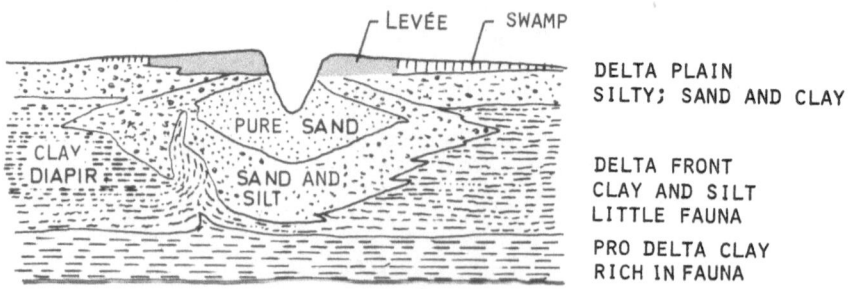

Fig. 5.15. Fluvial channel sand in clay-rich delta. The channel sand sinks into the clay (interdistributary mud) due to its greater density, and clay diapirs may form

Sedimentation at the Mouth of Channels

At the channel mouth the fluvial water rapidly loses its energy, and all the sand which has been transported along the bottom (the *bed load*) is deposited in the form of a *channel mouth bar* (Fig. 5.16). Most of the suspended material is deposited fairly rapidly, at a relatively short distance from the mouth. While the river water, which is lightest, flows out of the channel, salt and brackish water flow back in along the bottom, and may reach a long way up the channel when the rate of fluvial flow is low. Marine fossils can therefore be transported quite a way into the fluvial environment. During floods the salt-water wedge is forced back over the bank at the mouth. The wedge reduces the cross-section of the river, and the velocity increases somewhat in consequence (Fig. 5. 17). Spring tides will force the river water up the channel and may cause it to overflow its banks.

The channel mouth bar itself is a very characteristic deposit which grades upwards from delta mud to well-sorted sand on top. Because of brackish water and the high rate of sedimentation, there are few organisms which live right at the mouth of the channel, but bioturbation may occur in the surrounding sediments.

Delta Front Sedimentation

The delta front is the area where fluvial and marine forces meet, and we can virtually quantify the marine influence on the delta. Tidal forces are a function of the tidal range, and from wave measurements we can also estimate the *wave power*. While wave power near the beach is estimated to be $0.0341 0^7$ erg/s on average for the Mississippi delta, it is 10.10^7 erg/s for the Nile, and 20.6×10^7 erg/s for the Magdalene River (Wright 1978). In rivers like the Mississippi, which has a shallow

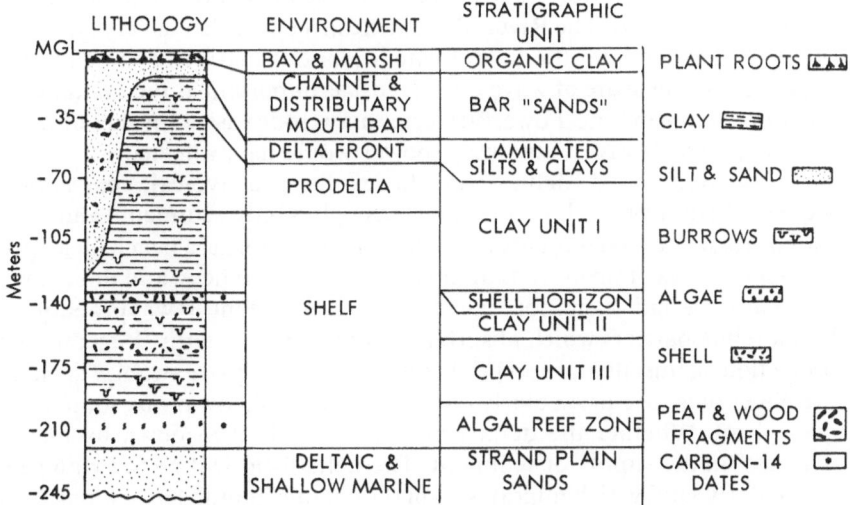

Fig. 5.16. Vertical section through deposits in the modern Mississippi delta. (Coleman and Prior 1980)

WATER FLOW OUT THROUGH A CHANNEL

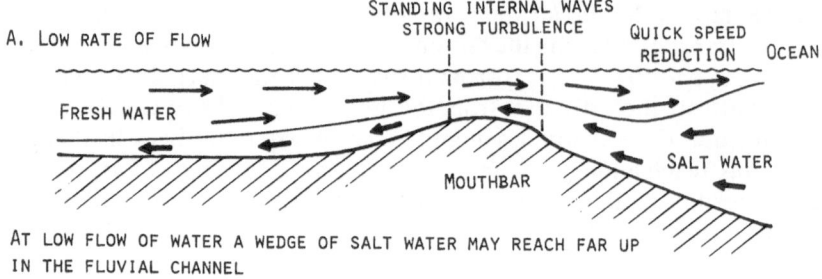

AT LOW FLOW OF WATER A WEDGE OF SALT WATER MAY REACH FAR UP
IN THE FLUVIAL CHANNEL

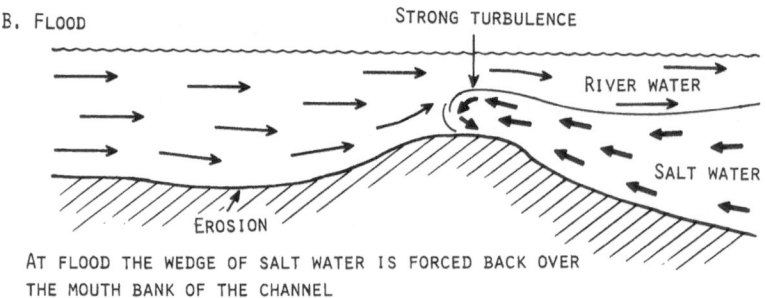

AT FLOOD THE WEDGE OF SALT WATER IS FORCED BACK OVER
THE MOUTH BANK OF THE CHANNEL

Fig. 5.17. Water flow through a fluvial channel in a delta. (After Wright and Coleman 1974)

profile beyond the delta front, most of the wave power dissipates before it reaches the delta front. Deltas which develop into deep water right by the continental slope are subject to the greatest wave power, because waves are not damped before they reach the delta front. The River Magdalene in Venezuela is a very typical example.

The thickness and the lateral extent along the strike of the delta-front sand deposited is a measure of wave power. Just as important as the *overall* wave power is its *distribution* in time. Powerful storms may erode vast quantities of sediment, far more than evenly distributed wave power would have affected.

Wave-dominated delta front facies have a typical coarsening-upward sequence: from pro-delta clay to increasingly sandy sediments and finally well-sorted crossbedded sediments with a low angle of dip, i.e. a beach facies profile. We often find traces of plant roots at the top of such sequences, and this shows that the sand bank has had vegetation right down to the shoreline. It was probably protected by shoreline barriers which absorb most of the wave energy. Vegetation can offer protection against fluvial and tidal erosion, and mangrove swamps such as those in the Niger delta are particularly effective. There will always be some erosion on a delta front. Whether the delta progrades or is broken down by marine forces depends on the supply of sediment. The delta front will be fed with sediment — particularly sand which migrates from the channel mouth bar — along the beach in the wave zone. We often call this "strike feeding" because sediment transport is parallel with the strike of the shore line, i.e. along a horizontal line. This is in contrast to the transport in channels, which is parallel with the dip of the deposit.

If there is little erosion of the channel mouth bar, the channels will extend far into the sea, and little sediment, least of all sand, will be supplied to the rest of the delta front. As a result we have muddy coastlines, with *interdistributary bays* which consist virtually only of clay and silt, right up to the shoreline.

Stability in a Delta

Sediments which are deposited in a delta have a very high porosity. Clay and silt may contain up to 60–70% water. Clay minerals and other clay- and silt-sized grains have a very unstable structure immediately after deposition. Clay layers will gradually lose some water and undergo compaction due to the weight of overlying sediments. Clay and silt have a low permeability, however, and will expel water very slowly. If sedimentation is rapid the sediment load will increase faster than the water can flow out, and overpressure will develop in the pore water. This means that more of the overburden is carried by the porewater and this reduces the effective stresses between the sediment grains, and as a result also the compaction. The friction between grains, which is a function of the effective stresses, is greatly reduced, and in consequence so is the shear strength of the sediments. If the pore pressure attains the pressure exerted by the overlying sediments, the effective intergranular stresses will be equal to zero. This means there is no friction between the grains, and the sediments can flow like a liquid (*liquefaction*). Overpressure may develop in sediment beds near the surface due to rapid deposition of sand over silt and clay during progradation. Since sand has a greater density than mud immediately after deposition, this will result in increasing density up the profile, and hence increased stresses. The resulting instability may cause the sand bed to sink down into the mud, or diapirs of mud to be squeezed up into the sand bed. During floods there is rapid deposition of sediments, which may cause overpressure to develop. Loading due to the sediment deposited on a delta during a flood may therefore lead to synsedimentary deformation. The Mississippi delta is characterised by rapid sedimentation, and we find overpressurized clays developing into mud diapirs. Localised liquefaction often causes collapse depressions which may be circular and rimmed by listric fault scarps. Liquefied mud may flow down the delta front from shallow depths (about 10 m) down to about 100 m. Clay diapirs rise like salt diapirs because they are less dense than the more compact clay, silt or sand, which have lower water contents.

The stability of sediments also depends on the chemical composition of the porewater. In fresh water, clay mineral particles with negatively charged surfaces will repel one another. In salt water these surface charges will be neutralised by cations (Na^+, K^+ etc.) so that clay minerals flocculate. Clay sediments containing fresh porewater are therefore more unstable, and artificial addition of salt may stabilise clay locally.

Stability of Delta Fronts and Slopes

Deltas which prograde out into deep water will develop a slope which may vary greatly. The force of gravity acts on the sediments in the delta front and on the slope so that shear stresses develop in the sediments. When these stresses exceed the shear strength of the sediments, the sediments will be deformed by some sort of gravity-governed process. This may take place through sliding and slumping, which in turn may generate turbidity currents, or steep fault planes may develop, i.e. *growth faults* (Fig. 5.18). The growth fault plane gradually deflects and flattens out with depth.

The growth fault may form near the delta front and lead to this part of the delta subsiding rapidly resulting in very thick deposits with delta front facies stacked on top of one another (Fig. 5.19a,b).

Tide-Dominated Deltas

A high tidal range affects delta sedimentation very strongly and in a number of ways. When the sea level varies by 6–8 m, it leads to a great difference in the gradient of the lower part of the river. Since the lower part of major rivers has a very low gradient, tidal range often affects the flow of the river as much as 50–100 km up-river. During high tide a great deal of sand (bedload) and biological material will be transported up-river. The difference between ebb and flow drops off gradually up the river, and the periods between rising water (when the tide comes in) will be shorter and shorter, right down to 2–3 h, while the low tides grow correspondingly longer. This leads to the rate of flow along the bottom at high tide being much greater than at low tide. Consequently it is often only the flow up the river course during high tide which gives rise to shear forces against the bottom strong enough to transport a bedload. Sand deposited in such rivers has structures (*bedforms*) which indicate transport up the river. It is very important to remember this when studying older rocks.

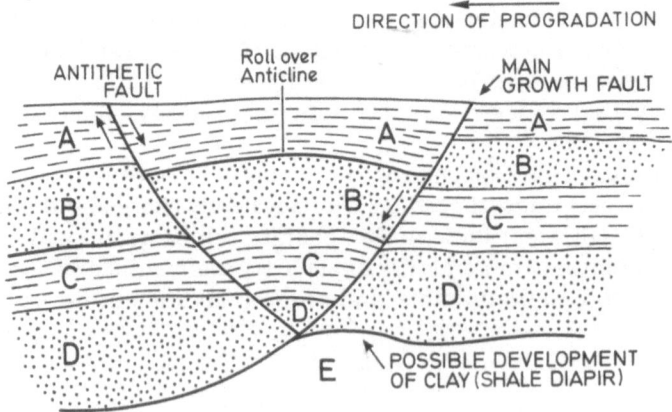

Fig. 5.18. Main features of growth faults

Rivers in areas with high tidal ranges broaden markedly as they approach the sea because of the ever greater volumes of water to be transported out and in. At the outermost point a broad estuary forms, with clay and sand banks cut by channels which transport tidal water.

The Thames and the Rhine are examples of modern rivers with well-developed estuaries.

The mixing of fresh and marine water also causes flocculation of mud and deposition in the more protected parts of the estuary between sand bars. Estuarine deposits are therefore characterised by a mixture of sandstone and mudstone, often with marine fossils which may be transported several kilometres upstream by the salt wedges along the bottom. Tidal flats are often found in the inner parts of the estuary. In the outer part tidal currents will produce elongated bars which are oriented perpendicular to the coastline. The geometry and orientation of the sandbars is very different from that of wave-dominated deltas and coastlines, where the sand bars are parallel to the coast.

Tide-dominated deltas are characterised by the fact that the erosion due to tidal currents is very strong compared to erosion by waves. The fluvial channels have cross-sections which are too small to allow transport of water out and in, and the result is the development of a broad belt of tidal channels, which furthest out are separated only by thin ridges of sand and clay. Very often special channels develop for low tide and high tide respectively, and in each channel one transport direction will dominate. Signs of transportation in two opposite directions have often been used as a criterion for tidal deposits. This will apply to the whole area, but we cannot always expect to find cross-bedding with opposite current directions in the same channel deposit. Cross-bedding with opposite directions may easily develop in a wave and current-dominated environment, however.

Wave-Dominated Deltas

Wave-dominated deltas have a vertical sequence very like that of a beach. Little of the fluvial part of the delta is preserved, and instead of distinct deposits at each channel mouth (distributory mouth bars), the waves distribute the sediments in a continuous *beach ridge* along the delta. The result is a coarsening-upward profile which is not always easy to distinguish from a beach deposit outside a delta. In a wave-dominated delta, however, beach ridges form right in front of the delta top deposits, which often contain coal beds from vegetation, whereas beach profiles formed by barrier islands have a lagoon behind them. The first unambiguous proof that one is dealing with a delta, however, is if the beach ridges are found to be cut by fluvial channels. In addition to the Rhône and the Niger deltas, the Nile delta is also a good example of a wave-dominated delta (Fig. 5.20). Here sediment supply to the delta has been greatly reduced through the building of the Aswan Dam, which traps sediments, and the balance between sediment supply and wave erosion has been disturbed so that the delta is now retreating. This is a modern example of how progradation or destruction (transgression) of a delta depends on a very sensitive balance between sediment supply and erosion.

Further reading: Coleman and Prior 1980.

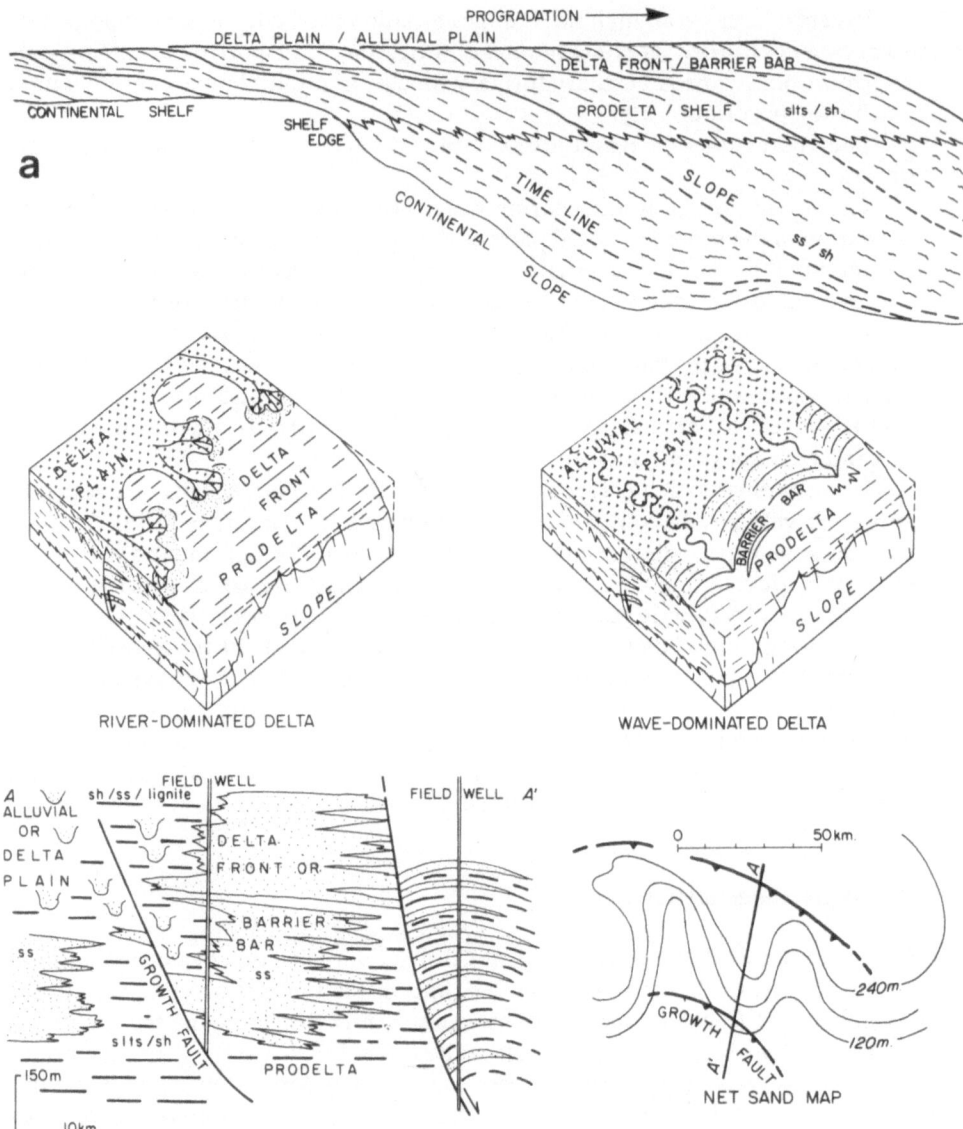

Fig. 5.19. a Section through a prograding delta system. Significance of growth faults for traps in delta systems. (Brown and Fisher 1977)

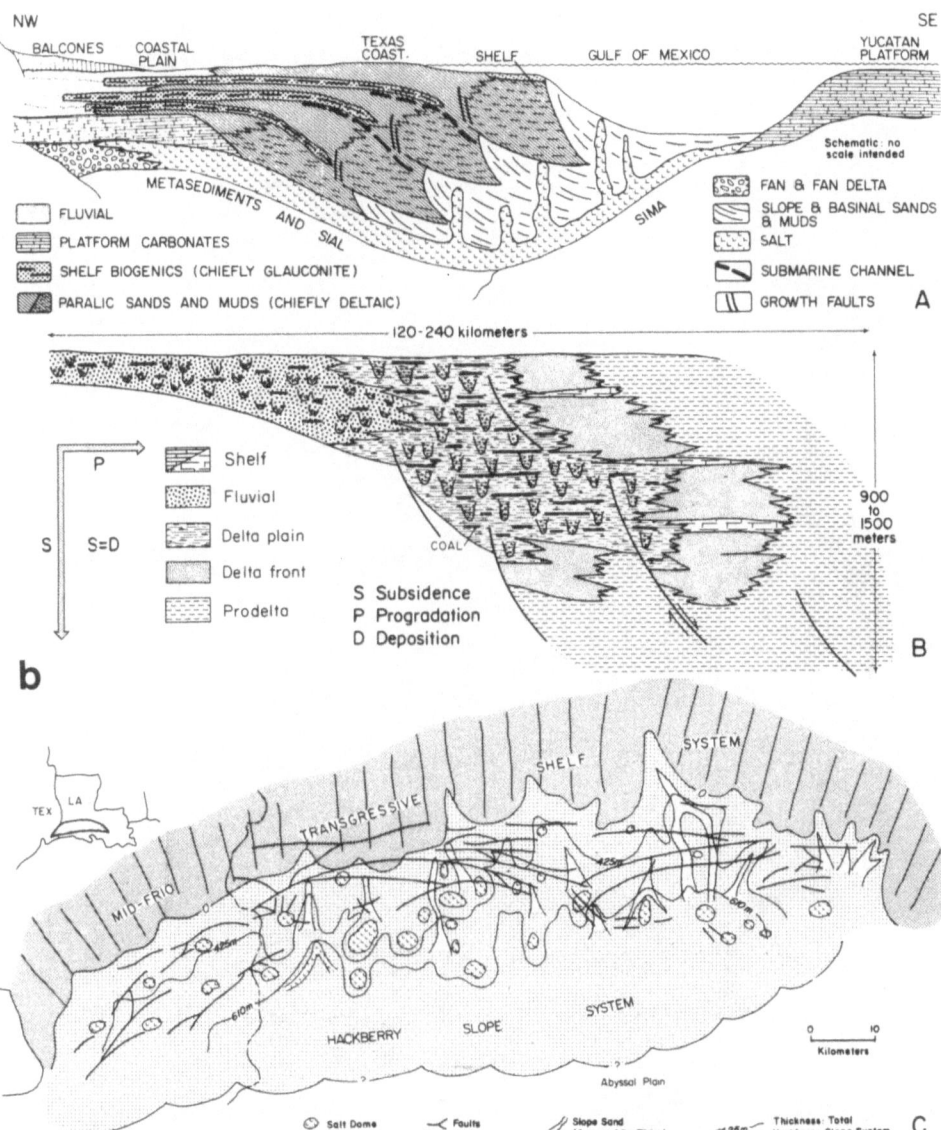

Fig. 5.19. b *A* A general section through the Gulf of Mexico. Mesozoic salt has a decisive influence on sedimentation. *B* A section through the Mississippi delta in Eocene times. *C* Palaeogeographic reconstruction of Oligocene shelf deposits. (Brown and Fisher 1977)

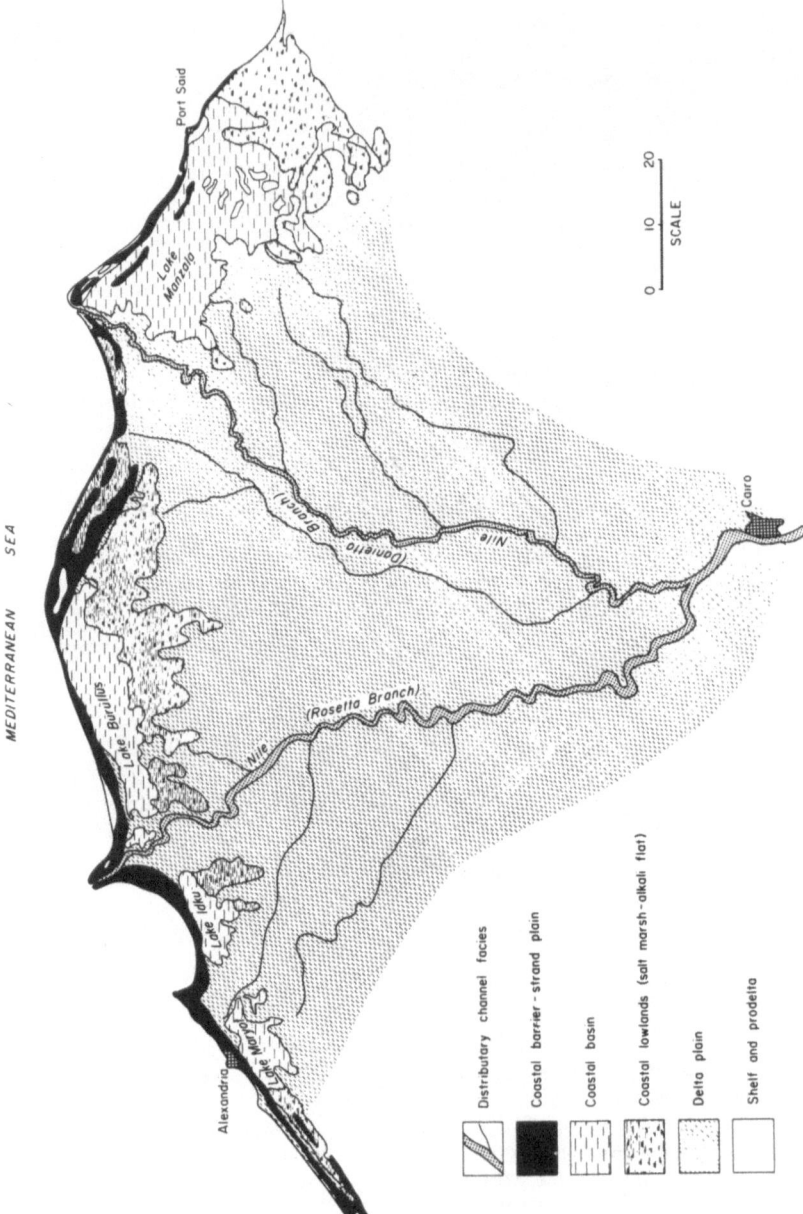

Fig. 5.20. The Nile delta system is an example of a wave-dominated, destructive delta system. (Fisher et al. 1974)

Coastal Sedimentation Outside Deltas

Most sediment transport to the sea takes place via rivers which drain into deltas. In fluvially dominated deltas, the greater part of the sediment is deposited there.

On destructive deltas, a great deal of sediment is eroded and transported out into the sea or along the coast. Coastal areas between deltas are very largely supplied with sediment which has been eroded from deltas and transported along the coast by waves and coastal currents. Since the surface of the coast slopes outwards, the direction parallel with the shoreline can be defined as the *strike* of the deposit, and the direction normal to the coast is called the *dip*.

Sediment supply parallel with the coast is often called "strike feeding". If the supply of sediment along the coast is greater than the erosion rate, the coast will build out into the sea and we will get regression through coastal progradation. If sediment supply is less than the rate of erosion, the coast will retreat, and we will get transgression.

Because there is a finite amount of sediment being transported along any stretch of the coast, engineering constructions which are sometimes built to accumulate sediment and prevent erosion at a particular spot (groynes) will as a rule lead to increased erosion along other parts of the coast. This has great practical significance in areas with high coastal erosion. Beach sedimentation must therefore be seen, not in isolation, but in the context of the overall sediment budget along the coast.

The Mississippi delta is the greatest source of sediment to the Gulf of Mexico, although there are a number of minor rivers in Texas. Here sediment transport takes place along the coast from east to west. Because the active part of the Mississippi delta lies as far east as it does today, less sediment arrives at the Texas coast. The fact that this coastline is subsiding at the same time leads to transgression of the coastline. This example shows that if the supply of sediment due to strike feeding cannot match subsidence, the result will be a transgression.

The Shore Zone

The shore zone is where land and sea meet, and we can distinguish between different types.

1. *Rocky Beach*. The beach is covered with pebbles and blocks or solid rock representing ancient, resistant bedrock. Much of the coast of Norway is of this type, but only 2% of the coastline of North America. Rocky beaches form in areas which have been uplifted tectonically, and where the coast erodes metamorphic or eruptive rocks.
2. *Sand or Gravel Beach*. This is the most common form of beach zone where we have active sedimentation or erosion of older sediments (makes up 33% of the coastline of North America.
3. *Barrier Island*. This is a beach which is separated from the rest of the continent by a lagoon. Common in North America (22%), but less common in other areas.

4. *Muddy Coastlines*. Beach zones which consist basically of clay are formed where we have sediments with a very low sand content, and where there is little wave activity to wash out or enrich any sand the sediments might contain. We find this in parts of clay-rich deltas and estuaries and where there is abundant vegetation along the beach zone which protects the clay sediments against erosion (e.g. mangrove forest).
5. *Cheniers* are isolated sandy beach ridges on coastal mud flats. They are abundant along the coastline around major deltas like the Mississippi and the Amazon. They require low tidal ranges, moderate wave energy and abundant mud, and a limited amount of sand.

The composition of beach sediments varies greatly, depending on the materials available and the grade of mechanical and chemical breakdown of the minerals. Quartz sand is the most widespread because quartz is the most stable of the minerals we find in most beach sediments. But on volcanic islands, which consist largely of basalt, there is no quartz, and we get sand derived from basalt or volcanic glass. Carbonate sand is formed locally from the broken skeletons of carbonate-secreting organisms. This does not only happen in tropical regions, e.g. the Bahamas, but also along coasts with cold climates. In many parts of Norway, for example, the beach sand consists very largely of carbonate sand from molluscs (barnacles, bryozoans and calcareous red algae).

Sand profiles are formed by wave power acting on the coast and depend on a number of different factors:

1. The composition of the available sediments.
2. Wave power
   a) Average wave height
   b) Size and frequency of storms
   c) Angle between the commonest orientation of the waves and the beach.
3. Tidal range.
4. Vertical profile off the beach.
5. Supply of sediment from land or along the coast.

The beach zone and the near-shore areas have a morphology which is the result of interaction of various features related to wave activity. As waves approach land, they will be affected by friction against the bottom. The depth at which this starts depends on wave height and length. When the depth becomes less than about half the wave length, friction against the bottom will be great, the circular wave motion (orbit) will be distorted and oscillatory sediment transport will affect the sea bed, producing ripple marks. The depth at which this occurs is called the *wave base*. Sand dunes are formed of sediments which are deposited when the waves break, and are at the same time the reason for the waves breaking precisely there. There is thus an interaction between the wave regime and the bottom geometry in beach sediments. In addition to breaking on the *foreshore* itself, we often find that waves break at two or three places offshore, and at each of these places we find a sand bar.

Prograding Beach and Barrier Sequences

The progradation of a beach will produce a characteristic coarsening-up sequence
(Fig. 5.21) which will obey Walther's Law of facies succession. The vertical
sequence will represent the environments from the shelf to the shoreline. The
transition between shelf mud and fine-grained sandstones with ripples may
represent the wave base. Isolated sand layers in mud may have been deposited near
the storm wave base, whereas the transition to continuous fine-grained sand may
represent the "fair-weather wave base". The thickness of the sequence from the

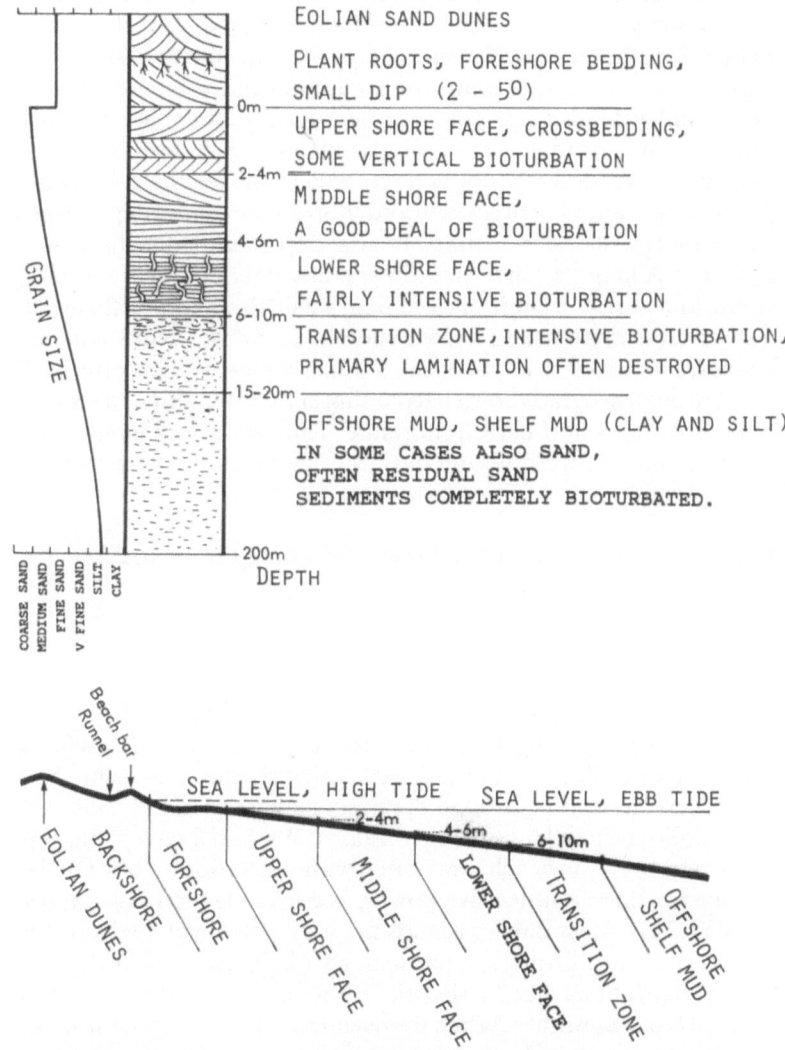

Fig. 5.21. Diagrammatic section through a sand deposit with some typical sedimentary structures. Just
under normal fair-weather wave base, 5–20 m, "hummocky" stratification occurs

wave base to the foreshore is an expression of the wave energy at the coastline when the sequence was deposited. Bioturbation occurs in the lower part of this sequence. Below the wave base it usually takes the form of horizontal feeding traces, and in the lower and middle shore faces, where there is relatively high wave energy as vertical traces (Scolithos facies). As erosion and reworking intensify, the preservation potential of bioturbation is reduced, and it becomes less frequent. The formation of sand bars and erosion surfaces on the upper shore face results in cross-bedding, usually trough cross-bedding, representing flow in the upper part of the lower flow regime.

In the breaker zone there is an upper flow regime, producing a planar facies which in vertical section will appear as very low-angle cross-bedding. On the beach we often have a beach bar which is flooded only during storms, and a depression behind it called a runnel, which helps to drain the backshore area. Because the exposed beach is a rich source of sand, the wind will tend to blow sand from the beach and redeposit it as eolian dunes, usually where it is trapped by vegetation. Eolian sediments therefore often cap ancient beach profiles, but the eolian dunes may also be eroded and not be preserved in the geological record.

The development of a beach and shore face does not depend only on the general wave energy of the coastline. It also depends on the bathymetry outside the coastline. A long stretch of shallow water will reduce the wave energy reaching the shore, and barrier islands or sand banks will also dampen the waves considerably. The result is often called barred shorelines. Storms will frequently set up strong rip-currents which may produce channels which dissect some of the bars.

Through progradation a barred shoreline will produce a more complex vertical sequence than a non-barred shoreline. The former will include important erosion surfaces and the foreshore and the upper shore face may repeat themselves because of the development of these bars.

Further reading: Davis 1978; Davis and Ethington 1976; Fischer and Dolan 1977.

Barrier Islands

Barrier islands are beach deposits which are separated from the mainland by a lagoon. They form long, thin sand ridges which are often only a few hundred metres to a couple of kilometres broad, and which rise up to 5-10 m above sea level.

A vertical section through the part of a barrier island facing the sea resembles an ordinary beach deposit (Fig. 5.22a,b). We find a coarsening-upward sequence from marine clay to beach sand, often with vegetation on top. On the lee side, facing the lagoon, there is little wave power, and in the lagoon clay, mud and often oyster reefs are deposited. Barrier islands are very well developed along long stretches of the coast of North America, particularly off Texas and North Carolina. There are several indications that barrier islands are due to transgressive conditions such as those of Holocene times. When the ocean rises in relation to the land, beach deposits can continue to grow through gradual deposition of sand in the beach zone, but the areas behind them sink below sea level and form a protected lagoon with clay sedimentation. More localised transgressions may be caused by compaction and subsidence of sediments along a coastline with a sediment supply which is insufficient to keep pace with subsidence.

One prerequisite for forming such islands is that there is an adequate supply of sand, so that the island can grow in pace with the transgression. This sand cannot be transported across the lagoon, and must be added along the length of the islands parallel to the coast (strike feeding from deltas or eroding coastlines).

During storms or hurricanes the sea level may rise due to wind stress, and waves may break over and through the barrier island. An erosion channel is then formed through the island, and at the rear, out in the lagoon a *washover fan delta* develops. A delta of this type can form in a matter of hours during a hurricane. Barrier islands may extend for tens of kilometres, but they will not form a continuous belt along the coast. Water has to circulate between the lagoons and the ocean through gaps between barrier islands. The gaps are called tidal inlets, and the distance between them will be a function of the tidal range. Both seaward and landward of the inlets small sandy deltas may develop in response to ebb and flood currents respectively (Fig. 5.23). Flow at ebb tide will normally be stronger than that at flood tide. This is due to the profile of the lagoons. At high water in the lagoon the volume of water which must flow out to compensate for a specific lowering of sea level is greater than the volume needed to raise the water level in the lagoon correspondingly at low tide. The tidal inlets are therefore capable of transporting more sediment out during the ebb, and structures indicating this flow direction may predominate (Fig. 5.23). Inlets are characterised by an erosional base and lateral migration of inlets produces a characteristic fining upwards sequence. The strong currents in tidal inlets often generate sand waves which tend to migrate in the ebb direction, but they may also be modified by flood currents.

Ebb-tidal deltas consist of a channel dominated by ebb currents with smaller flood-tide channels on the sides. At the ocean end of the channel sediment is deposited in a sand ridge which is similar to a channel mouth bar in an ordinary delta. This sand ridge, which is called a *terminal lobe*, is subject to wave erosion, and smaller *swash bars* may form, which reach above sea level. In areas with strong wave power ebb-tidal deltas will be less obvious because of erosion and further transport along the barrier ridges. Ebb-tidal deltas will be characterised by deeper submarine channels.

Flood tidal deltas form inside the lagoon and are well protected against wave erosion. Here the water flows into flow channels which branch inwards in a flood-tidal delta, where the sediments are deposited on a tidal flat. Ebb currents move back along the edges of the outer side of this delta and may form small *spillover lobes* when ebb-currents penetrate over the edge of the flood-tidal delta. Flood-tidal deltas are associated with shallower channels than ebb-tidal deltas and are not much eroded by waves.

Tidal channels fill with sand which forms an upward-fining sequence overlain by tidal flat sediments (Fig. 5.25). In areas with carbonate sediments or cohesive clays, erosion due to sideways migration of tidal channels has caused the formation of intraformational breccias (Fig. 5.26).

Barrier island deposits thus consist of a long, thin body of sand. The thickness of the sand layer will correspond to the depth of the wave base plus a few metres which correspond to the height above sea level.

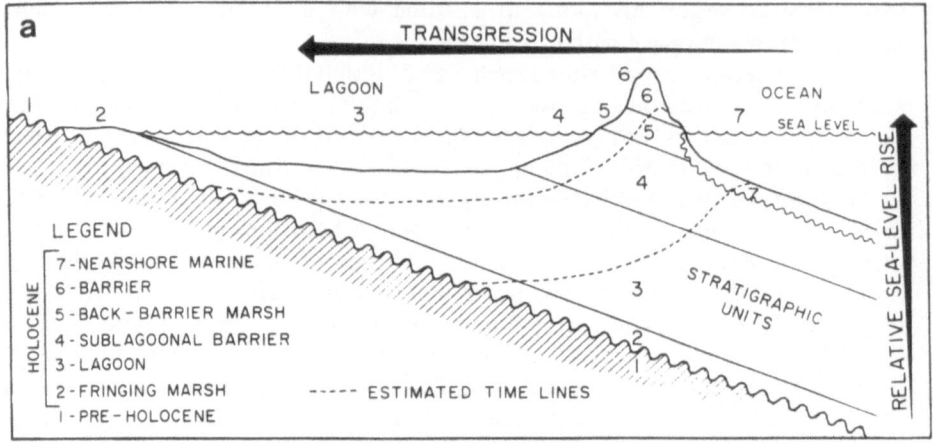

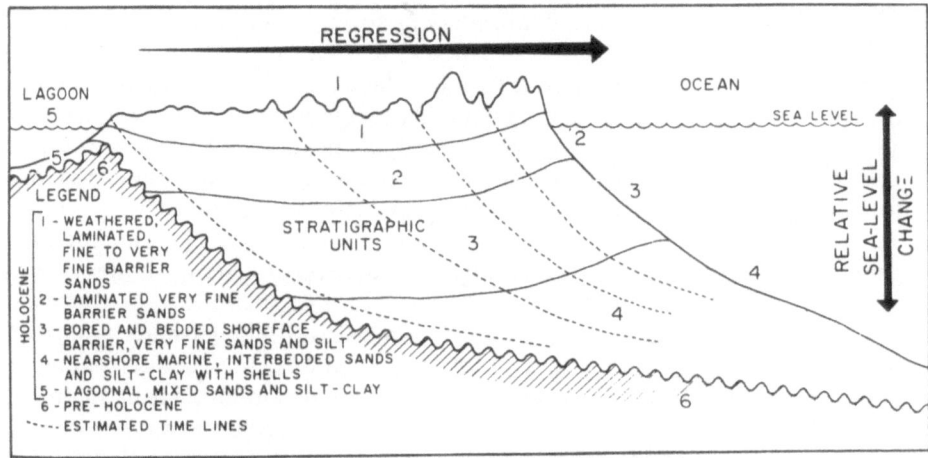

Fig. 5.22. a Transgressive and regressive barrier islands. (Kraf and John 1979).

In areas with a higher tidal range, lateral migration of tidal channels will be rather pronounced, and fining up sequences will also be common.

If the barrier islands are drowned by a transgression, a carpet of clay and silt will be deposited over these sandstone deposits. This represents the ideal stratigraphic trap. Compaction or tectonic tilting will cause the sandstone deposits to interfinger with mud from the lagoon deposits, which is a good source rock. Oil will be able to collect in the top of the barrier ridge sand or in flood-tidal delta deposits (or *washover fans*) which represent *pinch-outs* in the muddy lagoon sediments.

Further reading: Leatherman 1979; Howard and Scott 1983.

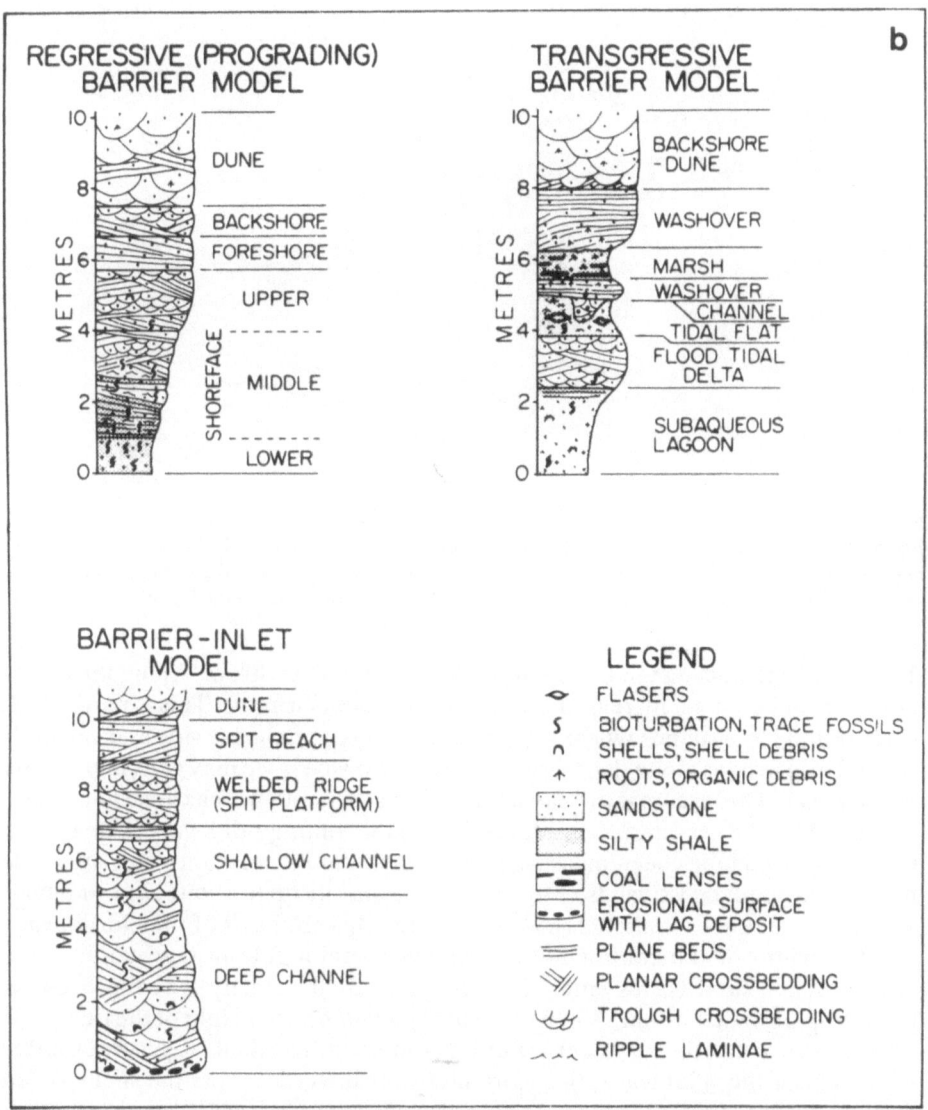

Fig. 5.22. b Sections through different types of barrier island deposits. (Walker 1979)

Tidal Sedimentation

Tidal range is an important factor in coastal sedimentation. We distinguish between

1. Micro-tidal environment (tidal range less than 2 m).
2. Meso-tidal environment (tidal range 2–4 m).
3. Macro-tidal environment (tidal range greater than 4 m).

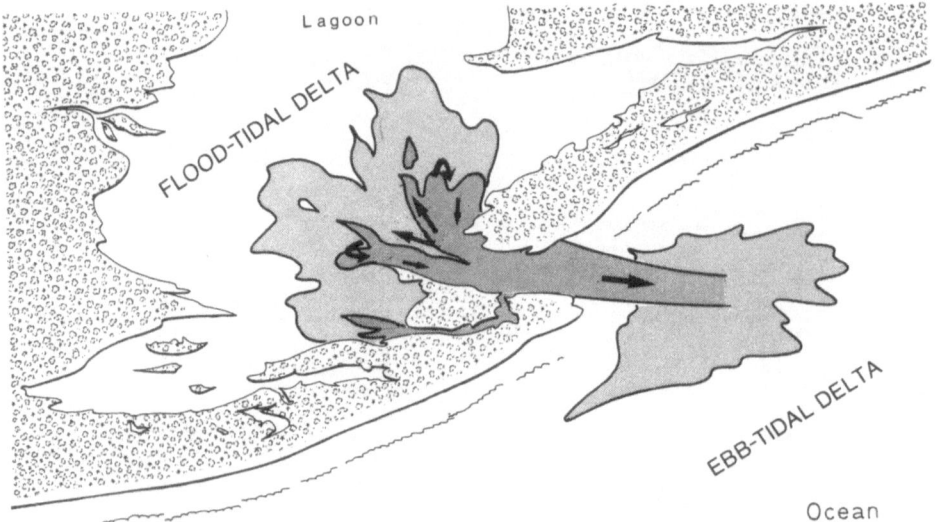

Fig. 5.23. Tidal channels with tidal delta forming between barrier islands. The barrier islands and the channels will migrate laterally and deposit channel facies sediments by lateral accretion. Note that the ebb-tidal delta is much more exposed to waves than the flood-tidal delta in the lagoon

The average tidal range in the open sea is only about 30–40 cm. Along the coasts, however, we often get increased interference by tidal currents. This is particularly true around large islands where tidal waves converge on the lee side and can build up, and also in bays along the coasts. In long, narrow bays we may get a high degree of *resonance*. This occurs if the bay has a length and depth which cause tidal waves which are on the rebound to reinforce the next incoming tidal wave. The highest tidal range which has been measured is 16.3 m in the Bay of Fundy in Canada. In the areas around the British Isles the tidal range may be up to about 12 m, and there are also large ranges in the German Bight and adjacent parts of the North Sea.

The width of the continental shelf plays a major role in determining tidal ranges. When tidal waves enter shallow water, their velocity is reduced due to friction against the bottom. When the velocity declines, the height of the tidal wave will increase so that the total energy flux is maintained. The tidal range, which is thus the height of the tidal wave, therefore increases inwards across the shelf. Where there are embayments along the coast, tidal waves become focussed, so that the tidal range increases towards the middle of the bay. This is true of the east coast of the USA, off Georgia, and of the German Bight in Europe.

If the shelf becomes even wider, as in an epicontinental sea, tidal power will gradually be exhausted in overcoming the high frictional resistance. This is the case on the Siberian continental shelf, and there is much to indicate that the great epicontinental seas, or Ordovician and Cretaceous times for example, were characterised by low tidal ranges. When the water is very shallow, as on the Bahamas Bank, friction damping takes place over much shorter distances.

Tidal waves also move around centres without tidal ranges (amphidromic points) and the tidal range increases radially with the distance from the centre. This is typical of the tidal pattern in the North Sea.

About 1/3 of the world's coasts have tidal ranges greater than 4 m (macro-tides), 1/3 have meso-tides (2–4 m) and 1/3 micro-tides (less than 2 m). Small inland seas, like the Mediterranean and the Black Sea, are too small to keep pace with the attraction of the moon and the sun, and have small tidal ranges. This is also true of lakes.

Further reading: Ginsburg 1975; Nummedal and Fischer 1978.

The most important features for recognising tidal deposits are:

1. Tidal channels
2. Flaser bedding
3. Tidal burdles.

None of these criteria is entirely unambiguous.

1. *Tidal Channels.* May resemble fluvial channels or submarine channels in that they form fining-upwards sequences. Channels formed in estuaries in fact often connect fluvial channel systems. Tidal channels which are not part of a river delta, however, tend to fill with sandy sediments from the surrounding tidal flat, because they have no supply from land. Channels will often erode the riverbank and cause it to collapse, and this may result in the formation of intra-formational conglomerates if the sediments are slightly lithified.

Tidal currents are also greatly influenced by wind stresses and waves, and we may get large-scale sediment transport and deposition during occasional storms and spring tides.

Lateral migration of tidal channels may produce typical epsilon cross-bedding which is the result of lateral accretion of point bars in the tidal channel (Fig. 5.24).

Tidal channels often contain a bed of marine fossils at the base and marine trace fossils. Channels on tidal flats which are not associated with deltas (estuaries) will not receive much clastic material from land. Conglomerates and breccias in these tidal channels will therefore typically be of the intraformational type, derived by local reworking of tidal flat sediment. Because of their early lithification, carbonate beds in particular can be reworked to form intra-formational conglomerates and breccias.

Tidal currents switch direction every six hours and we sometimes find good examples of cross-bedding with opposite current directions. This is not always the case in tidal environments, however. Some tidal channels are dominated by ebb flow and others by flood currents. This is because the ebb and flood often find different dominant pathways. Bipolar cross-bedding is therefore not an essential feature of tidal channels. Storms with opposing wind directions may also produce some from of bipolar cross-bedding.

In a regressive sequence tidal channels will be overlain by lagoonal sediments (Fig. 5.25).

2. *Flaser Bedding.* Consists of clay laminae in a matrix of sandstone with ripple cross-lamination. The clay occurs mainly as infill in ripple troughs, but may also drape over the ripples. Flaser bedding forms as a result of alternating periods of

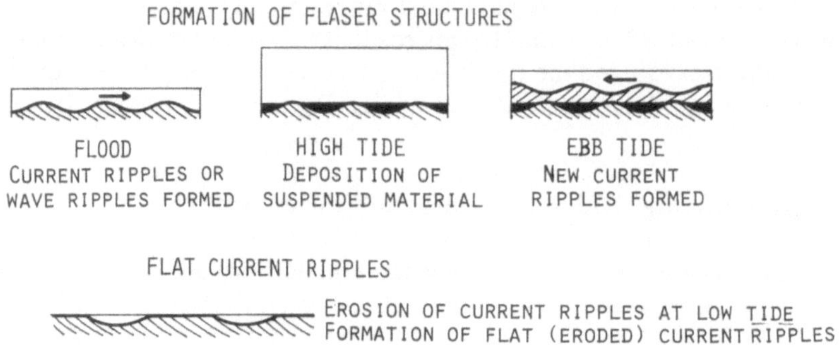

FORMATION OF FLASER STRUCTURES

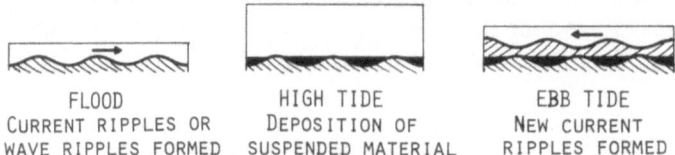

FLOOD	HIGH TIDE	EBB TIDE
CURRENT RIPPLES OR WAVE RIPPLES FORMED	DEPOSITION OF SUSPENDED MATERIAL	NEW CURRENT RIPPLES FORMED

FLAT CURRENT RIPPLES

EROSION OF CURRENT RIPPLES AT LOW TIDE
FORMATION OF FLAT (ERODED) CURRENT RIPPLES

Fig. 5.24. Formation of flaser structures and truncated current ripples which are typical of tidal environments

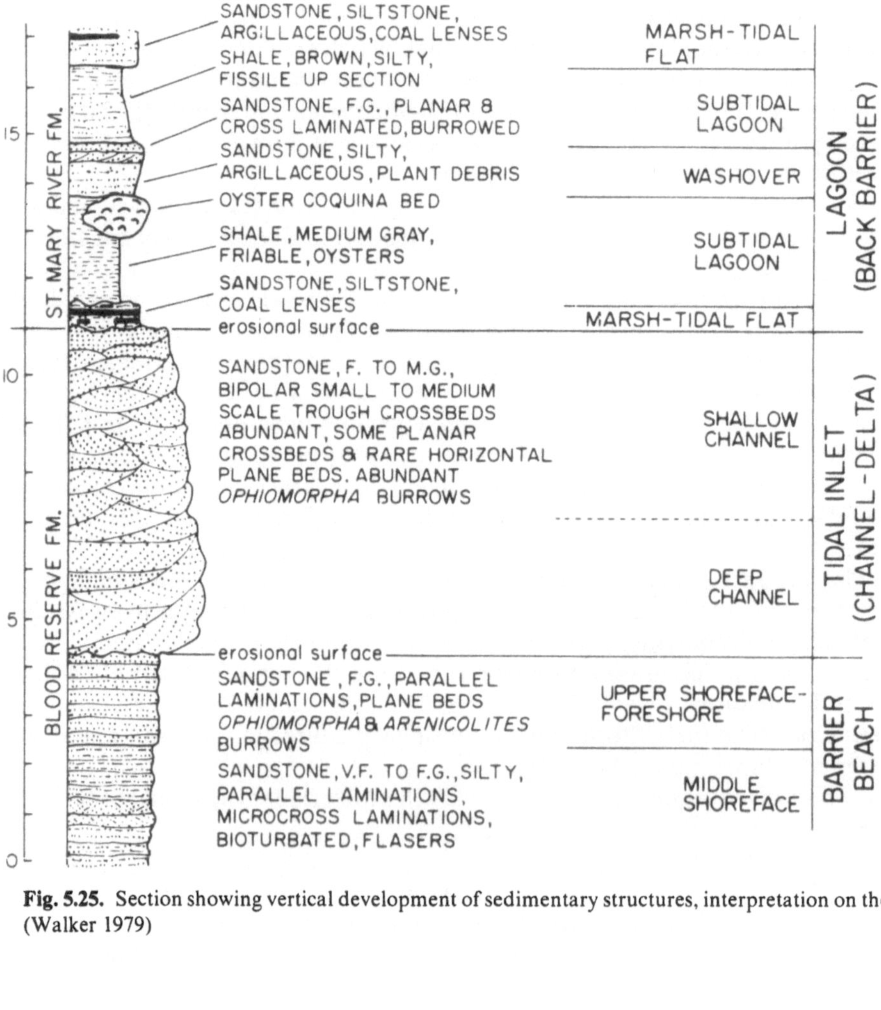

SANDSTONE, SILTSTONE, ARGILLACEOUS, COAL LENSES	MARSH-TIDAL FLAT	
SHALE, BROWN, SILTY, FISSILE UP SECTION		
SANDSTONE, F.G., PLANAR 8 CROSS LAMINATED, BURROWED	SUBTIDAL LAGOON	
SANDSTONE, SILTY, ARGILLACEOUS, PLANT DEBRIS	WASHOVER	LAGOON (BACK BARRIER)
OYSTER COQUINA BED		
SHALE, MEDIUM GRAY, FRIABLE, OYSTERS	SUBTIDAL LAGOON	
SANDSTONE, SILTSTONE, COAL LENSES		
erosional surface	MARSH-TIDAL FLAT	
SANDSTONE, F. TO M.G., BIPOLAR SMALL TO MEDIUM SCALE TROUGH CROSSBEDS ABUNDANT, SOME PLANAR CROSSBEDS & RARE HORIZONTAL PLANE BEDS. ABUNDANT *OPHIOMORPHA* BURROWS	SHALLOW CHANNEL	TIDAL INLET (CHANNEL-DELTA)
	DEEP CHANNEL	
erosional surface		
SANDSTONE, F.G., PARALLEL LAMINATIONS, PLANE BEDS *OPHIOMORPHA & ARENICOLITES* BURROWS	UPPER SHOREFACE-FORESHORE	BARRIER BEACH
SANDSTONE, V.F. TO F.G., SILTY, PARALLEL LAMINATIONS, MICROCROSS LAMINATIONS, BIOTURBATED, FLASERS	MIDDLE SHOREFACE	

ST. MARY RIVER FM.   BLOOD RESERVE FM.

metres   0   5   10   15

Fig. 5.25. Section showing vertical development of sedimentary structures, interpretation on the right. (Walker 1979)

currents or wave activity and slack water. The clay settles out in the slack water periods in a tidal environment, but this type of bedding may also form in other environments where there is rhythmic sedimentation, such as in certain fluvial environments. On tidal flats clay and silt will settle out at high tide and will deposit between ripples formed due to the ebb and flood currents. It may consist partly of clay pellets (faecal pellets from marine organisms) which settle out faster than clay-sized particles. Flaser bedding belongs to a type of structure which we get with mixtures of sand and clay. Lenticular bedding represents isolated laminae or lenses of sand in mud.

The inner part of a tidal shelf consists of very muddy sediments usually with abundant bioturbation. Mollusc shells in the mud (Fig. 5.26) are often eroded and deposited as shell lag.

3. *Tidal Bundles.* The best identifying feature for tidal environments is regular lamination consisting of fine sand and mud, making up tidal couplets representing a tidal cycle. Both modern and ancient tidal sediments have shown a regular variation in the thickness of such couplets, reflecting the spring and neap cycles. Regular laminations reflecting tidal cycles are often called tidal bundles (Fig. 5.27).

Shallow Marine Shelves

The shelf extends from the nearshore environment to the shelf edge, where there is a rather abrupt increase in slope, usually at a depth of 200–500 m. The width of the shelf varies considerably, and may exceed 1000 km. Continental shelves are generally very flat areas which may be cut by deeper channels transporting sediments across the shelf from nearshore or deltaic environments.

The study of sedimentation on modern continental shelves is complicated by the fact that sea level was more than 100 m lower only 10,000 years ago. This means that most shelf areas have not yet reached an equilibrium with respect to the modern environment. Sandy shelf deposits are much more difficult to core than muddy sediments and the commonly used gravity corer has to be replaced by vibro-core equipment. For this reason we have few vertical cores through sandy shelf environments.

Most shelf areas are below the wave base for normal waves (fair-weather wave base) and sedimentation is governed largely by storms and tidal currents.

When the wind is landward, waves will usually approach the beach obliquely, resulting in wave refraction effects, particularly on relatively steep beaches with high wave energy. This will produce rip currents, where the water piling up against the beach may flow seawards. Rip currents are often rich in suspended material and transport material towards the shelf. Another component of wave energy is transmitted parallel with the beach as longshore currents, contributing to the longshore drift of sediment transport. During onshore storms the sea level near the coast may be raised by several metres due to the combined effect of wind stress, tides and the low barometric pressure associated with storms.

Fig. 5.26. a Tidal channel on a tidal flat. Along the channel bank lag deposits of mollusc shells are formed. From Jade Bay, near Wilhelmshaven, Germany. **b** Mollusc in living position in the muddy facies of the inner part of a tidal flat. Erosion by tidal channels in this mud produce a lag of mollusc shells. Locality as in **a**

The increased potential of the coastal water will result in strong bottom currents (storm surges) which are capable of transporting sediment further out onto the shelf. Storm surges may transport fine sand and mud in suspension, but they are not true turbidity currents. In this case the increased potential (elevation) of the coastal water is the driving force and not the density difference between the current and the surrounding water, as with turbidites. Hummocky cross-bedding is a characteristic sedimentary structure produced by the deposition of coarse particles from suspension during storms, and rounded, undulating sand surfaces tend to form. Finer particles will be transported further into areas with lower energy.

Fig. 5.27. Tidal bundles. These are bundles of laminae which reflect the tidal cycle between spring tides. From the Late Precambrian Wonoka Fm, Patsy Spring, Flinders Ranges, Australia

In many modern shelf areas tidal currents are important transport mechanisms, often in combination with storms. Since shelf sedimentation is mostly governed by the rare action of very strong currents, the sediments deposited in this environment are characterised by rather abrupt transitions from well sorted sand to mud and from bioturbated to non-bioturbated strata. However, more continuous, well-sorted shallow marine sandstones showing no evidence of a near-beach environment are often found in ancient sandstones. We do not fully understand how these form, but in some cases they may have been formed through progradation of sheets of sand which have built up to fair-weather wave base, where the rate of erosion has kept pace with the rate of accumulation, preventing build up to sea level.

In shelf areas with relatively strong tidal currents ($>$ 150 cm/s) we may get *furrows* and *gravel waves*. At velocities of less than 1 m/s *sand ribbons* may be deposited - longitudinal bedforms developed parallel to the currents. Sand waves are large-scale transverse bedforms, perhaps 2–15 m high, with a wavelength of 150–500 m. Sand waves require current velocities exceeding 60–70 cm/s. Their internal structure has not been thoroughly studied, but they probably have large-scale, low-angle cross-stratification with smaller-scale internal cross-stratification. Cross-stratification may be symmetrical or asymmetrical on both sides of the sandwave, depending on the relative strength of the opposing currents.

Relatively low-energy environments ($<$ 50 cm/s) are characterised by *sand patches* and mud. The sand forming the patches probably only moves during storms. Lateral structures are typically current ripples.

   Shelf sediments characteristically deposit at relatively lower overall sedimentation rates. Short periods of progradation of sandy bedforms are separated by longer periods of non-deposition or erosion. Sedimentary sequences
through shelf sediments may often be divided into depositional sequences separated by sequence boundaries representing major unconformities. These sequences
may reflect changes in sea level, and if they are thick enough ($> 30\text{--}40$ m) they may
be recognised in seismic sections.

Continental Slopes

Continental slopes are areas between the continental shelf and the *continental rise,*
where the ocean deep begins. The edge of the continental shelf commonly lies at a
depth of 2-300 m. The slope continues down to a depth of 2-4000 m, while the
continental shelf is very flat for the most part. The continental slope is typically 2-6°,
and it is 20-100 km broad. The average gradient of continental slopes all over the
world is 4.20° (Shephard 1967). The gradient is a function of a number of different
factors, but the stability of the shelf edge constitutes a major control factor. The
steepest slopes are therefore to be found off carbonate banks with well-cemented
coral reefs and carbonate beds which therefore have a high shear strength. In areas
with rapid sedimentation, particularly outside deltas, the slope will be least because
loose sediments have little shear strength and submarine slides, slumping and
formation of turbidites occurs on relatively gentle slopes (1-2°). Where sedimentation is slower the sediments have more time to consolidate, and will be more
stable. The steepest submarine slopes (greater than 10°) in clastic sediments are
therefore found in submarine canyons, where erosion cuts into older, well-consolidated sedimentary rocks.
   Along passive continental margins the continental slope is associated with the
transition from continental crust to oceanic crust. In areas with a large supply of
sediment, the shelf may have prograded beyond this boundary.

Organic Sedimentation on the Slope of the Continental Shelf

Analyses of the organic content of the basal sediments in a section from the
continental shelf to the ocean deep show that we normally find most organic
material in continental slope sediments. This is because it is in this area that we have
the greatest upwelling of nutrients from the deep. On slopes off deltas there will also
be high productivity because of the supply of nutrients from river water. Thereafter,
however, organic matter is greatly diluted by rapid clastic sedimentation. We also
find minimum oxygen content in the water column on the slopes, and much of the
organic matter produced is therefore retained in the sediments. On the shelf we
often find that the supply of nutrients is small, but this will depend on water
circulation. The shallow areas, however, commonly have stronger currents and
turbulence, which will increase the oxygen supply in the water. A far smaller
proportion of the organic production on the shelf will therefore be retained in the

sediments. The North Sea, for example, has high biological productivity, but the Holocene sediments now being deposited are not particularly rich in organic matter.

Out in the ocean basin organic production at the surface is less, due to a limited supply of nutrients, and basal currents with cold, oxygenated water from the polar areas contribute to oxidation of the organic matter sinking down from the surface (Fig. 5.28).

Sediments deposited on the continental slope are therefore more promising as source rocks for oil than shelf and deep-water facies (Fig. 5.29).

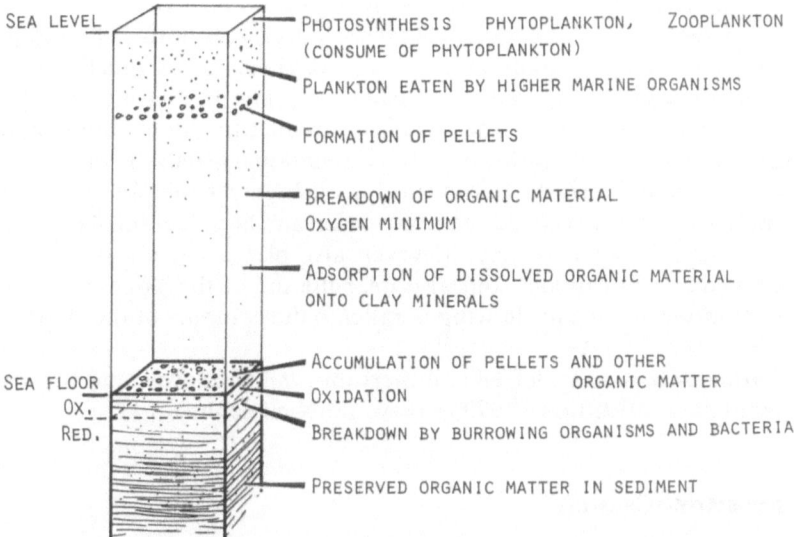

Fig. 5.28. Only a small fraction of the organic matter produced in the photic zone in the ocean becomes trapped in sediments. Oxidation of organic matter occurs in the water column and on the sea floor above the red/ox boundary

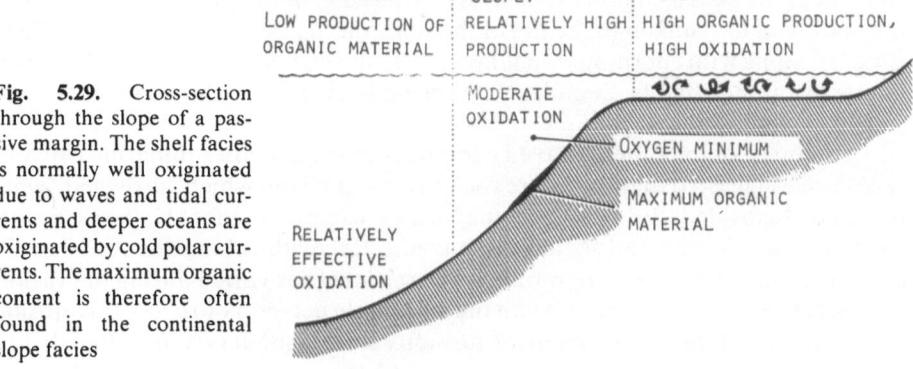

Fig. 5.29. Cross-section through the slope of a passive margin. The shelf facies is normally well oxiginated due to waves and tidal currents and deeper oceans are oxiginated by cold polar currents. The maximum organic content is therefore often found in the continental slope facies

As regards suitable reservoir rocks, the possibilities are more limited. The greater part of the continental slope sediments are fine-grained clay and silt which have deposited from suspension. Sand is transported mostly in submarine canyons. We find deposition of purer sand deposits in submarine fans. The sand here is also often sorted by strong traction currents, and may form good reservoir sandstones surrounded by source rocks. The sand will often pinch out upslope, and may form a natural stratigraphic trap (Fig. 5.30).

Sediment Transport on Submarine Slopes

Gravitational processes are naturally important on submarine slopes.

Gravity forces can be decomposed into forces normal to and parallel to the slope. The parallel component consists of shear forces which may overcome the shear strength of the sediment, causing slumping. Sliding of large volumes of sediments downslope produces extensional faulting in the upper part of the slope and compression in the lower part. Gravitational instability on the slopes may also develop into debris flows and turbidity currents. Collapse of the sediment grain framework may cause sudden compaction and liquefaction of the slope sediments.

Traction currents may, however, also play a part on the submarine slopes, particularly in canyons, but also near the toe of the slope, where we may have contourites — currents flowing parallel to the contours of the slope.

Further reading: Andersen and Bjerrum 1968; Galloway and Brown 1972; Doyle and Pilkey 1979; Lowe 1979; Saxov and Nieuwenhuis 1982.

Submarine Canyons

Surveys of the sea bed have resulted in detailed bathymetric charts. We have also gradually gained a better idea of the distribution of sediments on the sea bed. One of the most interesting features revealed by sea-bed surveys were submarine canyons. These are valley-shaped depressions which extend from the top to the bottom of the slopes, down to 2000–4000 m. In some cases they may start in shallow water near the beach, and in others close to the edge of the shelf. The height from the bottom of the canyon to the top of the slope on each side may be up to 2000 m. We are dealing with enormous topographical features, which would have been very impressive if they had been on land, towering structures on the same scale as the Grand Canyon.

A number of hypotheses have been presented to explain submarine canyons. It has been suggested that they were river canyons on land which "drowned" when the sea bed subsided. We now know that in most cases this cannot be the case, since we do not find continental sediments in connection with submarine canyons. We also have a fairly good idea now of how the sea level has varied during the various geological periods, and it has not had the amplitude necessary to cause this erosion.

At the end of the 1930's, when the turbidity current model was developed, Daly (1936) suggested that submarine canyons had been formed through erosion by

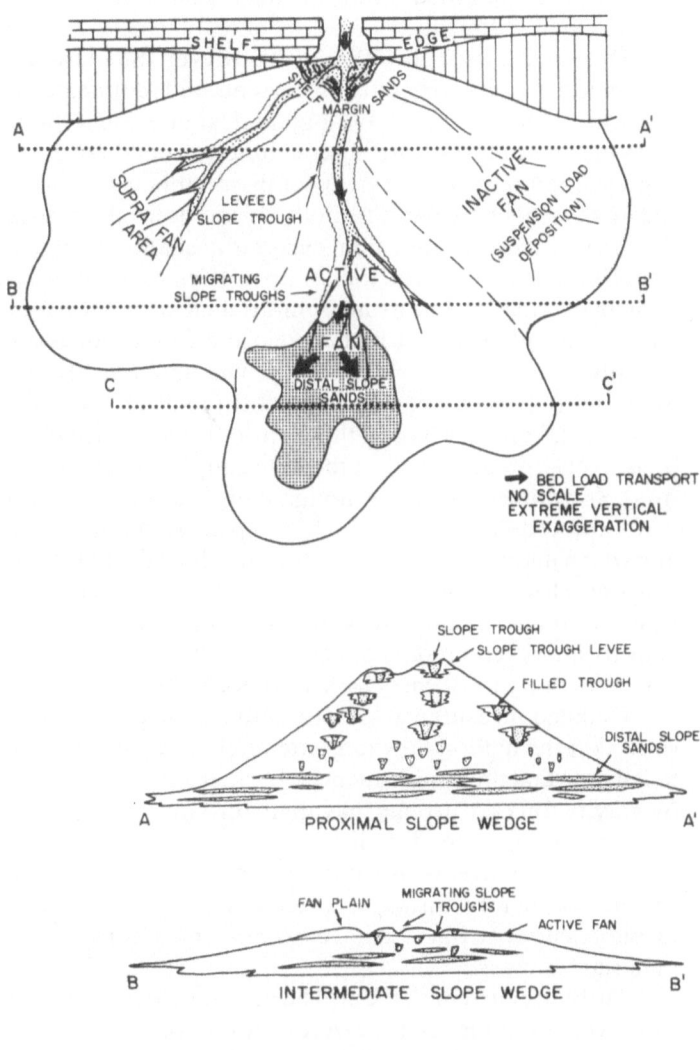

Fig. 5.30. Model of submarine fans deposited at the foot of submarine slopes. Submarine fans have channels which sometimes meander. On the sides of the channels we find fine-grained levée deposits resembling fluvial channels. The pattern of shifting depositional lobes resembles that of deltas. (Brown and Fisher 1977)

turbidity currents. This theory appeared more satisfying than the previous ones. Some workers doubted, however, that such currents could erode relatively well-consolidated sediments.

During the past 20-30 years, however, Shepard et al. (1979) have been systematically gathering data on currents and sedimentary transport in submarine canyons, so that we now have a great deal of evidence to base our theories on.

One of the most surprising conclusions from these data is that tidal currents are of great importance *also* at great depths in submarine canyons. Current meters have shown that there are currents both up and down the submarine canyons, and that they switch every 6 h, like tidal currents in shallow water. Current velocity is often only 10-20 cm/s, but velocities of up to 40 cm/s occur in many canyons, and this is enough to transport fine to medium-grained sand. Measurements taken over a long period will often record higher velocities. Occasionally very powerful currents actually carry off the current meters. The flow velocity tends to be greatest in the upper part of the canyon (see Fig. 3.10) and fall off downwards. Most sediment transport takes place during these episodic and unusually high flow velocities. We do not yet know the causes of the unusually high flow velocities (as much as several m/s). Sometimes it looks as though there is a connection with storms which create wind-induced shear forces which sweep the water up against the coast (*storm tides*) and which may cause currents to develop along the bottom and down the submarine canyons. However, high flow velocities have also been measured without it being possible to associate them with storms or wind stress. The most powerful flow velocity is most commonly directed downwards, but upward-directed streams with velocities of up to 90 cm/s have also been observed.

Detailed measurements in the submarine canyons off the coast of California show that the highest flow rates are oriented up the canyon in a way which seems to indicate that they are influenced by internal waves from the ocean basin, and not by gravity forces. Currents may then develop when the waves "break" against the coast or the continental shelf.

Powerful currents in submarine canyons are capable of transporting sand, sometimes in rare instances even coarser material. These are frequently not turbidity currents, but traction currents, which transport and deposit better-sorted material.

Turbidity currents have also been observed in submarine canyons. But definite (observed) examples of these were only low-velocity, low-density turbidity currents with a maximum velocity of 70-100 cm/s. These are most common at the upper end of submarine canyons, in the neighbourhood of a delta, where there is a good supply of suspended material. Outside deltas, slides or deformation of unstable sediment may be necessary before turbidity currents form. This happens relatively seldom, and may be the reason that we still nave no good direct measurements or observations of high-velocity turbidites.

We thus have both traction currents, which are controlled partly by tidal forces, and turbidity currents in submarine canyons.

Downward-moving currents driven by tidal forces may, if they contain much suspended material, also turn into turbidity currents. We can imagine that downward-moving currents have both a traction component and a gravitation component if the density, due to the sediment in suspension, is greater than that of the surrounding water.

The relief of submarine canyons is due partly to erosion down into the underlying sediments, and partly to lack of deposition in the canyon while the adjacent beds were being deposited.

At low sea level stand rivers may prograde closer to the shelf edge and hence supply more sediment to the submarine canyons, thus feeding submarine fans. At sea level high stands currents in the submarine canyons and the shelf may erode shelf and slope sediments and deposit pure sand onlapping an erosional unconformity at the toe of the canyon.

Most of the canyon itself is an area of sediment transport and erosion. Deposition takes place where there is a change of slope near the basin floor. Here the channel defined by the canyons splits up into several channels which build depositional lobes called suprafan lobes (Fig. 5.31).

As the lobes build up due to deposition, the gradient of the slope is reduced and a new channel will form in a part of the fan where there is a steeper slope. This produces lobe-shifting similar to that we observe in fluvially dominated deltas. Each lobe will tend to build a fining-upward sequence, with conglomerate and coarse sand near the base. In the main channels the coarsest material is sometimes deposited with inverse to normal grading. On the sides of the channels fine-grained material in suspension deposits as thin-bedded turbidites. The levée builds up on both sides of the channel, and resembles a river levée.

The distal fan is also dominated by fine-grained sediments deposited as thin, graded fine sand, silt and clay.

Further reading: Shepard et al. 1979; Howell and Newmark, 1982.

Sedimentation Along Island Arcs and Submarine Trenches

Submarine trenches form along converging plate boundaries where oceanic lithospheres are disappearing into a subduction zone. Along these converging plate boundaries sediment basins with very special deposition environments are formed. There are three main sources of sediment:

1. From the continent
2. From island arcs, which may consist of continental crust, oceanic crust and/or volcanic rocks
3. Pelagic sediment, including biogenic sediment and wind-blown volcanic ash.

These have quite different compositions. Sediment which is added from the continent is deposited in deltas and in turn fills the basin behind the arc (*back-arc basin*). Sediments which are formed on the island arcs are rich in volcanic material, and this will characterise deposits in small basins on island arcs and *fore-arc basins*. Back-arc basins may also receive volcanic sediments from the island arcs. During the initial subduction phase fore-arc basins will tend to be characterised by turbidites deposited in relatively deep water, but sediment filling may take them

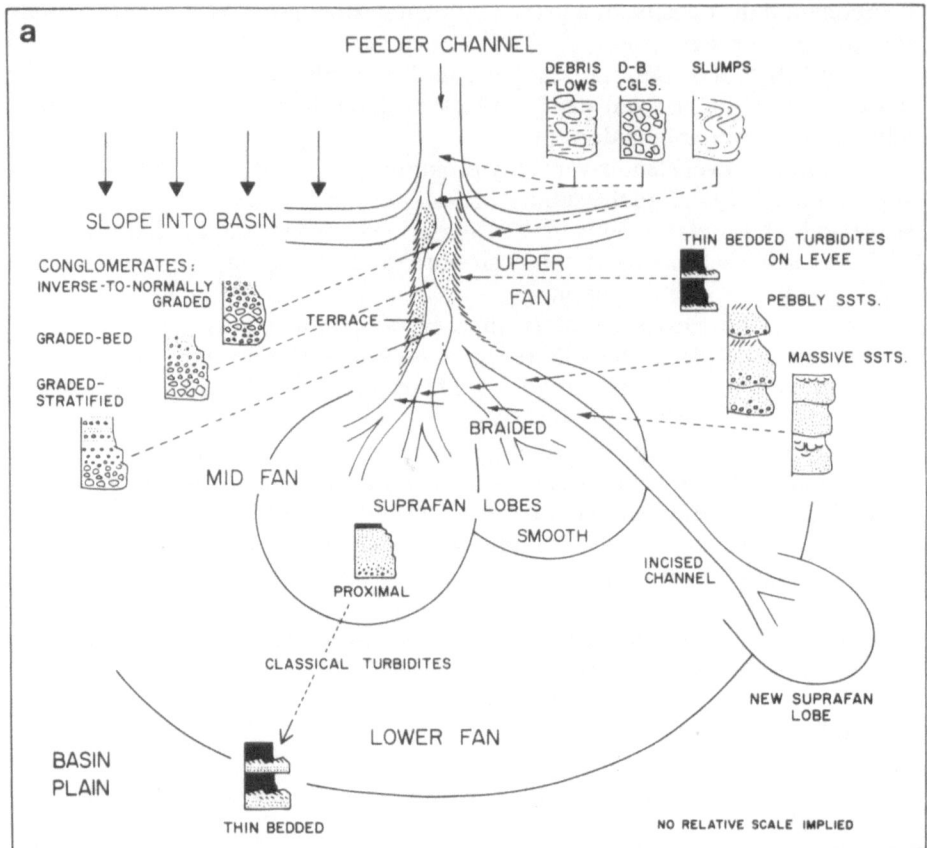

Fig. 5.31. a Submarine fan model showing progradation and shifting of lobes similar to delta lobe shifting. One of the main differences between submarine fan and delta facies is the absence of wave reworking. (Walker 1984) C.T. Classical turbidite, M.S. Massive sand, PS Pebbly sandstone, C U coarsening upward sequences, F U fining upwards sequences.

into a shallow water environment, which may include carbonate sediments. A structural high separates the fore-arc basin from the actual slope down to the submarine trench (Fig. 5.32).

Off the deep sea trenches a significant amount of pelagic sedimentation takes place on a relatively flat ocean crust. Some of the sediments get scraped off the subducting volcanic crust and stacked up in what are called "accretionary prisms" (Fig. 5.32). Some sediment may also be carried down with the subducting plates. The supply of sediment to the deep-sea trenches themselves is often very limited, so that they do not fill up with sediment. They represent the greatest ocean depths (up to 10 km) and this can be explained isostatically by the fact that the oceanic plate which is undergoing *subduction* is cold, and therefore heavy. The downward movement acts against the direction of heat flow, resulting in low geothermal gradients and therefore dense crust.

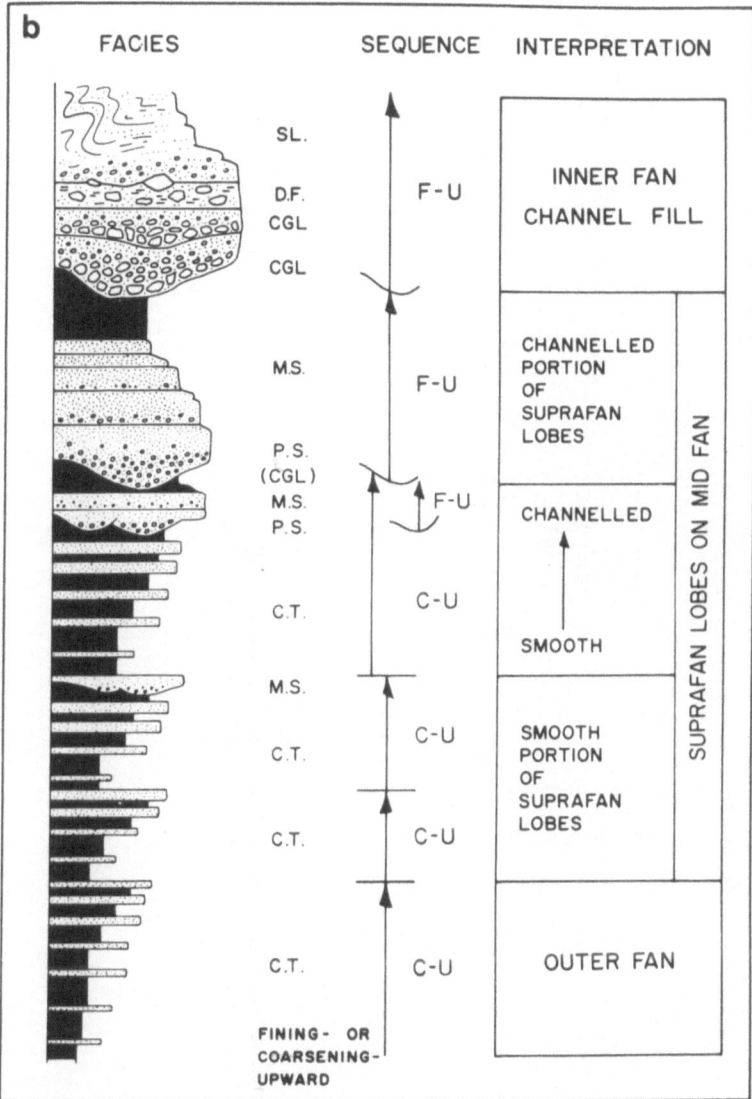

Fig. 5.31. b Vertical sequence through a submarine fan. (Walker 1984)

The sediments may be pelagic oozes or distal turbidites. Since the oceanic plate is moving towards the island arc, the sediments on the oceanic crust have been deposited further away from the sediment source, and in consequence we do not normally have very thick sedimentary sequences in the subducting plate. The slopes represent a very special sedimentary environment. The accretionary prism consists of a series of sliding faults which are steepest near the surface and have a lower gradient downwards. They are often draped with a blanket of pelagic sediments.

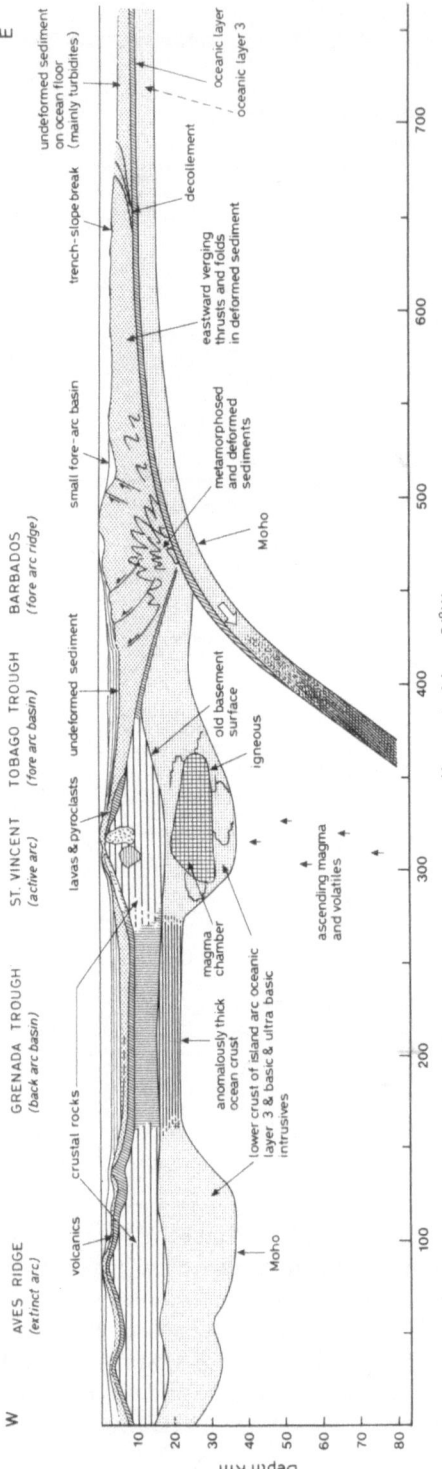

Fig. 5.32. Section through an area with converging plate boundaries in the Caribbean sea. Note that sediments deposited on oceanic crust are scraped off the descending plate and imbricated in the accretionary prisms. (Westbrook 1982, cited in Leggett 1982)

Listric faults of this type are similar to those we find in plate boundaries with tension (rifting) but the relative movements are in opposite directions (reverse faults). Sediments which are still not very consolidated tend to deform along the imbricated faults, and form various kinds of *drag folds*. Continued movements of the imbricated fault planes cause this slope to become very steep locally, and conglomerates and fan deposits may become unstable and slide. Lithified carbonates and sandstones will break up and form large blocks in a more clay-rich matrix. Volcanic rocks may also be included in this package of broken-up sediments and be incorporated into coarse conglomerates with large blocks called *olistostromes* (Fig. 5.33). The blocks, which lie in a matrix of clay sediments, may be as large as houses or small mountains. The result is called a tectonic *melange*.

The sediments in an accretionary prism are subjected to strong tectonic deformation prior to deeper burial, and soft sediment deformation is a very characteristic feature of such deposits. The sediment may be folded and overturned but within each wedge most of the sediments tend not to be inverted. If we look at the total package of imbricated wedges, there is a younging in the opposite direction, towards the subduction zone. This is a feature that can also be used to recognise this depositional environment.

Further Reading: Allen 1970; Fisher et al. 1972; 1974; Galloway and Brown 1972; Reineck and Singh 1975; Weimer 1975; Lowe 1976; Reading 1978; Walker 1979; Selley 1979; Hallam 1981; Leggett 1982.

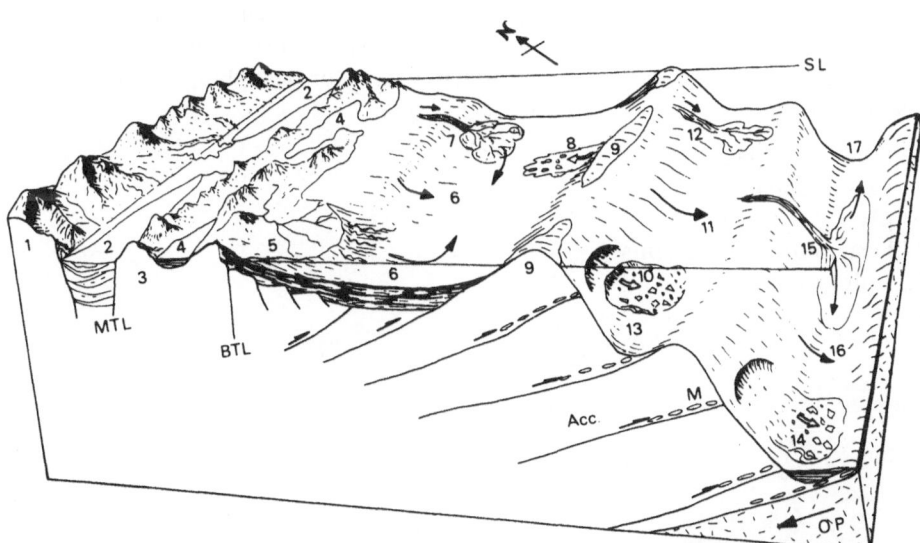

Fig. 5.33. Depositional environment in deep-sea trench off Japan. (Taira et al. 1982 in Leggett 1982). Diagram which shows environments of deposition during the deposition of the Lower Shimande Group during the Cretaceous. *2* Inter-arc basin; *4* small fore-arc shelf basin *5–8* accretionary fore-arc basin; *7* submarine fan; *8* slump and olistostromes; *9* partly exposed trench-slope break; *10, 14* slump and olistostromes: *11, 16* turbidites *12* and *15* submarine fans

6 Chemical and Mineralogical Factors Which Influence Sedimentological Processes

Geological processes in sediments and sedimentary rocks consist primarily of reactions between minerals and aqueous solutions at moderate temperatures (0–200°C). Naturally occurring aqueous solutions are very complex systems which are extremely difficult to study in the laboratory. This is because many reactions are so slow that they may require millions of years to take place or to attain anything approaching equilibrium. In addition, biological and microbiological processes often take place in conjunction with the purely chemical processes. It is useful, nevertheless, to look at processes involving water and sediments in terms of physical chemistry. A detailed treatment of this topic falls outside the scope of this book, but a short introduction to geochemical processes may be useful.

Water (H_2O) consists of one oxygen atom linked to two hydrogen atoms, with the H-O-H bonds forming an angle of 105° (Fig. 6.1). The distance between the O and the H atoms is 0.96 Å, and between the hydrogen atoms 1.51 Å. Water molecules therefore have a strong dipole, and are held together by hydrogen bonding. This gives water a relatively high boiling point and viscosity, and makes it a good solvent for polar substances. Another consequence of the molecular structure of water is its high surface tension, which enables water to transport particles and organisms on its surface. The capillary forces which cause water to be drawn up through fine-grained soils are also a result of this high surface tension.

A number of concepts are particularly useful for describing and explaining geochemical processes:

1. Ionic potential
2. Redox potential Eh
3. pH
4. Distribution coefficients
5. Isotopes.

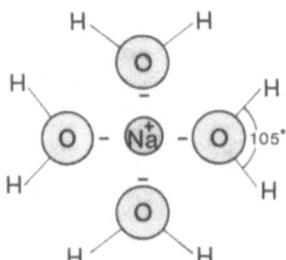

Fig. 6.1. Water molecules have a strong dipole and are attracted to cations, which thereby become hydrated

Ionic Potential

Ionic potential is a term introduced by Goldschmidt to explain the distribution of the elements in sediments and aqueous systems. It must not be confused with ionisation potential. Recent authors have proposed the term "hydropotential" for the concept, in order to prevent confusion.

Ionic potential (I.P.) $= Z/R$, where Z is the valency of an ion in solution and R is its radius. Ionic potential is an expression of the charge on the surface of an ion, i.e. its capacity for absorbing electrons. Small ions with a high charge have a high ionic potential and large ions with a small charge have a low ionic potential (see Fig. 6.2).

In aqueous solutions ions with a small ionic potential (e.g. K^+) will not cause the O-H bonds in water molecules to break, and will therefore remain in solution as hydrated cations, i.e. K^+ ions, surrounded by water molecules with the negative charge of the dipole towards the positive ion. This is because the O-H bond is stronger than the bond which the cation forms with oxygen (M-O bonding), and is true particularly of alkali metal ions (Group I) and most alkaline earth elements (Group II) (I.P. less than 3). If the M-O bond is approximately as strong as the O-H bond (I.P. 3–12), hydroxides are formed as a result of the metal ion replacing one of the hydrogen atoms, and we get compounds of the type $M(OH)_2$ or $M(OH)_3$ (see Fig. 6.2). Examples of these hydroxides are $Fe(OH)_3$, $Al(OH)_3$, $Mn(OH)_4$, etc. They tend to have a very low solubility.

Ions with a high ionic potential (greater than 12) form a stronger bond with oxygen than the H-O bond, and form soluble anion complexes, e.g. SO_4^{2-}, CO_3^{2-} etc. and release H^+ ions into solution. This approach fits well for those elements on both sides (electropositive and electronegative) of the Periodic Table which form ionic

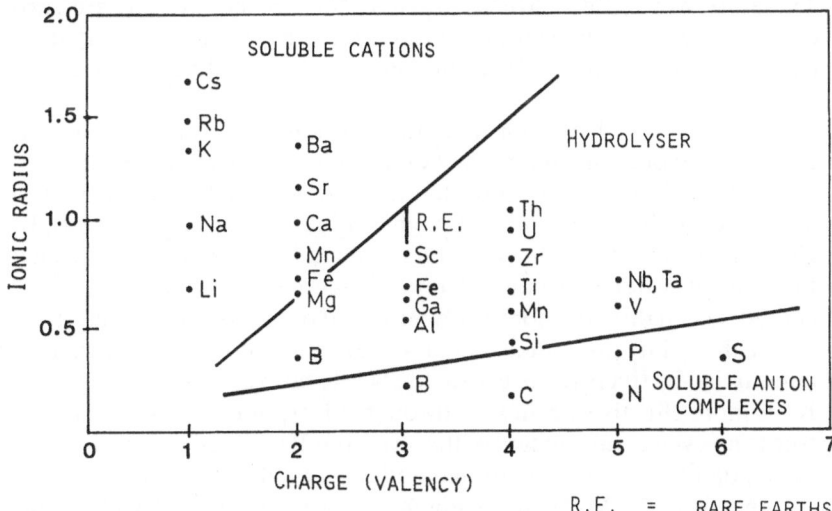

Fig. 6.2. Classification of the geochemical behaviour of some elements as a function of ionic potential, i.e. ionic radius and valency

bonds. The elements in the middle of the periodic table, however, have a greater tendency to form covalent bonds, in which the strength of the M-O bond is not merely a function of valency and radius, and the picture becomes far more complex. The concept of ionic potential is still useful, however. We see that during weathering elements with a small ionic potential remain in solution along with anionic complexes of metals and non-metals with a high ionic potential.

Hydrolysates

The composition of sea water reflects the solubility of different compounds, but so does biological precipitation. Weathering of rocks leaves residues of the least-soluble compounds, which are enriched in elements like Al^{3+}, Fe^{3+}, Mn^{4+}, Ti^{4+}, U^{4+} etc. in the form of oxides or hydroxides. Sediments rich in these weathering products are often called *hydrolysates*. Note also that Fe^{2+} and Mn^{2+}, which have lower ionic potentials than in the oxidized state are more soluble. Since ocean water is normally oxidising it contains practically no iron, aluminium or manganese in solution. The ions which are precipitated in salt deposits (evaporites) are for the most part cations with a low ionic potential and anions with a high ionic potential, e.g. $CaSO_4 \cdot 2H_2O$, Na_2CO_3.

Metallic ions with a low ionic potential will not form M-O bonds, and will remain in solution, but will be affected by the dipole of the water. There is attraction between the metal ion (e.g. K^+, Na^+) and the oxygen of the water molecule due to the water dipole (Fig. 6.1), so that each metal ion is surrounded by coordinated water molecules. This has a strong effect on the chemical properties of the ion and on its ability to form part of a crystal structure and form minerals. Metals which have an intermediate ionic potential (I.P. 3–12) will be most strongly hydrated. These are Mg^{2+}, Fe^{2+}, Mn^{2+}, Li^+, and Na^+. Since the ions are surrounded by coordinated water molecules, we can use the expression "hydrated radius" as an expression of the space which the ion with the coordinated water molecules will occupy (Fig. 6.3).

The different hydration potentials of ions and variations in ionic radii are capable of explaining many geochemical phenomena. Of the elements in Group I of the Periodic Table, we know that Li^+ and Na^+ are the most soluble in sea water and are only slightly adsorbed by other minerals, e.g. clay minerals. K^+, Rb^+ and Cs^+, on the other hand, have a larger ionic radius and are not as strongly attracted to adjacent water molecules. They will therefore have a more effective surface charge, so that they are more easily adsorbed onto clay minerals etc. In consequence we see that while just as much potassium as sodium is dissolved during weathering and carried by rivers to the ocean, the Na/K ratio in sea water is 30. This is because K^+ is more effectively removed through adsorption and various reactions. This is true to an even greater extent of the large ions Rb^+ and Cs^+, which have the smallest ionic potential, are least hydrated and most easily adsorbed onto clay etc.

In Group 2, Mg^{2+} has a higher ionic potential and will be more strongly hydrated than Ca^{2+}. As a result Mg^{2+} has a greater tendency to remain in solution. Despite the fact that the Mg/Ca ratio in sea water is 5, calcium carbonate is normally the first to precipitate. Part of the reason we do not get dolomite and

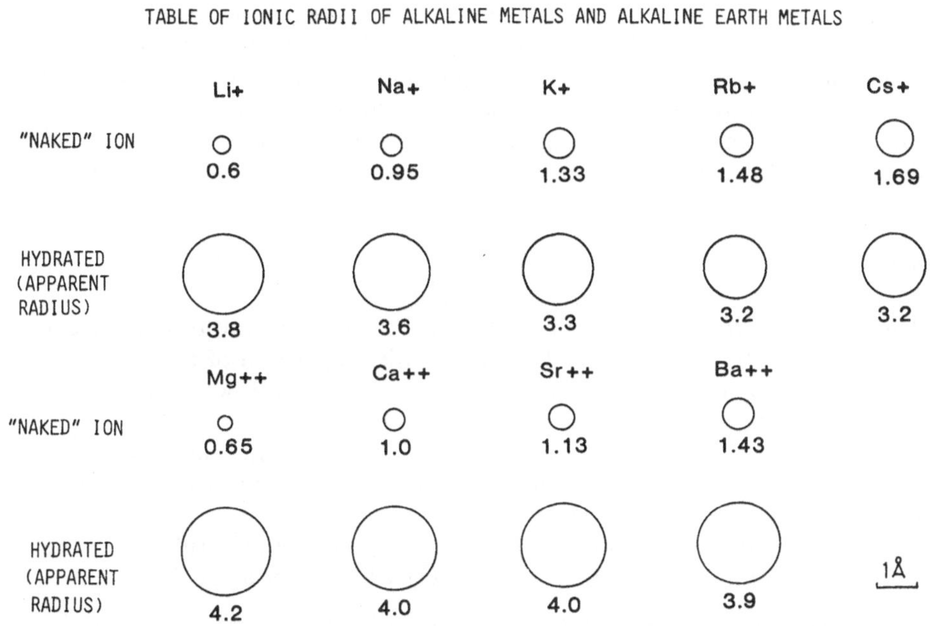

TABLE OF IONIC RADII OF ALKALINE METALS AND ALKALINE EARTH METALS

	Li+	Na+	K+	Rb+	Cs+
"NAKED" ION	0.6	0.95	1.33	1.48	1.69
HYDRATED (APPARENT RADIUS)	3.8	3.6	3.3	3.2	3.2

	Mg++	Ca++	Sr++	Ba++	
"NAKED" ION	0.65	1.0	1.13	1.43	
HYDRATED (APPARENT RADIUS)	4.2	4.0	4.0	3.9	1Å

Fig. 6.3. Ionic radius of hydrated and non-hydrated ("naked") ions of alkaline metals and alkaline earth metals

magnesite formed as primary minerals in sea water is the strong hydration of the Mg^{2+} ion. If we had naked, unhydrated ions, $MgCO_3$ and $FeCo_3$ would be more stable than $CaCO_3$, because Mg^{2+} and Fe^{2+} have greater ionic potentials and stronger bonding to the CO^{3-} ion. As the temperatures rises, the effects of hydration will decline, and the ions will become less strongly bonded to surrounding water molecules. During diagenetic processes at 80–100°C magnesium carbonates will be very stable, and may form even if the Mg^{2+}/Ca^{2+} and Fe^{2+}/Ca^{2+} ratios are low. When the ions are not hydrated, the smaller ions (which have the greatest ionic potential Mg^{2+}, and Fe^{2+}, will be preferred to Ca^{2+} in the carbonate structure.

Redox Potentials

Oxidation potential is an expression of the tendency of an element to be oxidised, i.e. give up electrons and remain with a positive charge. This potential can be measured by recording the potential difference (positive or negative) which arises when an element functions as one electrode in a galvanic element. The other electrode is a standard one, normally hydrogen. The oxidation potential of the reaction $H_2 = 2H^+ + 2e$ is defined as $E° = 0.0$ V at 1 atm and at H^+ concentration of 1 mol/l at 20°C. There have been different conventions used to assign plus and minus values. In geochemical literature, metals with a higher reducing potential than hydrogen are assigned negative values, e.g. $Na = Na^+ + e^- = -2.71$ V, while strongly oxidising elements are given a positive sign: $2F^- = F_2 + 2e = 2.87$ V (redox potential). A list of redox potentials shows which elements act as oxidising agents or reducing agents in relation to others. Reactions which result in a negative oxidation potential (E) will proceed spontaneously, while those which have positive voltage will require the addition of energy from an outside source to take place. We can predict whether a redox reaction will occur using Nernst's Law (see chemistry textbooks or Krauskopf 1969) (see Fig. 6.4).

pH

The ionisation product for water $H^+ \cdot OH^- = 10^{14}$. The concentration of H^+ in neutral water will be 10^{-7}. pH is defined as the negative logarithm of the hydrogen ion concentration, and is therefore 7 for neutral water at 25°C. However, it is important to remember that the ionisation constant (product) varies with temperature. At 125°C the ionisation constant for water $H^+ \cdot OH^- = 10^{-12}$. In other words, neutral water has a pH of 6. It is important to remember this when considering the pH of hot springs or in deep wells, e.g. oil wells.

In nature the pH of surface water varies from 4–9 as a rule. Rainwater is frequently slightly acid due to dissolved CO_2, which gives an acid reaction

$$H_2O + CO_2 \rightarrow H_2CO_3 \rightarrow H^+ + HCO_3^- \rightarrow CO_3^- + 2H^+.$$

Humic acids may give the water in lakes and rivers a low pH. Sulphur pollution from burning oil and coal gives SO_2, which is oxidised in water to sulphuric acid.

$$2SO_2 + O_2 + 2H_2O = 2H_2SO_4.$$

In areas with calcareous rocks or soils this sulphuric acid is neutralised immediately, while in areas with acid granitic rocks, as in the south of Norway and large areas of Sweden, the rock does not have enough buffer capacity to counteract acid rain or acidic water produced by vegetation (due to humic acids). Vegetation is an important producer of organic acids, and in many cases runoff is more acid than rainwater, suggesting that the cause of acidification is very complex, and also related to the type of vegetation.

Draining swamps or bogs may also cause acid reactions in runoff water, as H_2S from organic material is oxidised to sulphates when the water table falls and oxygen penetrates the system. Very high pH values (pH 9–10) may result where there is very high organic production (photosynthesis), because CO_2 (which forms acidic solutions) is involved in the reaction

$$CO_2 + H_2O = CH_2O + O_2.$$

Eh-pH Diagrams

Eh and pH are important parameters for describing natural geochemical environments, and the diagram obtained by combining these two parameters is particularly useful (Fig. 6.4).

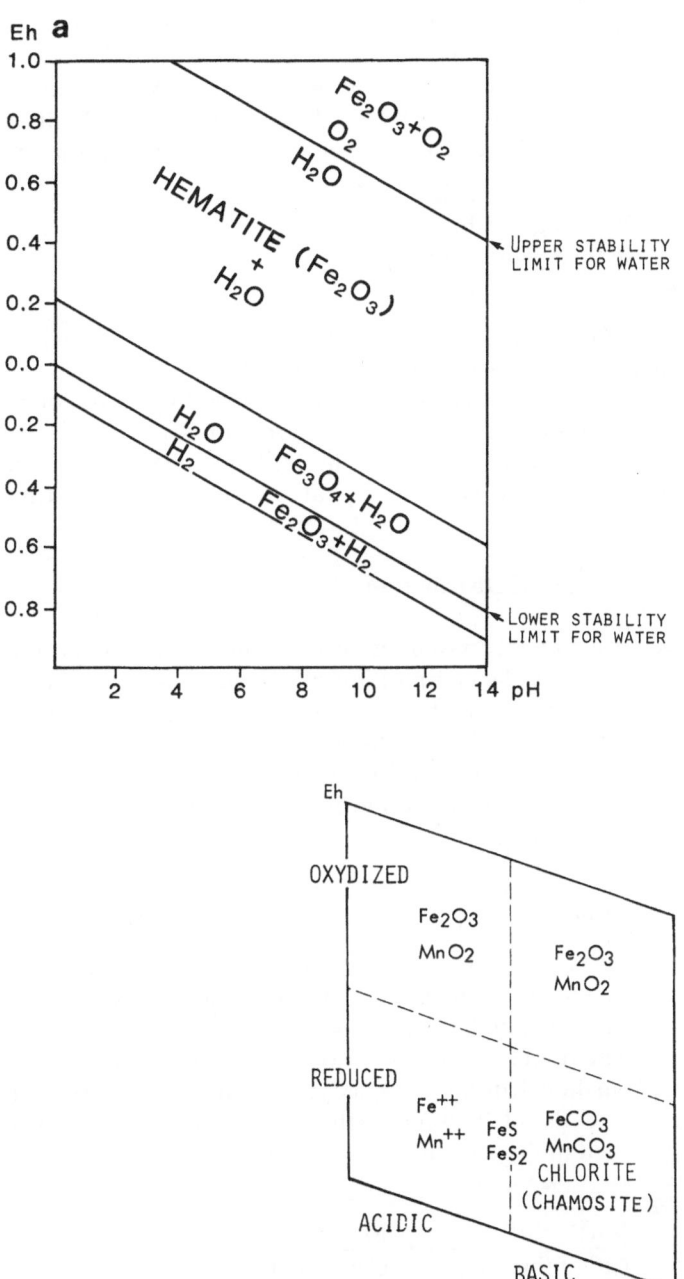

Fig. 6.4. a Eh-pH diagram showing the stability areas of iron oxides and water. The stability area of water lies between the lines corresponding to Eh = -0.059 pH and Eh = 1.22-0.059 pH. Naturally occurring water therefore has Eh-pH values which fall between these lines. **b** The occurrence of common iron and manganese minerals in an Eh-pH diagram

The lower Eh limit in non-marine aqueous natural environments is set by the line $Eh = -0.059$ pH, because otherwise we would have free hydrogen, and the upper limit corresponds to $Eh = 1.22-0.059$ pH, beyond which free oxygen would be released from the water. If we also set pH limits at 4 and 9 in natural environments, we can divide environments into four main categories:

1. Oxidising and acid
2. Oxidising and basic
3. Reducing and acid
4. Reducing and basic.

Variations of pH and Eh are the major factors involved in chemical precipitation mechanisms in sedimentary environments.

A number of elements are precipitated by oxidation and also increase of pH, e.g. iron and manganese, while uranium and vanadium, for example, are least soluble in the reduced state. The oxidation phases of sulphur are also important, as these determine whether sulphates or sulphides will form.

Coefficients of Distribution

When a mineral crystallises out of a solution, the composition of the mineral will be a function of the composition of the solution and the temperature and pressure. Trace elements which are incorporated in the mineral structure are particularly sensitive to variations of these factors.

At equilibrium, with constant temperature and pressure, the concentration of an element in the mineral is proportional to its concentration in solution:

$$\frac{x}{y} = k \cdot \frac{x}{y}.$$
Mineral   Solution

Here we are dealing with the concentration of a trace element, X, and one of the main elements in the mineral Y. The ratio between concentration in solution and concentration in the mineral is expressed through the coefficient of distribution, k.

The mineral calcite is a good example. A number of elements substitute for Ca^{2+} in the calcite lattice: Mn^{2+}, Fe^{2+} and Zn^{2+} have a distribution coefficient which is greater than 1. This means that they will be captured so that the mineral becomes enriched in these elements relative to the solution.

e.g.:

$$\frac{Mn^{2+}}{Ca^{2+}} \text{ mineral} = k \cdot \frac{Mn^{2+}}{Ca^{2+}} \text{ solution}$$

k here is about 17, i.e. the Mn^{2+}/Ca^{2+} ratio in calcite is about 17 times greater than that in solution.

At low temperatures ($25°$) Mg^{2+}, Sr^{2+}, Ba^{2+}, Na^+ have distribution coefficients which are less than 1. This means that the mineral phase will contain proportionately less of these elements than the aqueous phase. For Sr^{2+} k is about 0.1 (0.05–0.14) in calcite, but it is considerably higher in aragonite.

Isotopes

A number of elements occur in nature as different isotopes: elements which have the same atomic number (number of protons) but a different number of neutrons. They therefore have the same chemical properties although their masses are slightly different. Unstable radioactive isotopes break down at a certain rate (disintegration constant).

The methods most commonly used in determining the age of rocks are the $^{87}Rb \rightarrow ^{87}Sr$ and $^{40}K \rightarrow ^{40}Ar$ methods. Isotope ratios in the series $^{235}U \rightarrow ^{207}Pb$, $^{238}U \rightarrow ^{206}Pb$ and $^{238}Th \rightarrow ^{208}Pb$ can also be used.

Age-dating of sedimentary rocks is very complicated, and it is often difficult to interpret the results adequately. Clastic sediments consist of fragments and minerals from older rocks, as well as minerals formed authigenically, and their apparent age may be strongly affected by the source rocks. During early diagenesis the sediment will also be a relatively open system, i.e. the reaction products which are formed by breakdown of the unstable isotopes are mobile. With increasing compaction, diagenetic transformation, and finally metamorphosis, clay minerals gain a more ordered crystal structure, and new minerals are formed. The water content of the sediment decreases, and it becomes less permeable. Attempts have been made to determine definite temperatures at which the reaction products (e.g. ^{40}K and ^{87}Sr) of both individual minerals and the whole rock are not mobile (blocked) but the results are not always consistent.

The fact that isotopes have different masses causes fractionation to take place through both chemical and biological processes. The simplest example is water, H_2O, which may contain two oxygen isotopes and two hydrogen isotopes. The oxygen isotopes are fractionated through evaporation, with more $H_2{}^{16}O$ evaporating than $H_2{}^{18}O$. As a result, rainwater and ice both contain less ^{18}O than sea water.

Oxygen isotope ratios are expressed as ^{18}O values:

$$\delta^{18}O = \left(\frac{^{18}O/^{16}O_{sample}}{^{18}O/^{16}O_{standard}} - 1 \right) \cdot 1000.$$

$\delta^{18}O$ values are expressed as a deviation from standard mean ocean water (SMOW) or from a Cretaceous belemnite (Pee Dee Belemnite). The fact that two scales are used may cause some confusion, but $\delta^{18}O_{SMOW} = 1.031\ PDB + 30.8\permil$.

PDB values are generally used for carbonates while silicate minerals and water samples are related to the SMOW scale. Evaporation concentrates the lighter isotopes in the vapour phase, so rain water ($\delta^{18}O = -2$ to -15) is lighter than sea water. The most negative values are typical of high latitude precipitation, and near the poles even more negative values, up to $\delta^{18}O = -50$, may be found.

The isotope composition of sea water is affected by the amount of ice present on the earth, because ice forms from water molecules with light oxygen isotopes and separates out, thus increasing the content of heavy oxygen in the sea water. Oxygen and hydrogen isotopes are therefore important in regard to palaeoclimatic studies.

Hydrogen also has two stable isotopes, ^1H and ^2H, and deuterium (^2H) undergoes even stronger fractionation during evaporation. The hydrogen isotope composition of organic matter is also affected by latitude.

Precipitation from a limited volume of water (closed system) will enrich the heavy isotopes in the water phase and the concentration of heavier isotopes in the minerals precipitated will increase even if the distribution coefficient remains the same.

Precipitation of authigenic minerals from porewater will fractionate the oxygen isotopes, and given constant porewater composition, the oxygen isotope value of the precipitated mineral becomes more negative with increasing temperature of formation.

With carbonates we have the following relationship:

$$\text{Temp (T)} = 16.9 - 4.38(^{18}O_{carb} - {}^{18}O_{water}) + 0.10(^{18}O_{carb} - {}^{18}O_{water})^2 .$$

We see from this equation that $\delta^{18}O$ PDB $= 0$ corresponds to a temperature of 16.9°C. If the temperature at which the mineral precipitates is known, the porewater composition can be calculated.

Carbon has two stable isotopes ($^{12}C - 98.9\%$ and $^{13}C - 1.1\%$). During photosynthesis a greater proportion of $^{12}CO_2$ forms organic compounds than $^{13}CO_2$ because $^{12}CO_2$ has a smaller mass. Organic material is therefore enriched in ^{12}C relative to atmospheric CO_2 and HCO_3^- in sea water. The isotope composition of carbon is expressed as $\delta^{13}C$ values:

$$\delta^{13}C = \left(\frac{^{13}C/^{12}C \text{ sample}}{^{13}C/^{12}C \text{ standard}} - 1 \right) \cdot 1000.$$

As is the case for oxygen isotopes, the $\delta^{13}C$ values are expressed in terms of PDB for carbonates and organic matter. Atmospheric CO_2 has $\delta^{13}C = -7$. Land plants have an average $\delta^{13}C$ value of -24 ($-15\%_o - -30\%_o$) and marine organisms have $\delta^{13}C$ values between $-15\%_o$ and $-30\%_o$.

Freshwater contains CO_2 from the atmosphere, and in addition groundwater which has filtered through a soil profile will take up CO_2 from roots and organic matter which has been oxidised.

Bacterial conversion of organic matter under reducing conditions forms methane, which is strongly enriched in ^{12}C ($\delta^{13}C - 55\%_o - -85\%_o$). Early diagenetic concretions are enriched in light carbon isotopes ($\delta^{13}C -15$ to -30), probably due to CO_2 formed through bacterial breakdown of organic matter (Fig. 6.5).

The oxygen isotope ratio in water is largely a function of salinity and temperature. Cold freshwater gives strongly negative $\delta^{18}O$ values, while evaporites are enriched in ^{18}O isotopes (positive $\delta^{18}O$). Shallow marine carbonates which have been diagenetically transformed by fresh groundwater have lower ^{18}O values than marine carbonates precipitated in deeper water.

Analyses of stable oxygen isotopes were first used by Urey in 1951 as proof of temperature variation in sea water. By taking samples through a cross-section of a belemnite it was possible to register temperature variations which showed annual variation in sea water 150 million years ago (Fig. 6.6).

Further reading: Hudson 1977.

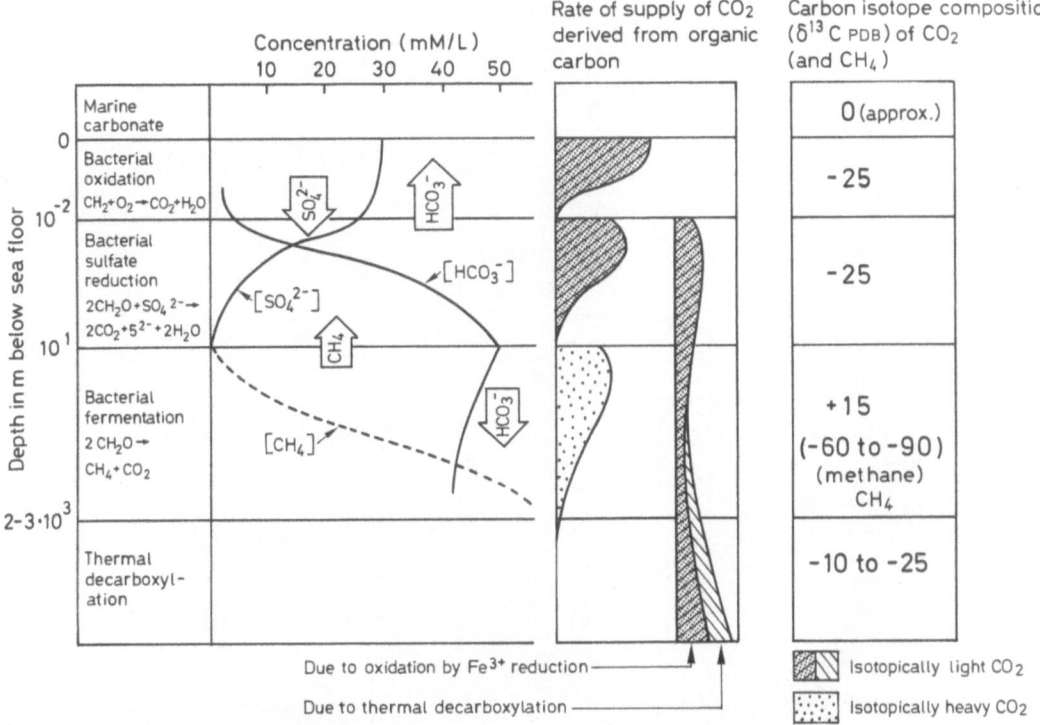

Fig. 6.5. Stable isotope composition of carbon dioxide in marine sediments. In the sulphate-reducing zone there is a strong biological fractionation towards negative $\delta^{13}C$ values. This is reflected in the composition of the carbonate minerals precipitated. Near the redox boundary there is a strong downwards diffusion of sulphate and an upwards diffusion of bicarbonate. At greater depth reduction of ferric iron to ferrous iron can supply electrons which may oxidise some organic matter. (Based on Irwin et al. 1977 and Gautier and Claypool 1984)

Clay Minerals

A number of minerals are referred to as clay minerals because they occur largely in the clay fraction of sediments and sedimentary rocks. However, this is not an accurate definition, because the clay fraction contains many minerals other than those we call clay minerals, and because clay minerals are often larger than 2μ (0.002 mm). By "clay minerals" we usually mean sheet silicate minerals which consist chiefly of oxygen, silicon, aluminium, magnesium, iron and water (H_2O, OH^-). These sheet minerals also occur as metamorphic and eruptive minerals (e.g. biotite, muscovite and chlorite) and they occur as *clastic* minerals in clay sediments. Clastic sheet silicate minerals are typically altered from their initial composition in metamorphic and igneous rocks. Micas (muscovite and biotite) lose some potassium which is replaced by water (H_2O, H_3O^+) to form illite. Clay minerals are also formed through weathering reactions or reactions between minerals and porewater after deposition (diagenesis), e.g. through the breakdown of felspar and mica. Clay

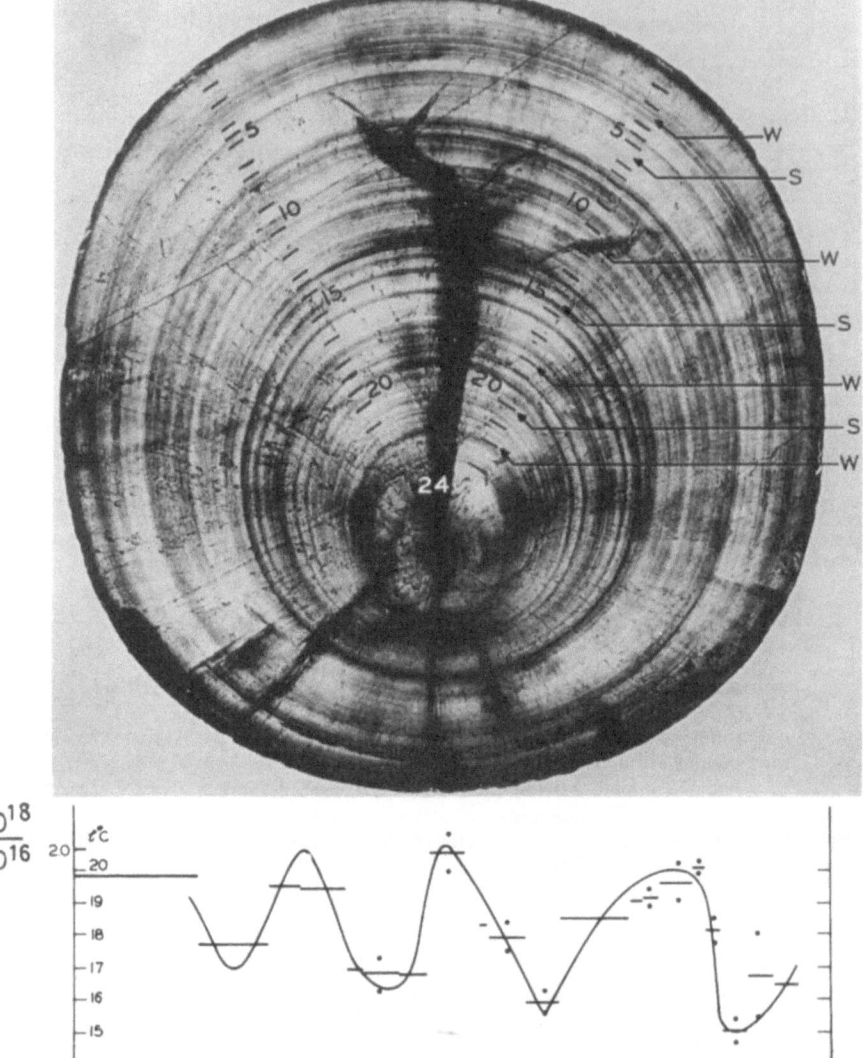

Fig. 6.6. Analyses of oxygen isotopes in samples from a section through a belemnite from the Jurassic (Urey et al. 1951). Fluctuations in the $^{18}O/^{16}O$ ratio show the annual variation in temperature in sea water 150 million years ago

minerals which are formed in sediments after deposition are called *authigenic* clay minerals.

Sheet silicates have a structure based on sheets of SiO_4^{-4} tetrahedra and octahedra alternating with layers of cations such as Al^{3+}, Fe^{2+} and Mg^{2+}. The tetrahedra layers contain silicon or aluminium surrounded by four oxygen atoms. In the octahedra layers the cation is surrounded by six oxygen or hydroxyl ions. Both bi- and trivalent ions can act as cations in the octahedral layer, and in sheet silicates with trivalent ions (e.g. Al^{3+}) only 2/3 of the positions are filled (resulting in a dioctrahedral form). With bivalent ions (Mg^{2+}, Fe^{2+}), however, all positions are filled (trioctahedral) (Fig. 6.7).

The main method of identifying clay minerals is now X-ray diffraction, by which the thickness of the sheet silicates (Fig. 6.7) is determined using X-rays which are diffracted according to Bragg's Law: $n\lambda = 2d \sin \alpha$. Here λ is the wavelength of the X-ray, α the angle of incidence and d the thickness of the reflecting silicate layers. d is thus a function of angle α.

For identification of clay minerals by means of X-ray diffraction, see Grim (1968), Carrol (1970), Thorez (1975), Brindley and Brown (1980).

Sheet silicates may also be identified by means of differential thermal analysis (DTA) (Grim 1968).

The scanning electron microscope (SEM) is very useful for studying clays, particularly the morphology and composition of authigenic clays in sandstones. The transmission electron microscope (TEM) is now becoming very important for high resolution pictures of clay minerals, and for reliable detailed analyses.

Figure 6.7 shows the structures of some of the main clay minerals. Illite consists of sheets with two layers of tetrahedra and one of octahedra bonded together by potassium. The ionic bonds between the potassium and oxygen in the two sheets (6 + 6) are relatively weak, so the mineral cleaves easily along this plane. The bonds within the tetrahedra and octahedra layers are more covalent and stronger. The potassium content in mica corresponds to the formula (about 9% K_2O), while illite, which is a clay mineral, has a deficit of potassium. Glauconite is an iron-rich mica mineral which forms authigenically on the sea bed. It contains both divalent and trivalent iron, and tends to form at the redox boundary in the sediment.

Smectite (montmorillonite) has the same structure, but most of the potassium is replaced by hydrogen and water (H_3O^+), other cations or organic compounds. In an atmosphere of glycol vapour, smectite will swell from 14Å to 17Å. Smectite has a very high ion-exchange capacity. The stability of smectite declines in aqueous solutions with high K^+/H^+ or Na^+/H^+ and with increasing temperature, and it is converted to illite. Vermiculite has a structure reminiscent of the smectites, but here the characteristic exchangeable cation is Mg^{2+} and the bonding between sheets becomes too strong for any real swelling to be possible. Vermiculite forms by weathering of mica, mainly biotite.

Kaolinite structures consist only of a tetrahedra layer and an octahedra layer, and are very stable at low temperatures. The structure has no positions where exchange proceeds easily, and kaolinite therefore has a far lower ion exchange capacity than smectite. At higher temperatures kaolinite becomes unstable and will be altered to illite at 120–130°C if K^+ is available.

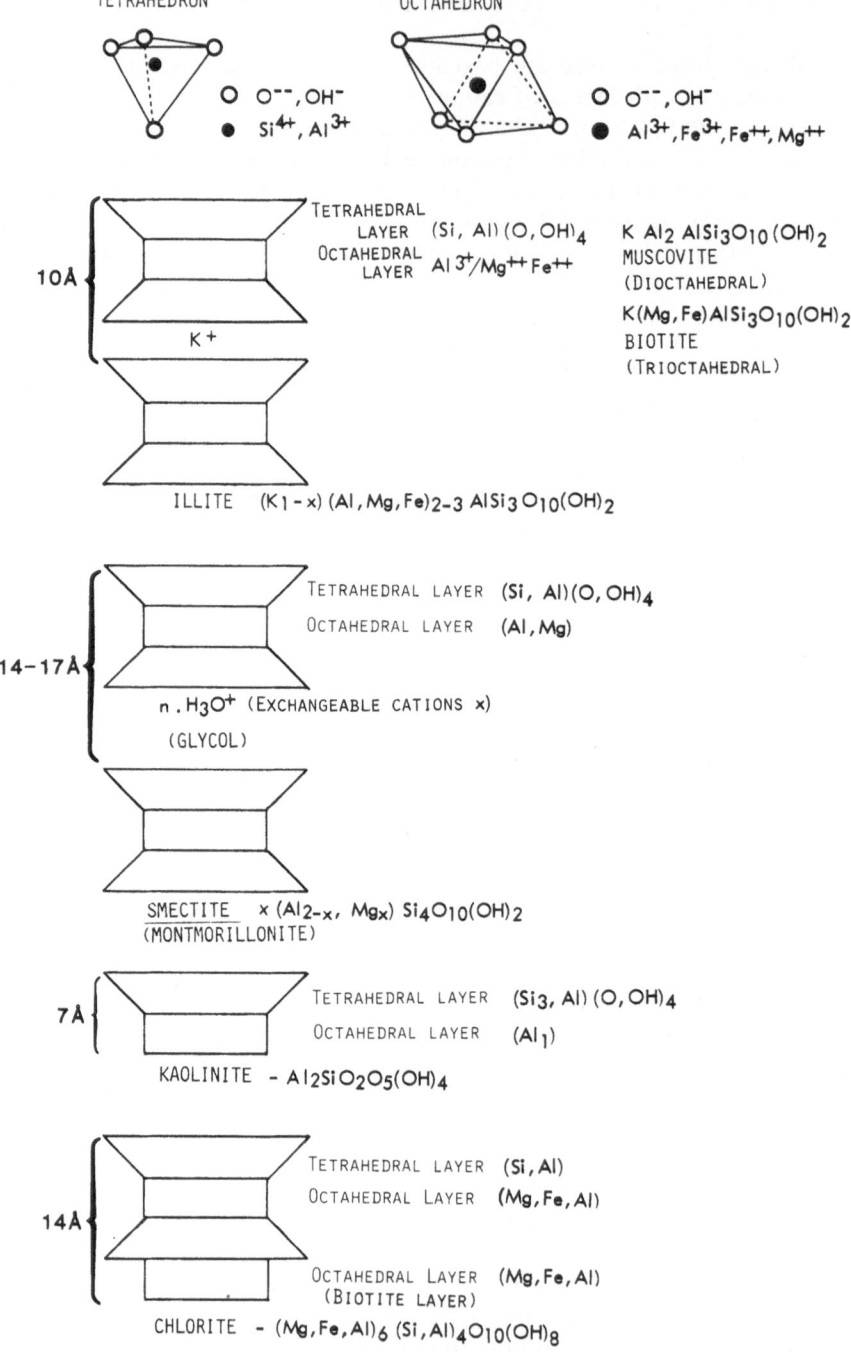

Fig. 6.7. Review of the major groups of clay minerals. See text

If there is no source of potassium, kaolinite will be stable at higher temperatures, and eventually form pyrophyllite ($Al_2Si_4O_{10}(OH)_2$).

Chlorite is a mineral which consists of two tetrahedra layers and two octahedra layers, totalling 14Å. The octahedra layer is filled with Mg^{2+} and Fe^{2+}. Magnesium-rich chlorites are stable only in metamorphic facies, while iron-rich chlorites may form as authigenic minerals in sediments. Chamosite (birthierine), is an iron-rich chlorite mineral which is often formed close to the sediment surface in reducing environments.

Clay minerals have many properties which distinguish them from most other minerals. Because of their small size and large specific surface area they have a great capacity for adsorbing ions, which is increased by the fact that clay minerals have negatively charged edges due to broken bonds. In water with a low electrolyte content, clay minerals will therefore repel each other. If cations are added, clay minerals therefore accumulate a layer of positive ions (a double layer), and repulsion between negatively charged clay minerals declines as the strength of the electrolyte increases. Van der Waal's forces will therefore more easily cause flocculation in salt water, where repulsion due to the negative charge is reduced. A colloid solution of clay in water is called a *sol*, which can be regarded as a Newtonian liquid. A flocculated clay is called a *gel* and has thixotropic properties: shear strength declines with increasing deformation, i.e. increasing shear stresses. This is the opposite of a *dilatant* system, which has a shear strength which is low at low shear stresses, but which increases with increased deformation. This is not usual in clay, but occurs often in water-saturated silt and sand.

Further reading:

Geochemistry — Goldschmidt 1954; Garrels and Christ 1965; Berner 1971, 1980; Garrels and MacKenzie 1974; Krauskopf 1979.

Clay mineralogy — Grim 1968; Carrol 1970; Millot 1970; Weaver and Pollard 1973; Thorez 1975; Velde 1977; van Olphen and Fripiat 1979; Brindley and Brown 1980.

7 Weathering and Geochemical Processes

How Are Sediments Formed?

Sediments which are deposited in the sea or in fresh-water basins derive chiefly from the continents, where erosion and weathering produce matter which is transferred to the sedimentary basins.

Mechanical weathering is the breakdown of a rock into smaller pieces which can then be transported as clastic sediment.

Chemical weathering involves solution of minerals and rocks, which allows parts of the rock to be carried away in aqueous solution.

As we shall see, biological processes are important, not only in connection with chemical weathering, but also in regard to mechanical weathering. Both mechanical and chemical weathering are due to rocks which are formed under other conditions (e.g. at greater depths, higher pressures or temperatures or different chemical environments) no longer being stable under the conditions which prevail at the surface: exposure to the atmosphere, water and biological activity.

Mechanical Weathering

By mechanical weathering we mean mechanical breakdown of rocks. A variety of forces may be involved. Perhaps the most important form of mechanical weathering is a result of the stresses which are released when the sediments overlying deeply buried rocks are removed by erosion and the rocks are subjected only to atmospheric pressure. The rocks then expand, giving rise to pressure release jointing (*exfoliation*) and horizontal scales (*sheeting*). We see this most clearly in granites, which are homogendeous, while in metamorphic and sedimentary rocks expansion will take place along bedding surfaces or along zones of tectonic weakness due to deformations or fractures formed at great depths. Water will then circulate in the cracks, and this will expose more surface area to chemical weathering.

In areas with frost, frost weathering is very important. When water in cracks in rocks freezes, it expands by 9%, and can develop a very great pressure, which in turn will widen cracks near the surface. Repeated periods of freezing and melting cause effective frost burst. Frost burst is an important factor responsible for the formation of blocks of rock in the mountains of Scandinavia.

Rocks at the surface are also subjected to daily temperature fluctuations which cause the outer layers to expand differently from the rest of the rock. In desert regions in particular, the daily temperature range may be great, but some doubt has been raised regarding the importance of this process for mechanical weathering.

The roots of plants and moss can also contribute to mechanical weathering as they grow into fractures, take up water and expand.

Biological Weathering

Rocks are a source of nutrient salts for plants, and plants are capable of dissolving and breaking down the major rock-forming minerals. Moss, which consists of algae and fungus living in symbiosis, produces organic compounds which are capable of dissolving silicate minerals. Even in the early stages of weathering, we see that fungus hyphae penetrate into microscopic cracks. Since moss can live on bare rock, it will also contribute to the weathering process by developing weathering products – soil, which paves the way for the establishment of higher plants later. Moss therefore probably played an important part in establishing the vegetation cover on the earth's surface during the Palaeozoic.

Plant roots produce CO_2 which helps to lower the pH and dissolve important minerals, such as felspar and mica, thus freeing potassium, which is an important plant nutrient. Plants also produce humic acids, which likewise have a strong effect on the solubility of silicate minerals, and on the stability of clay minerals. The production of humic acids is perhaps the major factor influencing the rate of weathering. In heavily vegetated areas, such as rain forest in the tropics, the weathering rate is very high because so much humic acid is produced.

Bacteria and fungi, which are found in almost all soil types, help break down minerals. Animals also contribute to weathering, and certain marine organisms in particular, such as mussels, are able to bore into rocks (see Chap. 8).

Chemical Weathering

It is difficult to draw a sharp distinction between biological and chemical weathering because we find biological activity in almost all soils and rocks near the surface. The chemical environment in water on the surface of the earth is very much affected by local biological activity, and in most cases it is biological processes which cause weathering to continue when rainwater is neutralised through reaction with minerals. We will therefore use the term "weathering" here for both chemical and biological processes.

Weathering Profiles (Soil Profiles)

Both chemical and biological weathering are a function of climate. The crucial factor is the ratio between precipitation and evaporation in an area. In areas where precipitation is far greater than evaporation, we get podsol profiles in which there is a net transport of ions down through the soil profile. In other words, we get weathering due to the fact that rainwater is slightly acid (CO_2 and H_2SO_4 content) and contains oxygen. Rainwater is initially undersaturated with respect to all

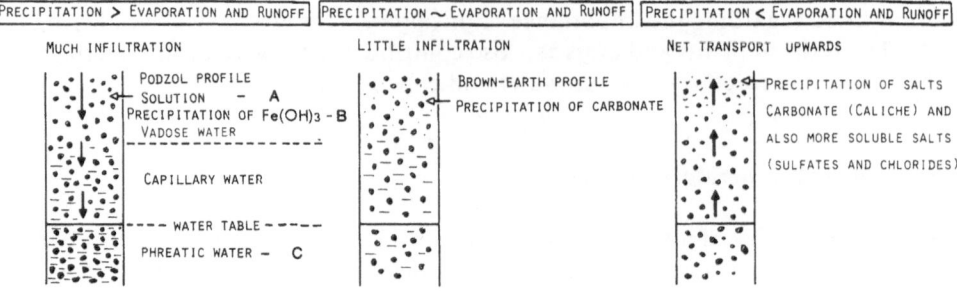

Fig. 7.1. Diagram showing different kinds of soil profiles

minerals. Some minerals are only very slightly soluble, others more soluble in this slightly acidic, oxidising water. Dissolved ions are transported down to the water table, but the ferrous iron in iron-bearing minerals will be oxidised and precipitated as $Fe(OH)_3$. Vegetation at the top of the soil profile produces CO_2 from roots and organic compounds, particularly humic acid which will increase the solubility of silicate minerals. The uppermost part of the soil profile, where dissolution due to undersaturated rainwater and organic acids dominates, is called the A-horizon. Some of the dissolved salts precipitate just below this horizon in the soil profile as the acids are neutralized. Horizons where ferric hydroxide is precipitated are called B-horizons (Fig. 7.1). These may form a solid rock cemented with iron and aluminium oxides and hydroxides (hard-pan).

Where precipitation is approximately equal to evaporation, there is less downward water transport and leaching of the soil profile, and at a certain depth (about 1 m) carbonate will be precipitated and form a hard layer. The organic content is also greater in the B-horizon due to less oxidation, and these factors give this soil a brown colour, hence the term *brown-earth profiles*. If evaporation is greater than precipitation, so that there is flow upward from the groundwater, there will be net transport of porewater upwards, and dissolved salts in the groundwater will therefore precipitate. The most common carbonate mineral in soil profiles is calcite, which occurs in "caliche". More soluble salts, such as chlorides and sulphates, occur in soils in areas where there is a very low rainfall and humidity. Precipitation of salts in the topsoil is a problem in areas with irrigation from wells or rivers, the water from which always contains some dissolved salts.

What Factors Control Weathering Rate and Products?

Because weathering is the most important sediment-producing process, we are interested in understanding how the rate of weathering depends on rocks, precipitation, vegetation, relief etc. We also try to establish a connection between weathering products, particularly clay minerals, and these factors. By studying sediments from older geological periods, we can learn something about weathering conditions at those times. Weathering products will also bear the stamp of the rocks undergoing weathering. We shall take as an example granite, which consists of

WEATHERING **=** ROCK **+** $H_2O + H^+$ → WEATHERING PRODUCT **+** IONS IN SOLUTION

Fig. 7.2. Weathering along exfoliation cracks in granite. Can be developed further into spheroidal weathering (see Fig. 7.3)

quartz, felspar and mica. The stability of a mineral during weathering is largely a function of the strength of the bonds holding the cations in the crystal lattice. Potassium in mica is held by weak bonds (low ionic potential) and Mg^{2+} and Fe^{2+} in the octahedral layer of biotite will also be weakly bonded. This means that these cations can be attacked by protons (H^+) which will replace them and send them into solution. Chain silicates, like hornblendes and pyroxenes will also be relatively unstable, and the minerals will weather rapidly. Dissolution will start on the surface, where we have broken bonds or along dislocations in the crystal structure (Fig. 7.2). The solubility of felspars during weathering depends on their composition. Stability is lowest in calcium-rich plagioclase, while pure albite and the potassium felspars are more stable. The breakdown of these silicate minerals will primarily liberate alkali cations to form new silicate minerals, largely clay minerals, but will also in many cases produce silicic acid (H_4SiO_4) in solution which may precipitate as quartz.

Some of the most important reactions are listed below:

1. $2KAlSi_3O_8 + 2H^+ + 9H_2O \quad Al_2Si_2O_5(OH)_4 + 4H_4SiO_4 + 2K^+$
   K-felspar   kaolinite

2. $2K\,Al_2AlSi_3O_{10}(OH)_2 + 2H^+ + 3H_2O = 3Al_2Si_2O_5(OH)_4 + 2K^+$
   muscovite   kaolinite

3. $2K(Mg{\cdot}Fe)_3AlSi_3O_{10}(OH)_2 + 12H^+ + 2H_2O + 2e = Al_2Si_2O_5(OH)_4 +$
   biotite   kaolinite
   $4SiO_2 + Fe_2O_3 + 4Mg^{++} + 6H_2O + 2K^+$
   quartz

4. $5KAlSi_3O_8 + 4H^+ = KAl_5Si_7O_{20}(OH)_4 + 8SiO_2 + 4K^+.$
   felspar   illite   quartz

We see that potassium has been replaced by hydrogen ions in the new silicate minerals. The same applies to sodium in albite. The extent of weathering depends on the solubility of the minerals of which the rock consists, and on the volume of water which flows through the rock, in other words the mineral surface area. Weathering occurs at low temperatures, when dissolution and precipitation, particularly of silicate minerals, require relatively high kinetic energy. Dissolution will therefore be slow, and the water in the soil is often supersaturated with respect to common minerals such as quartz. Weathering tends to begin in cracks and fractures, formed by exfoliation or tectonic deformation, for example, into which water penetrates (Fig. 7.2). The weathering process spreads outward from the

joints, and after a while isolated blocks of unweathered rock are formed. They have
rounded corners and may be entirely round (spheroidal weathering) (Fig. 7.3). In
desert areas where there is little precipitation, weathering proceeds much more
slowly. Illite and montmorillonite may be formed under higher K^+/H^+ and
Na^+/H^+ ratios than kaolinite, and they are frequently formed where there is less
water percolation and the removal of potassium or sodium is slower.

Granites undergoing weathering often develop a special topography.
Weathering proceeds fastest in fractures and fault zones, and valleys develop in
which groundwater collects so that weathering accelerates. The most massive
granite blocks will stand out in the terrain, and because precipitation runs swiftly
down into the depressions between the elevated portions, the topographic difference will become more and more pronounced. Granites surrounded by sedimentary rocks (into which they may have been intruded) will normally weather
faster than the surrounding sediments. Particularly if the sediments consist of
quartzites and shales, the granite will form a depression in the terrain. The
weathering products from granites will normally be quartz grains the same size as
those in the granite, and clay consisting of kaolinite, and possibly also some illite
and smectite, formed from felspars and micas. We have a bimodal grain-size
distribution from the outset, since there is a very small silt fraction.

Fig. 7.3. Spheroidal weathering. Weathering of rocks which give round blocks of unweathered rock
surrounded by weathered matrix. To the *left*, weathered granite in a quarry outside Kampala, Uganda.
To the *right* a drawing by H. Reusch (1878) of weathered granite from Corsica

Basic rocks (e.g. gabbro) will weather far more rapidly than granites, because basic plagioclase (Na-Ca felspar), pyroxenes and hornblende are more soluble and therefore less stable. Magnesium and iron will dissolve readily out of these minerals. While Mg^{2+} will have a tendency to remain in solution, all the iron in a normal, oxidising weathering environment will precipitate out again as iron oxide. The other basic minerals (Si, Al) will form kaolinite if there is abundant porewater circulating. This will reduce the concentration of Mg^{2+} and Ca^{2+}. Where circulation of porewater is slower, we may get a higher build-up of Mg^{2+} and Ca^{2+} concentration in the water and formation of smectite (montmorillonite) or chlorite. Smectite also requires porewater with a relatively high silica concentration (Fig. 7.5) and is therefore often found in sediments derived from volcanic rocks containing glass or soluble silicate minerals. Biogenic sources of silica (diatoms, radiolaria) will also increase the silica concentration in porewater. Figure 7.4 shows analyses of rocks at various stages of transformation due to weathering. Weathering proceeds particularly rapidly in amphibolites, and Na^+, Mg^{2+} and Ca^{2+} are quickly leached out, while Al^{3+}, Fe^{3+} and Ti^{4+} become enriched.

After the alkali cations have been dissolved out of the silicates, and kaolinite has formed, very slow leaching of quartz begins. When the concentration of silica in the porewater is sufficiently low (see Fig. 7.5), kaolinite will be unstable, and gibbsite $Al(OH)_3$ or $Al_2O_3 \cdot 3H_2O$ will form. We also see from the stability diagram that as long as the porewater is in equilibrium with quartz, gibbsite cannot form. All quartz must therefore first dissolve or become encapsulated (e.g. in a layer of iron oxides), before the silica content of the porewater drops low enough for gibbsite to form. Consequently gibbsite forms far more rapidly from basic rocks where the initial silica content is lower than that in granites. It will take a very long time to dissolve all the quartz in granite. The solubility of quartz at surface temperatures and pH 7–8 is about 5 ppm, increasing at higher pH values. Alkaline (basic) water can therefore speed up solution of quartz. *Laterite* consists of gibbsite and iron oxides or hydroxides. It is the end product of the weathering process, when all cations apart from Al^{3+} and Fe^{3+} have been removed. Under atmospheric conditions with a neutral pH, aluminium hydroxide and iron oxides may for practical purposes be regarded as insoluble.

At low pH values, e.g. under the influence of humic acids, in humid tropical climates, aluminium will be more soluble than iron, and selective leaching of Al^{3+} will leave iron-rich laterites. Aluminium hydroxide may then precipitate and form purer aluminium deposits with a lower iron content.

Laterisation is a very slow process, and takes millions of years, even in tropical regions with rapid weathering. It is therefore primarily in tropical areas that we find bauxite for the aluminium industry. Iron-rich laterites may have an iron content of over 50%, and in some areas (e.g. India) have been exploited as iron ore. Laterite forms a hard, cement-like crust over the weathering profile. Because it contains virtually no nutrients and is very hard, cultivation cannot take place in areas which are covered by laterite. In East Africa (especially Uganda) erosion has, however, cut through a layer of tertiary laterites, so that the laterite surface remains on flat, elevated surfaces, while the valleys have incised into fresher, more fertile rocks and weathering material. The vegetation in tropical areas may be more abundant, even if the soils are very poor in nutrients, and the vegetation recycles those nutrients

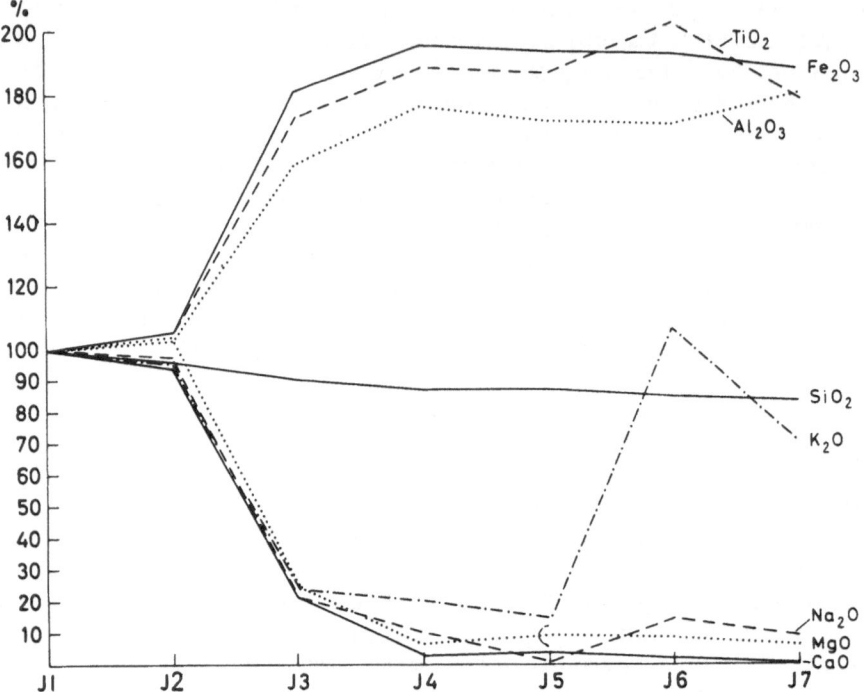

Fig. 7.4A. **A** and **B** show analyses of rocks which represent a gradual transition from fresh amphibolite (J1) to weathered amphibolite (J7). **A** is a graphic representation of the analysis in **B**, where the composition of sample J1 is set as 100%

which are available. If the vegetation is removed, and when organic material is no longer produced, oxidation and the absence of humic acids (increased pH) will lead to precipitation of oxides and hydroxides which make the soil hard and uncultivable.

Distribution of Clay Minerals and Other Authigenic Minerals as a Function of Erosion and Weathering

What Determines the Type of Clay Minerals We Find in Sediments and Sedimentary Rocks?

When rocks are subjected to erosion and weathering, clastic minerals are broken up and perhaps somewhat altered, compared to the minerals in the source rocks. We also get new minerals formed in the source rock, and precipitation of new minerals through weathering.

The more rapidly erosion and transport take place compared to the rate of weathering, the closer the composition of sediments is to the source rock. Glacial sediments represent one end of the scale in terms of sediment composition: because

Fig. 7.4B. Chemical analyses of weathered amphibolites from the buganda-Toro series east of Jinja, Uganda. Samples J1-J7 represent stages of increasing weathering

	J1	J2	J3	J4	J5	J6	J7
SiO_2	52.06	50.99	47.79	46.22	46.66	45.58	44.44
TiO_2	1.36	1.44	2.37	2.58	2.56	2.77	2.46
Al_2O_3	13.77	14.24	21.88	24.46	23.74	23.66	25.80
Fe_2O_3	13.28	14.01	24.16	26.03	25.74	25.86	25.07
MnO	0.17	0.18	0.12	0.07	0.10	034	0.22
MgO	6.25	6.54	1.53	0.43	0.58	0.51	0.45
CaO	11.28	10.71	2.47	0.35	.50	0.26	0.09
Na_2O	1.86	1.83	0.41	0.20	0.04	0.30	0.18
K_2O	0.22	0.05	0.05	0.05	0.24	0.24	0.16
P_2O_5	0.14	0.13	0.03	0.08	0.04	0.18	0.15
Total dry rock	100.22	100.10	100.69	100.40	99.99	99.90	99.01
Loss on ignition	1.15	1.98	10.66	11.30	11.18	12.36	10.84

Fig. 7.4C. Chemical analyses of weathered granitic gneisses from Mutoga Quarry, Uganda (Samples M1-M4 represent stages of increasing weathering.)

	M1	M2	M3	M4
SiO_2	64.16	65.69	68.10	64.89
TiO_2	0.62	0.60	0.46	0.64
Al_2O_3	17.37	17.66	17.59	20.34
Fe_2O_3	4.87	4.99	4.12	5.27
MnO	0.10	0.08	0.06	0.06
CaO	3.26	2.64	1.12	0.97
Na_2O	4.20	4.03	2.46	2.21
K_2O	3.33	3.32	3.78	4.30
P_2O_5	0.31	0.29	0.01	0.07
Total dry rock	99.86	100.59	98.36	99.79
Loss on ignition	1.43	1.27	2.61	4.13

of the high rate of erosion, low temperature and absence of vegetation during glaciations, there will be very little chemical weathering, and Quaternary sediments – particularly clay – will have a composition which virtually represents the average of the rocks which have been eroded. Sediments deposited in fault-governed basins (e.g. rift basins), where there is a short distance from the erosion site to the sedimentation site, have little time to weather on the way. Chlorite and biotite are minerals which break down relatively rapidly during weathering. Clastic chlorite and biotite are therefore found almost exclusively in areas with rapid erosion and/or cold climates.

An analysis of the distribution of clay minerals in modern sediments shows that we find clastic chlorite almost only at high latitudes (Fig. 7.6), except around islands of basic volcanic rocks (e.g. basalts). Clastic chlorites from metamorphic

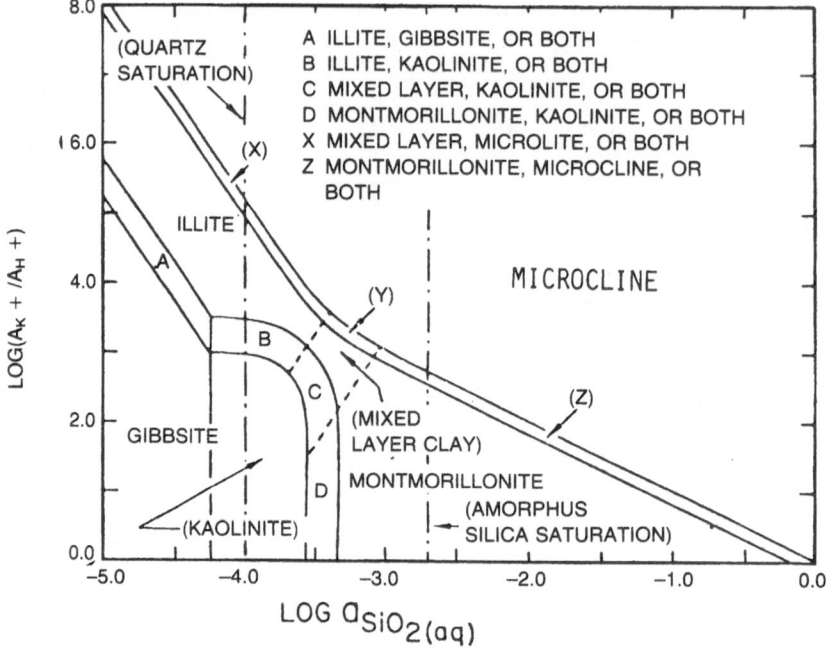

Fig. 7.5. Activity diagram showing the stability of various minerals as a function of silica (SiO_2) and K^+, H^+ concentrations. (After Aagaard and Helgeson 1982)

rocks or altered basic rocks are more magnesium-rich than authigenic chlorites, which are rich in iron. In temperate areas with moderate to high precipitation, weathering proceeds relatively rapidly kaolinite is formed from felspar and chlorite, and biotite is generally broken down if erosion is not too rapid. In desert areas weathering proceeds very slowly because all weathering reactions require water.

When felsper and other unstable minerals are altered gradually in a dry climate, alkali ions and alkaline earth ions such as K^+, Na^+, Ca^{2+}, Mg^{2+} will not be removed rapidly enough due to little percolation of fresh rainwater. As a result we will usually get illite or smectite formed as authigenic minerals because they are stable in the presence of high K^+/H^+ ratios in the porewater.

Smectite (montmorillonite) is thus a common clay mineral in desert areas, and its ability to swell makes sediments very plastic during floods. The expansion of the smectite also contributes to reducing the permeability of surface sediments during floods, so that water can flow over the surface for a long time before sinking into the ground. In addition capillary forces will prevent rapid percolation of water through dry soil. On the ocean floor near desert regions we find illite and smectite as typical minerals added through eolian transport.

In tropical areas where precipitation is relatively high, the rate of weathering will be very rapid. This is not only because weathering processes accelerate with temperature, but also because vegetation produces large amounts of organic acids (humic acids) which are very effective in breaking down silicate minerals.

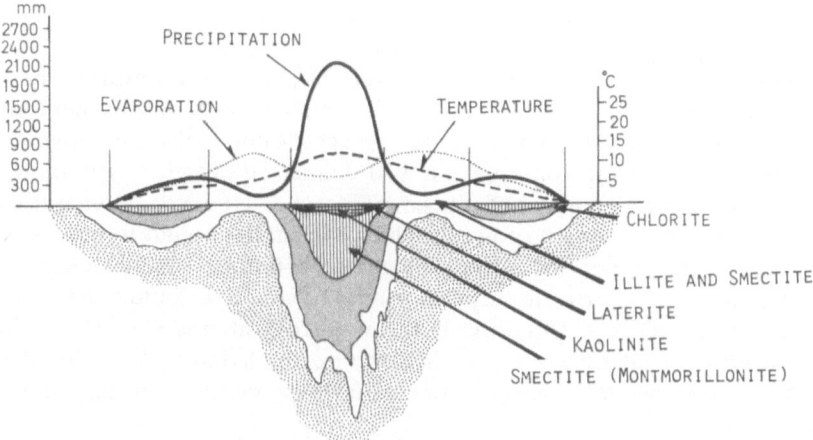

Fig. 7.6. Schematic diagram showing variations in climate and clay mineral composition as a function of latitude. (After Lisitzyn 1972)

Microbiological organisms such as fungi and bacteria also help in the breakdown process by producing CO_2 which forms carbonic acid, H_2CO_3.

Gibbsite ($Al_2O_3 \cdot 3H_2O$) and iron oxides (haematite, goethite, $Fe_2O_3 \cdot 3H_2O$) are constituents of the laterite which we find only in tropical areas with rapid weathering and slow erosion. Whereas iron oxides are also found at higher latitudes, gibbsite occurs almost exclusively in tropical, humid areas.

Volcanic rocks, particularly amorphous material (volcanic glass) will often form zeolites. These require a high concentration of both silicate and alkalis. Zeolites, particularly phillipsite, are formed authigenically on the Pacific Ocean bed and are also found in lakes (e.g. in East Africa).

In summary we can say that the factors which determine which types of clay minerals are "produced" in the various areas are:

1. The rocks which are eroded/weathered (source rocks)
2. Rate of erosion
3. Temperature
4. Precipitation
5. Vegetation
6. Permeability of source rocks and sediments (percolation of water).

Typical distribution of various minerals:

1. Chlorite and biotite — high latitudes (cold climate) — rapid erosion
2. Kaolinite — humid temperate and humid tropical regions — good drainage
3. Smectite (montmorillonite) — low precipitation or poor drainage (dense rocks). Typical of desert environments, but also formed in dense, e.g. basaltic, rocks in more humid environments. Typically formed from volcanic rocks
4. Gibbsite — tropical humid climate — long weathering period
5. Zeolites — formed in areas with volcanic material. Require a high concentration of silica and alkali ions.

Sandstones

Sandstones are rocks which consist very largely of sand grains, i.e. sedimentary particles between 1/16 and 2 mm in diameter. However, sandstones contain greater or lesser amounts of other grain sizes, and grade into rocks with a higher silt and clay content. Most sandstones have a well-defined upper grain-size limit, but if they have a significant content of coarser grains we call them conglomeratic sandstones.

Sandstones consist of grains which are small enough to be transported relatively easily at moderate current velocities (25–60 cm/s) and which are normally too large to be transported in suspension or to be cohesive. Sand grains are transported largely as bed load, and differ in this way from silt and clay. Only when we have powerful turbulent flow such as in turbidites, high density flows (mud flows, debris flows) or transport by ice (moraines) do we get mixtures of sand and clay.

Classification of Sandstones

There are a great many different ways of classifying sandstones. Most classification systems are based on the relationship between the relative quantity of sand-sized grains, the composition of the sand grains and the clay and silt content (matrix). Sandstones with more than 15% matrix are called *greywackes*.

If we use a four-component diagram, in addition to the silt and clay content we can distinguish between sand grains which consist of quartz (+ chert), felspar, rock fragments (or unstable rock fragments) and mud (30 μ) (Fig. 7.7).

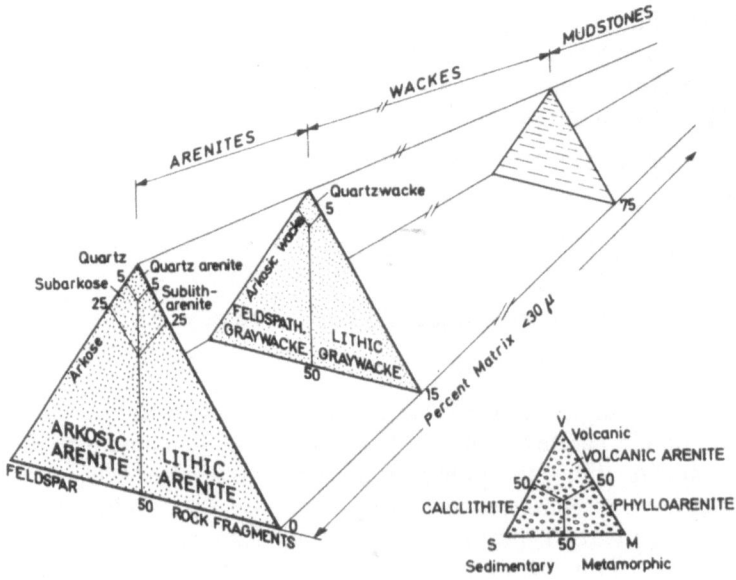

Fig. 7.7. Classification of sandstones. (After Dott 1964)

According to the classification of Dott (1964) we then have: sandstones with more than 25% felspar and a low rock fragment content, called *arkose*. If the percentage of rock fragments increases, we speak of *lithic sandstones*. *Quartz arenite* or *orthoquartzite* are the terms for sandstones which contain less than 5% felspar or rock fragments. Sandstones with a moderate felspar content (5–25%) are called subarkoses.

When granitic rocks and other coarse- to medium-grained rocks break down, they form sand grains which consist for the most part of a single mineral. Sandstone from these rocks will therefore consist largely of quartz or felspar with a small fraction of lithic fragments, e.g. arkose. The prerequisite for forming arkose, however, is that not too much of the felspar in the source rock is broken down by weathering to form minerals (e.g. kaolinite or illite). Arkose is therefore formed if there is rapid erosion in relation to weathering and a short transport path for sediments. This is the case in most rift valleys (grabens). Arkose is therefore in the great majority of cases associated with sedimentary basins formed by faults in gneisses and granitic rocks, i.e. the continental crust.

If we have a more mature relief in a tectonically stable area, a greater amount of felspar will break down during transport and weathering, so that subarkoses or orthoquartzites are deposited. Orthoquartzites (or quartzitic arenites) are formed in tectonically stable areas, where there is a high enough rate of weathering to break down the unstable grains, thus concentrating quartz. In Norway, the upper part of the Vangsås Formation (Lowermost Cambrian) is a good example of an ortho-quartzite deposited in a stable area on the Baltic Shield, and indicates the end of the faulting which led to the deposition of the underlying arkoses (sparagmites). Similar transgressive quartzites are found on other cratons (e.g. in N. America).

Sand which is transported in suspension or through mass flow (e.g. turbidites) will have poorer sorting and therefore a high matrix content, and form greywackes. In recent years we have also become aware that many occurrences of clay minerals in sandstones are not primary, but have formed after deposition. In sandstones which contain significant amounts of volcanic material, or in basic rocks with many unstable minerals, the primary sand grains will tend to form a "matrix" after deposition. This may occur through chemical breakdown of unstable clastic grains and formation of diagenetic clay minerals which form a matrix. We may also find mechanical breakdown of weak sand grains, e.g. chloritic basalt or basic minerals may be compressed and deformed to form a mass which is difficult to distinguish from a primary matrix. Greywacke therefore typically contains sand grains formed in areas of volcanic or basin rocks. This sort of rock is typical of sedimentary basins associated with island arcs along converging plate boundaries (fore-arc, inter-arc, back-arc).

Lithic sandstones are most easily formed from very fine-grained rocks, and one sand grain will often consist of several minerals. Sandstone formed through the erosion of basalts, rhyolites, intrusives or fine-grained metamorphic rocks therefore tends to produce lithic sandstones. Fine-grained metamorphic rocks, e.g. flint (chert) also form lithic sandstones.

We have seen that the various types of sandstone reflect different source rocks and varying tectonic stability in the area. Studies of different types of sandstone and their mineralogical maturity are therefore important indicators of relief, climate and also deformation in older geological periods.

Mineralogical and Chemical Composition of Sandstones

The mineralogical composition of sandstones can be analysed by means of a microscope, and quantitative mineral analyses can be carried out. In order to distinguish easily between potassium felspar and plagioclase, we can etch thin sections slightly with hydrofluoric acid (under a suction hood!) and then colour them with sodium-cobalt nitrite, so that potassium minerals turn yellow (due to formation of potassium cobalt nitrite). We can also analyse the mineral composition of sandstones by means of X-ray diffraction. It is difficult to obtain good quantitative results from these analyses, but the advantage is that we can analyse the average of a large crushed sample, and that we obtain a surer identification of the clay matrix than we do under a microscope. Although it is often difficult to obtain good quantitative mineral analysis results by means of XRD, the relative intensities of characteristic peaks for the different minerals are more reproduceable. Chemical analyses of sandstones can be carried out by means of X-ray fluorescence or atomic absorption. The aluminium content will be an expression of the amount of felspar and clay minerals, and the Na^+/K^+ ratio is a measure of the ratio of potassium felspar to albite, if there are not too many sheet silicates like mica (illite) in the matrix.

With a microcobe we can analyse the composition of individual minerals and variations in composition within a mineral grain. This is often useful for studying alteration due to weathering or diagenetic processes.

The scanning electron microscope has proved to be a very useful aid in studies of sandstones. With it we can study the three-dimensional geometry of minerals and pores in sandstones. By means of an energy-dispersion analyser linked to a scanning electron microscope it is also possible to analyse the chemical composition of minerals.

Geochemical Processes in the Ocean

The ocean can be regarded as a reservoir of chemicals dissolved in water. It looks as though the composition of sea water has not altered radically throughout the geological ages from the late Precambrian and Palaeozoic until the present day, although there have certainly been variations. The picture in the early Precambrian is less clear, because oxygen content and the conditions for organic life were so different 2,500 million years ago.

Let us look first at sea-water composition. The rate of additon of elements must be just as rapid as the rate of removal of the same elements from sea water. How are salts and nutrients produced in and added to the ocean? The most important source of supply to sea water is river water, which contains dissolved ions liberated from minerals and rocks through chemical and biological decomposition. Evaporation from the sea consists of pure water. It returns to the sea via rivers along with dissolved ions. There is thus a net addition of salts to the sea, even though fresh river water has a low salt content. The annual addition of salts dissolved in river water is about 2×10^9 tonnes per year. It is possible that the supply has been less in the past. The development of land plants which could produce humic acids, which in turn caused more rapid weathering, has probably increased the supply of salts (since Devonian times). Periods with worldwide rises in sea level, transgressions and limited land areas (e.g. the Upper Cretaceous) probably also led to reduced addition of salts and nutrients to the sea.

How are elements dissolved in sea water removed? As we have seen, the total removal of elements must correspond to the amounts added. The most important processes are presented in Fig. 7.8:

1. Adsorption onto clay minerals supplied by rivers (B in Fig. 7.8). When clay minerals are carried into the ocean by rivers, they will react with the sea water, which is a stronger electrolyte, by adsorption and exchange of ions, for example. Cations become bonded to clay minerals and thereby removed from sea water. Such ion exchange or adsorption varies considerably from one mineral or element to the next, and depends largely on ionic potential and degree of hydration. Sodium is so strongly hydrated that it has a tendency to remain in solution, while potassium will be far less hydrated and can be more easily adsorbed onto clay minerals and rapidly removed from sea water. It is no easy matter to quantify removal of elements from sea water through adsorption and ion exchange. The composition of clay minerals can vary considerably, and it is difficult to determine how much removal is due to adsorption.

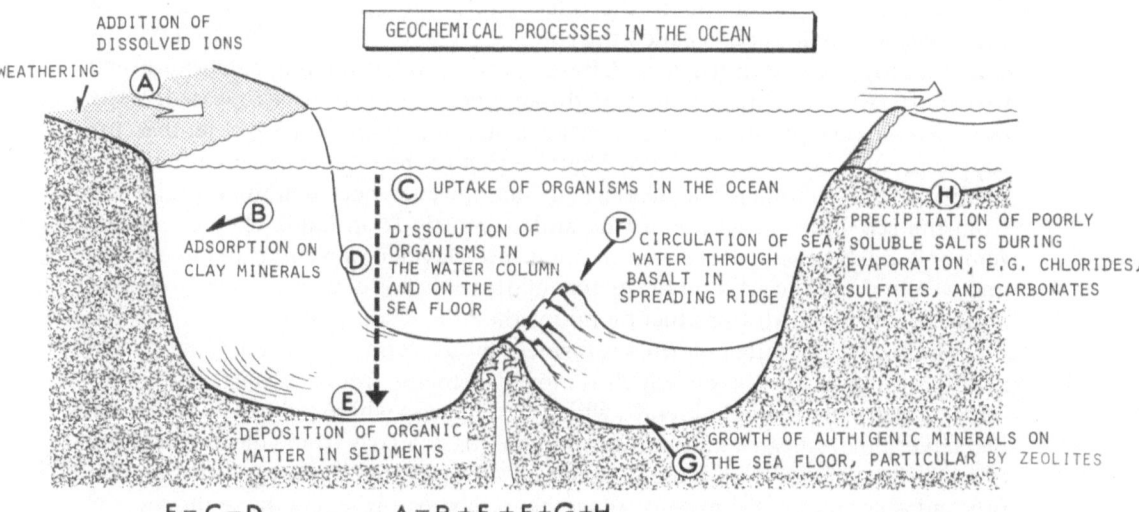

Fig. 7.8. Geochemical processes in the ocean. The addition of ions from the continents, plus possible addition from spreading ridges, must correspond to what is removed from the sea water by biological and chemical precipitation and adsorption. This is a requirement for the composition of the ocean to remain constant over geological time

The major felling mechanism for elements in the ocean is organic precipitation. Organisms can build their own internal chemical environment, and use their energy to precipitate minerals which are not normally stable in sea water. Carbonate-secreting organisms, e.g. foraminifera, molluscs etc., will precipitate aragonite or calcite even when the water is cold and undersaturated with respect to these minerals. Diatoms are so effective in precipitating silica, amorphous silicon dioxide (SiO_2), that the sea water near the surface of the photic zone (photosynthetic zone) is generally highly depleted with respect to silica.

The breakdown of organisms (D in Fig. 7.8) will begin as soon as they are dead. Organic compounds will start undergoing oxidation if the water is not reducing. Shells will dissolve if the sea water is not saturated in the minerals they are built up of. The more efficient organisms are at building skeletons despite undersaturation, the more rapidly they will dissolve. Diatoms, for example, dissolve to the extent of 99–99.9% before they sink to the sea bed. Only a very small amount is therefore preserved in sediments. Photosynthesising organisms in the surface water use CO_2 and produce oxygen and organic matter.

$$CO_2 + H_2O + \text{nutrients} \rightarrow CH_2O + O_2.$$

This helps to keep the pH high so that carbonates are stable or only dissolve slowly in warm water. But the temperature drops with depth, and CO_2 pressure increases so that solubility of carbonates increases with depth. Here, below the photic zone, oxygen is consumed and CO_2 released

$$O_2 + CH_2O \rightarrow H_2O + CO_2 + \text{nutrient}.$$

The depth at which the solubility of carbonate increases relatively rapidly is called the *lysocline*. The depth where the rate of solution is greater than the rate of carbonate sedimentation is called the *carbonate compensation depth* (CCD). This solution process causes nutrients (N, P etc.) to be liberated and they can be returned to the surface through upward flow. In this manner they are re-used repeatedly. The annual biological production in the ocean is therefore many times greater than the supply of nutrients from the land. That fraction of the organic products which is preserved in the sediments on the sea bed, has, however, been definitively removed from the sea, and nutrients must be added, mostly from land. They cannot be returned to the ocean before the sediments are elevated and subjected to erosion and weathering. The amount of organic matter which is deposited in sediments is a function of the rate of production minus the rate of solution.

The growth of authigenic (newly formed) minerals on the sea bed (C in Fig. 7.8) is also an important process which removes elements from sea water. The most important are the zeolite minerals, which can develop where sediments on the sea bed have a high silicate or aluminium content, particularly from volcanic material (glass). They may take up Na^+, K^+ and Ca^{2+} from sea water, but growth can also proceed very much at the expense of elements already present in the sediment (e.g. in the Pacific Ocean). This applies particularly to phillipsite, heulandite, clinoptilite and analcite.

Apart from this there is little direct chemical precipitation from sea water with normal salinity. This is because biological precipitation is more efficient in many cases and prevents the build-up of concentrations of the elements needed for chemical precipitation. Bacteria are active in the uppermost centimetres of the sediment, however, and sulphate-reducing bacteria remove sulphur from sea water in the form of sulphate, and reduce it to sulphides which are precipitated (e.g. iron sulphides, FeS and FeS_2).

In enclosed areas with greater evaporation than precipitation, however, we find even the very soluble salts being precipitated. In such areas there is enrichment of those elements which are otherwise precipitated only to a limited degree through biological or chemical processes, such as Na, Cl, S, Mg and trace elements such as B and Br. The amount of salt precipitated in evaporites through evaporation has probably varied very considerably throughout the geological ages.

In recent years it has become clear that biological and chemical precipitation, as described above, are not sufficient to explain the geochemical balance in the ocean. It has recently been discovered that important geochemical reactions take place in the spreading ridges in the oceans (F in Fig. 7.8). The heat from the basalt, which flows up along the spreading ridge, drives convection cells which cause ocean water to flow through the basalts and up along the ridge. The sea water reacts with hot basalt, and since sea water contains sulphur (as SO_4^{2-}) this leads to precipitation of metals which have been dissolved from the basalt as sulphides, e.g. iron sulphides and copper sulphides. When hot water flows up pipe-like *chimneys* in the ocean floor near spreading ridges, sulphides are also precipitated. When water emerging from the basalt onto the sea floor is oxidised, iron oxides and manganese oxides are precipitated around the spreading ridges.

Residence Periods for Different Elements in the Ocean

How long does an element spend in the sea after being brought by rivers before being chemically or biologically precipitated? The *residence* period is an expression of how rapidly an element is removed compared to its concentration in the ocean. Sodium has the longest period of residence (about 200 million years). That of potassium is about 1 million years, while the rare earths have periods of only a few hundred years. This is because sodium is the most abundant element in sea water and consequently little sodium is removed through biological or chemical precipitation outside evaporite basins. However, potassium and the rare earth elements are more rapidly adsorbed onto clay minerals and thereby removed.

Circulation of Water in the Oceans

Ocean currents are driven by

1. The rotation of the earth (Coriolis force)
2. Tidal forces
3. Differences in water density due to variations in salt content and temperature
4. Wind forces due to atmospheric circulation.

Ocean currents are very important for distributing heat from low latitudes to higher latitudes. This circulation is very dependent on the topography of the ocean floor and distribution of ocean and continents. In the ocean basins bottom currents are very different from those at the surface, and often flow in opposite directions. While warm surface water flows from the equator to the poles, cold surface water sinks down at the poles and flows along the floor to the equator (Fig. 7.9). Both currents are deflected by Coriolis forces, towards the right in the northern hemisphere and towards the left in the southern hemisphere. While surface currents (like the Gulf Stream) will be deflected eastwards, deeper currents will be deflected towards the west of the oceans (e.g. the Atlantic Ocean). These deep-sea currents may be strong enough to transport fine-grained sediments (silt and fine sand), and sediments deposited by such currents, which follow the depth contours, are called *contourites*.

The vertical circulation of sea water is very sensitive to variations in temperature and salt concentration. In periods with glaciation at the poles the temperature gradient in the surface water flowing from the equator to the poles is far greater than in non-glacial periods (e.g. the Mesozoic). The oxygen-rich cold water flowing down from the polar regions into the ocean basins is important for maintaining oxidising conditions in the deep ocean basins.

Animals and bacteria use oxygen from sea water continuously (for respiration), and oxidation of dead organic material also requires oxygen. If we did not have this downward flow of cold surface water, the water in the ocean basins would be reducing. Some ocean basins are isolated from this circulation, and we may then have a more permanent layering of water based on temperature and salt content. Warm surface water with low density can flow over heavier, colder basal water without the water masses mixing to any major extent. The boundary between warm

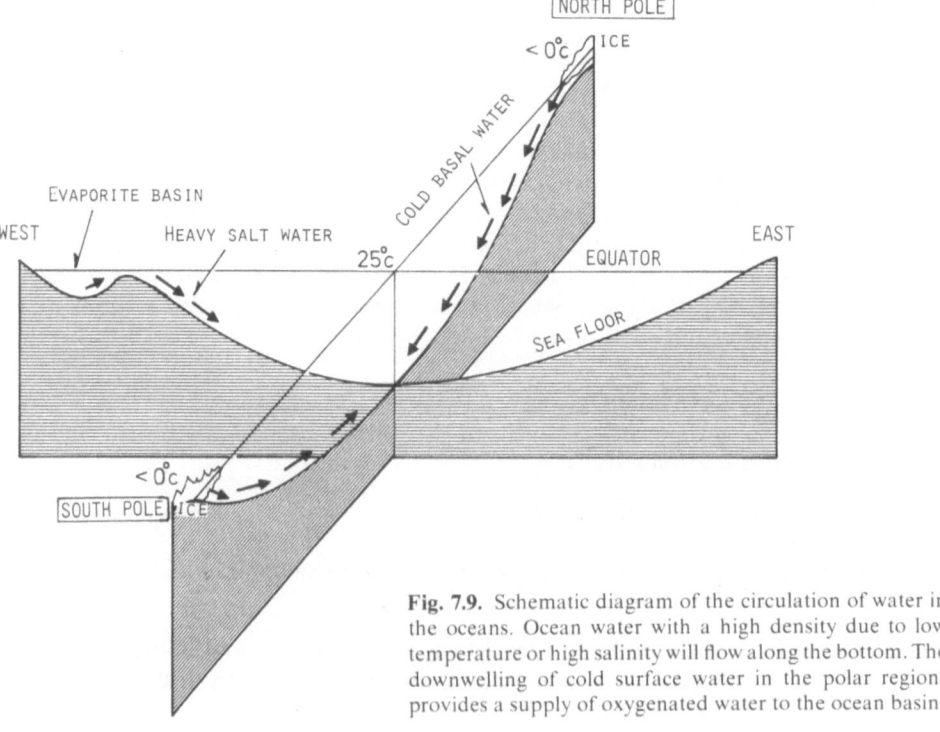

Fig. 7.9. Schematic diagram of the circulation of water in the oceans. Ocean water with a high density due to low temperature or high salinity will flow along the bottom. The downwelling of cold surface water in the polar regions provides a supply of oxygenated water to the ocean basins

and cold water masses is called a *thermocline*. If the density difference is largely due to salt content, we call the boundary a *halocline*. A *pycnocline* is the boundary between two water masses with different densities, mostly due to the combined effect of temperature and salinity. In lakes in temperate and cold regions there is good circulation, due to the inversion of water density at 4°, when it reaches a minimum density.

The addition of freshwater to a basin (e.g. the Baltic Sea and the Black Sea) leads to stratification of the water due to salt concentration. Evaporite basins give rise to water which is heavy due to its high salt content and therefore forms a layer which flows along the bottom. Such a layer will lead to reduced circulation and reducing conditions in the bottom layer. If the density contrast due to salt concentration is greater than that due to temperature, this will reduce or prevent the downward flow of cold, oxygen-rich surface water (Fig. 7.10).

There are factors which indicate that in previous geological periods (e.g. the Cretaceous) there was warm, salty water (about 15°) in the ocean basins instead of the present situation with cold basal water (2–3°) and a normal salt content. A higher average temperature in the oceans leads to reduced CO_2 solubility and a lower carbonate compensation depth (CCD). The volume of water welling up from deeper water layers to the surface corresponds to the amount of downflow. If we have basal water with high salinity, we have less downward flow and consequently less upwelling, so less nutrients are added to the surface water.

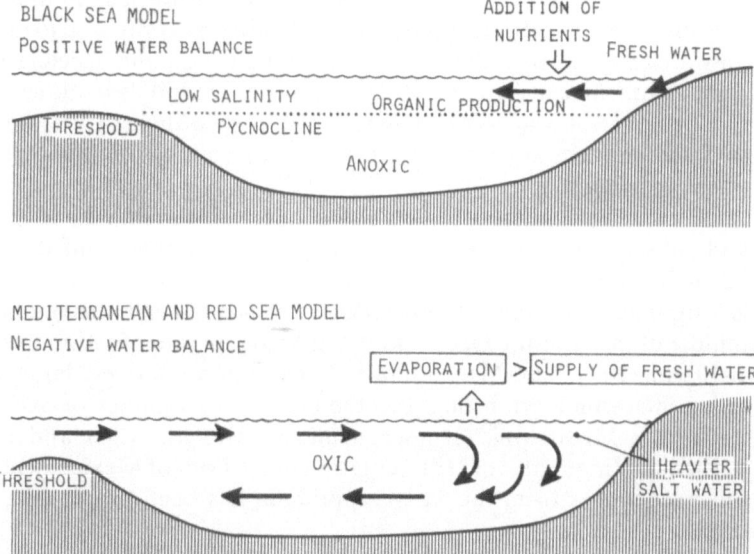

Fig. 7.10. Two different circulatory patterns in virtually isolated ocean basins. When the addition of fresh water is greater than evaporation (positive water balance) the lighter surface water will not mix with the deeper water masses and organic matter which sinks down to the bottom uses up the oxygen and causes reducing conditions. This is the case in basins like the Black Sea, and also in many glacially eroded fjords (e.g. in Norway). In desert climates the surface layer will become heavier through evaporation, and sink to the bottom, thus providing a supply of oxygen to the bottom water. This will cause increased oxidation of organic matter in bottom sediments

Conclusion

We have seen that a number of different processes affect the geochemical equilibrium of the ocean. There must also be an equilibrium between addition and
precipitation of chemical and biological sediments in order for the composition of
the ocean water to remain constant. The prerequisite for maintaining this equilibrium, however, has varied throughout geological time. During our earliest
geological history, up to about 2.10^9 years ago, when the atmosphere was reducing,
most geochemical processes acted very differently from the way they do now.
Weathering was less efficient then because of the low oxygen concentration in
the atmosphere and limited biological weathering. Consequently, less ions were
added to the sea via rivers. The spread of the ocean floor, on the other hand, was
probably more rapid, and more sea water circulated through the spreading
ridge. Isotope surveys ($^{87}Sr/^{86}Sr$) of sea water in early Precambrian rocks indicate
that at that time the composition of sea water was more strongly controlled by
circulation through the basalt on the spreading ridge. We can say that the composition of the sea water was buffered by material from the spreading ridge, i.e. the
mantel (Veizer 1982).

Clastic Sedimentation in the Oceans

Clastic sediments are produced chiefly on the continents, and brought to ocean
areas either through fluvial or eolian transportation. Island arcs associated with
volcanism may produce large amounts of sediment compared to their area, because
they are tectonically active, which causes elevation and accelerated erosion. Volcanic sediments are formed in large quantities around volcanic island groups, while
fine-grained volcanic ash becomes spread over wide areas. Submarine volcanism
may also produce some sediment, for example along the Mid-Atlantic Ridge, but
this is very limited.

The greatest quantities of clastic sediment are fed into the ocean through deltas,
and subsequently transported further along the coast and down the continental shelf, to the abyssal plains at the foot of the slopes. Around Antarctica there
is a significant amount of deposition of clastic, glacial sediments. In areas in the
middle of the Atlantic Ocean, far from land, the rate of sedimentation is as low as
1-10 mm/1000 years. The Atlantic Ocean has a relatively large supply of clastic
sediments, which are brought to the ocean by a number of major rivers: the St.
Lawrence, Mississippi, Orinoco, Amazon, Congo, Niger and Rhine. We have
particularly high rates of sedimentation in the Gulf of Mexico, where deposition of
thick sequences from the Mississippi delta has been proceeding since Mesozoic
times.

The South American and African continents drain mainly into the Atlantic.
The water divide between the Atlantic and the Pacific Oceans is very far to the west
in South America, and that with the Indian Ocean in Africa is very far to the east.
(Table 7.1).

The Pacific Ocean is surrounded by a belt of volcanic regions and island arcs.
There are relatively few rivers and these carry large amounts of clastic sediment

Table 7.1. A comparison between sedimentary conditions in the Atlantic Ocean and the Pacific Ocean

	Atlantic ocean	Pacific ocean
Plate tectonic situation	Ocean-floor spread (Mid-Atlantic ridge)	Ocean-floor shrinkage (subduction)
Margins	Passive	Active
Volcanism	Largely submarine (apart from Iceland)	Mostly above sea level
Extension of volcanic sediments	Limited	Great
Drainage area on continents	$6,710^7$ km^2	$1,810^7$ km^2
Relation between area of sea bed and drainage area	0.72	0.10
Clastic sedimentation	High	Low
Annual sediment added	1.95×10^9 t	2.98×10^9 t[a]
Area	9.3×10^7 km^2	18×10^7 km^2

[a] 2.38×10^9 tonnes of these sediments are carried into the Yellow Sea in China by the Yellow River and the Yangtze River and deposited there.

right out into the Pacific Ocean, as they do in the Atlantic Ocean. Sediment which is eroded, for example on the Asian continent, is deposited in shallow marine areas (marginal seas) such as the Yellow Sea and the China Sea. The sediments are cut off from further transport by the island arc running from Japan and southwards. The Pacific Ocean is therefore dominated by volcanic sediments.

Volcanic sedimentation takes the form of volcanic dust and glass, which may be transported aerially over long distances. After sedimentation, volcanic glass will turn into *palagonite*, an amorphous compound formed by hydration of basaltic tuff. Palagonite may then be further converted into montmorillonite or zeolite minerals. The zeolite phillipsite is very widely found in the Pacific Ocean, but is far rarer in other ocean areas. Pumice is also a volcanic product, and may be transported by water over large areas. The eruption of volcanoes in the Pacific Ocean area in historic times shows that large eruptions produce 10^9–10^{10} tonnes of ash, and much the same amount of pumice and agglomerates.

Submarine volcanism, by contrast, gives very little sediment. The lava which flows out onto the sea-bed will form an insulating crust on contact with the water (often forming pillow lava) so that little volcanic matter goes into suspension.

Weathering and erosion processes produce the entire volume of sediment which can be deposited in sedimentary basins. Material added by rivers takes the form of clastic and dissolved matter. The ratio between the quantities of these two forms of sediment addition is a function of precipitation, temperature and relief. Dry areas, like Australia, produce mainly clastic material, while the African continent produces mainly dissolved material because of the intensive weathering in some places on the continent (Table 7.2).

It is important to remember that the total deposition corresponds closely to the total production of sediment if we assume the sea to maintain approximately

Table 7.2. Review of the ratio between mechanical and chemical denudation of the different continents. (After Garrels and MacKenzie 1971)

Continent	Annual chemical denudation, tonnes/km²	Annual mechanical denudation, tonnes/km²	Ratio mechanical chemical denudation
North America	33	86	2.6
South America	28	56	2.0
Asia	32	310	9.7
Africa	24	17	0.7
Europe	42	27	0.65
Australia	2	27	10.0

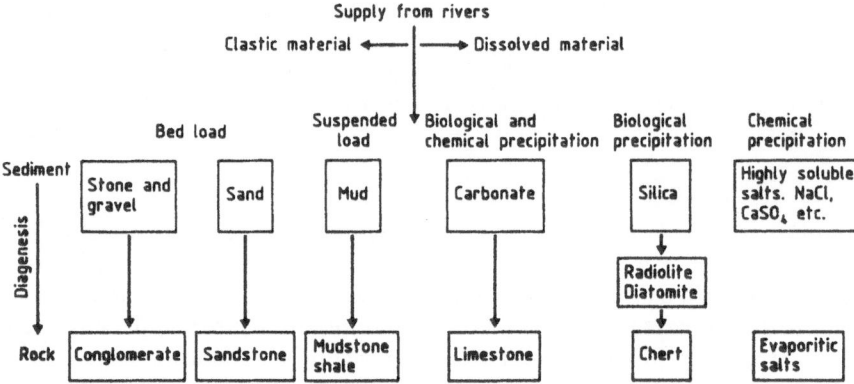

Fig. 7.11. Revue of different kinds of sediments and sedimentary rocks

constant composition. Sediments produced are deposited as a variety of different clastic sediments, or biochemically or chemically precipitated sediments which all derive from erosion and weathering (Fig. 7.11).

Further Reading: Garrels and MacKenzie 1971; Lisitzyn 1972; Anikouchine and Sternberg 1973; Veizer 1973, 1980; Stowe 1979; Hay et al. 1982; Kenneth 1982.

8 Carbonate Sediments

The carbonate sediments constitute a large group of varied origin. Common to them all is that they consist largely of carbonate minerals. They may be deposited as biogenic sediments, as chemical precipitates, or less commonly as clastic sediments eroded from other carbonate rocks. We often find clastic carbonate grains forming a minor component of sandstone and conglomerates. Because carbonates are less resistant to transport and weathering than quartz and felspar, the clastic carbonate content of sediments will tend to decline with increasing transport and weathering.

We distinguish between carbonate fragments (clasts) from older carbonate beds which have become exposed to form an external source for a basin (*extrabasinal clasts*) and clasts formed locally within the basin through mechanical breakdown or biological fragmentation of sediment crusts or fossil aggregates. These are intrabasinal clasts (*intraclasts*) which are about the same age as the matrix.

The first condition for obtaining relatively pure carbonate deposits is that there must be very little supply of clastic material. This puts an important limitation on the occurrence of carbonate sediments.

Carbonate deposits occur in the following environments:

1. *Carbonate Platforms and Reef Environments Along the Coast.* They require clear water and can form only outside the range of clastic sediment supplies from land. In dry areas, in particular, there will be very little runoff from land which could add clastic sediments. The drainage pattern also plays a major role.
2. *Tidal and Supratidal Environments Along the Coast.* Form first and foremost in dry areas.
3. *Reef and Carbonate Platforms Surrounded by Deep Ocean* (e.g. the Bahamas) or atolls — carbonate reefs topping volcanic seamounts in the ocean basins. The ocean basin on the landward side will function as a trap for clastic sediments (Fig. 8.1). In such an environment carbonate sedimentation may continue for long geological periods. Carbonate production has to keep pace with the subsidence of the sea floor or the rise in sea level, so that the area does not sink below the photic zone. The Bahamas platform represents the build-up of carbonate from Jurassic-Cretaceous times right up to the present, and forms a very steep slope out to the ocean basins.

Warm water carbonates of this type contain a varied fauna of carbonate-secreting organisms, including hermatypic corals and green algae, and are often referred to as *chlorozoan* associations (Lees and Buller 1972).

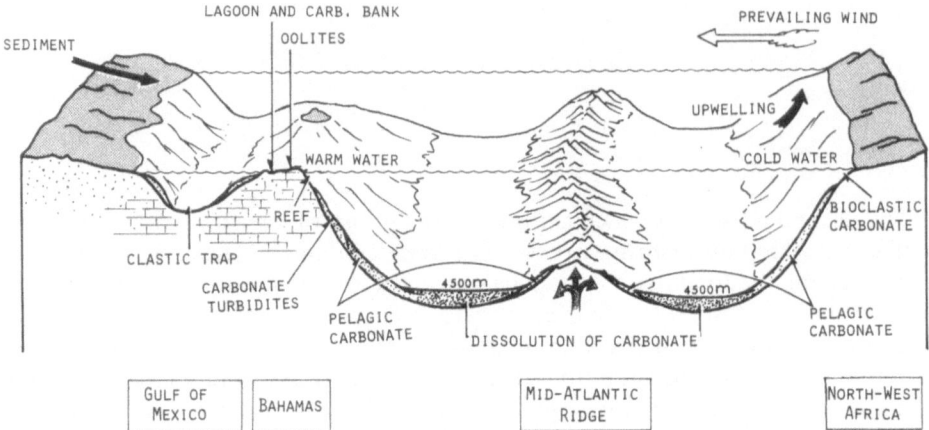

SCHEMATIC DIAGRAM OF THE VARIOUS CARBONATE ENVIRONMENTS
IN THE ATLANTIC OCEAN BETWEEN THE MEXICAN GULF AND N.W. AFRICA.

Fig. 8.1. Distribution of carbonate sediments across the Atlantic Ocean

4. *Carbonate Turbidites on Slopes Skirting Carbonate Platforms.* They may be formed from carbonate sand and mud which is stirred up during storms over shallow water and transported down the slopes. Mud flows, debris flows and submarine breccias may also form, as carbonate platforms often have very steep slopes.

5. *Carbonate Sediment Formed from the Remains of Carbonate-Producing Pelagic Organisms.* The most important of these are foraminifera and coccolithophores, a type of algae which secrete small calcite shells. Since the planktonic carbonate organisms are small, pelagic carbonate deposits form a fine-grained ooze of clay and silt size, with occasional larger fossil fragments. Large areas of the South Atlantic and the Pacific are covered by sediment which contains more than 50% $CaCO_3$ from planktonic calcareous organisms. At depths greater than 4000–5000 m, the solubility of $CaCO_3$ becomes too great, and the carbonate percentage declines. Planktonic carbonates may also be deposited in shallow ocean areas, and the chalk deposits of Northwest Europe from Cretaceous times are a good example of this. The depth of water here was only a few hundred metres below the photic zone. The major carbonate-producing planktonic organisms, such as foraminifera and coccolithophores, have only existed since mid-Mesozoic times. Consequently we do not have chalk deposits or deep-water pelagic carbonates from older periods, when most carbonate production took place in shallow water.

6. *Bioclastic Carbonate Deposits in Temperate and Cold Ocean Areas.* The biological precipitation of carbonate does not depend on whether the water is warm or saturated with calcium carbonate. On the contrary, we often find highest productivity in cold areas because the water there is richer in nutrients, particularly in areas of upwelling. Even if the water is undersaturated with

respect to carbonate, so that carbonate begins to dissolve as soon as the organisms die, carbonate will be deposited if the rate of production of carbonate is faster than the rate of solution. Carbonate will also be deposited in shallow areas which do not receive clastic sediments. The Spitsbergen Bank northwest of Bear Island in the Barents Sea is a good example of this (Bjørlykke et al. 1978). This area, which is from 30–100 m deep, is surrounded by deeper channels, and therefore receives little clastic matter. The area lies on a cold front where cold currents from the north and east mix with the warmer water of the Atlantic Ocean. We find similar carbonate deposits on other banks off North America, for example Grand Bank, and along the coasts of western Scotland and Norway we find Holocene carbonate deposits which form a shell gravel.

Cold-water carbonates consist mainly of molluscs, benthic foraminifera, barnacles, bryozoa and calcareous red algae, and are called a *foraminiferal* association (Lees and Buller 1972).

7. *Evaporite Basins.* In ocean areas with normal salinity, practically all deposition of carbonate takes place by way of biological precipitation. In ocean areas with somewhat higher salinity, for example in the Persian Gulf, however, chemical precipitation of calcium carbonate may occur. Here too this precipitation may nevertheless be linked to biological factors. Periods when algae flourish in the surface water entail photosynthesis and consumption of CO_2. This raises the pH, creating oversaturation and thus favourable conditions for chemical precipitation. Carbonates make up a small percentage of the salt precipitated when sea water evaporates to dryness. However, they are among the least soluble of the common salts in restricted ocean basins where the salinity is too low for the more soluble salts to precipitate (e.g. NaCl). Carbonates and sulphates often form thick evaporite sequences.

8. *Lakes and Inland Seas.* Lacustrine sediments may also contain considerable amounts of carbonate, and at lower latitudes in particular we find pure carbonate beds. In cold lakes the solubility of carbonate will be relatively high, and the carbonate content will tend mostly to be biogenic (e.g. molluscs). In lakes in temperate areas we also get calcareous sediments (*marls*) deposited. Algae often play an important role in carbonate production in lakes, and certain higher plants which grow in lakes can also precipitate carbonate.

The Dead Sea is a good example of an inland sea where carbonate is precipitated chemically due to strong evaporation. In Africa and other tropical areas, lakes will be subject to seasonal evaporation to dryness, and we may find alternation between biogenic carbonate layers and chemically precipitated carbonate. Here again, the production of diatoms and algae in the surface water layer consumes CO_2 (increasing the pH) and plays a major role in the precipitation of carbonate. Molluscs (bivalves and gastropods) and algae are important components of freshwater carbonates as well as chemically precipitated carbonate.

9. *Calcareous Tufa Deposits and Travertine.* Tufa is the name of a porous limestone which is precipitated where groundwater flows out at the surface or comes into contact with the atmosphere, as in limestone caves or springs. When it is inside rock, groundwater will have a higher partial pressure of CO_2, which can

remain in solution due to the lower temperature and higher pressure than at the surface. When water flows out of the rock, it will give up CO_2, and in summer at least, will warm up so that carbonate is precipitated mainly as calcite. Exposure to light will also cause biogenic precipitation (photosynthesis). Evaporation may also enrich the water and lead to precipitation. Calcareous tufa is common in areas with limestone, and deposits often create good moulds of plants and occasionally even of animals. Travertine is a more massive, banded (laminated) carbonate which has been used as a decorative stone, particularly in Mediterranean countries.

Carbonate-CO_2 Systems in the Sea

Carbon dioxide concentration is the factor which has greatest influence on pH and the solubility of carbonates in the ocean. CO_2 dissolves to form carbonic acid, which dissociates into bicarbonate (HCO_3^-) and carbonate (CO_3^{2-}) ions.

$$CO_2 + H_2O \rightleftharpoons H^+ + HCO_3^- \rightleftharpoons CO_3^{2-} + 2H^+.$$

The CO_3^{2-} concentration controls solubility, and thus the CO_2-carbonate system is self-buffering.

The dissociation constant for H_2CO_3 and HCO_3^- is:

$$K_1 = \frac{aH^+ \cdot aHCO_3^-}{aH_2CO_3} \quad \text{and } K_2 = \frac{aH^+ \cdot aCO_3^{2-}}{aHCO_3^-}.$$

We see that at high pHs, i.e. low H^+ activity, the reaction will be driven to the right and the CO_3^- concentration will have to be higher. At low pH values the equilibrium will shift to the left, giving more free CO_2. CO_2 is found in both water and in the atmosphere, and is exchanged between them (Fig. 8.2). The solubility of CO_2 in water is greatest at low temperatures, and decreases as the temperature rises. Since it is largely CO_2 concentration which determines the pH of water, the pH is highest (8–8.5) in the warm surface layer at low latitudes, and lowest in polar areas (7.5–8). Whereas photosynthesis consumes CO_2, removing it from the surface layer of the water so that the pH rises, respiration adds CO_2, causing the pH to fall.

$$H_2O + CO_2 \underset{\text{Photosynthesis}}{\overset{\text{Respiration}}{\rightleftharpoons}} CH_2O + O_2.$$

The water below the photic zone will gain CO_2 from the respiration of zooplankton and the breakdown (i.e. oxidation) of organic matter which sinks down through the water column will also produce CO_2 and lower the pH. In shallow areas, a daily variation in pH has been registered as a result of the fact that photosynthesis takes place only during the day, while respiration continues at night. Respiration in the water below the photic zone and oxidation of organic matter also contribute to the pH declining downwards through the water column. Microbiological breakdown of organic matter will also liberate CO_2 and lower the pH (Fig. 8.2).

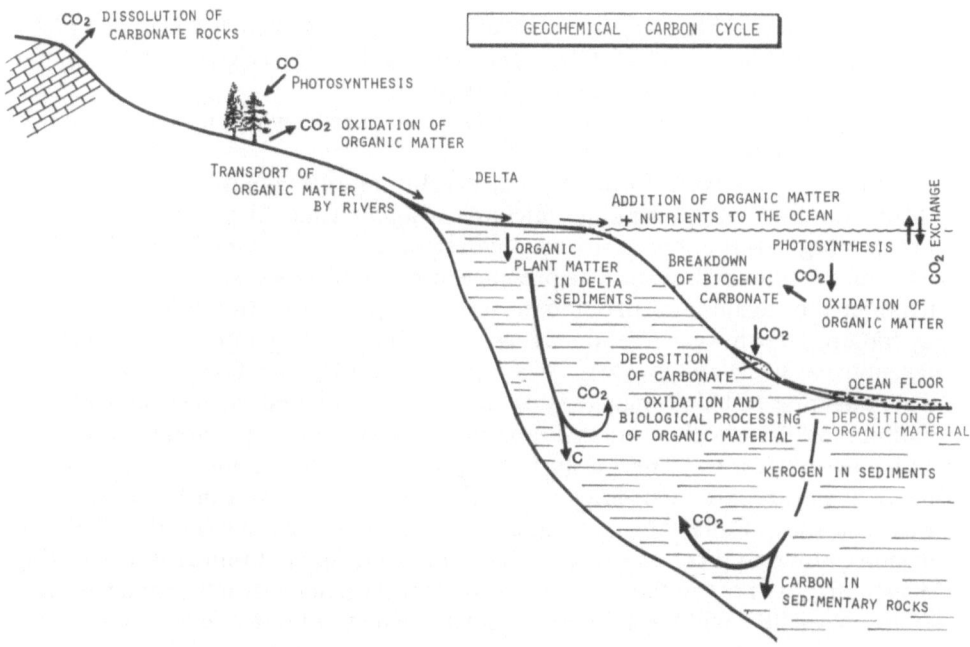

CO2 DISSOLUTION OF
         CARBONATE ROCKS

GEOCHEMICAL   CARBON   CYCLE

CO
    PHOTOSYNTHESIS

CO2 OXIDATION OF
        ORGANIC MATTER

TRANSPORT OF
ORGANIC MATTER
    BY RIVERS

DELTA

ADDITION OF ORGANIC MATTER
+ NUTRIENTS TO THE OCEAN

PHOTOSYNTHESIS

CO2 EXCHANGE

ORGANIC
PLANT MATTER
 IN DELTA
-SEDIMENTS-

BREAKDOWN
OF BIOGENIC   CO2
CARBONATE

OXIDATION OF
ORGANIC MATTER

CO2

DEPOSITION
OF CARBONATE

OCEAN FLOOR

CO2
 OXIDATION AND
BIOLOGICAL PROCESSING
 OF ORGANIC MATERIAL

DEPOSITION OF
ORGANIC MATERIAL

C

KEROGEN IN SEDIMENTS

CO2

CARBON IN
SEDIMENTARY ROCKS

Fig. 8.2. The carbon budget (CO_2) on land, in the ocean and in sediments

Geochemistry of Carbonate Minerals

The most important minerals in carbonate sediments are: calcite ($CaCO_3$), aragonite ($CaCO_3$), siderite ($FeCO_3$), magnesite ($MgCO_3$), dolomite ($CaMg(CO_3)_2$),
ankerite ($CaFe(CO_3)_2$).

Calcium carbonate occurs as two common polymorphs: calcite, which has
hexagonal symmetry, and aragonite, which is orthorhombic. The orthorhombic
lattice has an arrangement of CO_3^{2-} ions which requires cations that are 1 A or larger.
Analogous with aragonite we therefore have strontianite ($SrCO_3$), witherite
($BaCO_3$) and cerrusite ($PbCO_3$), which are all orthorhombic. In addition to Ca^{2+},
the aragonite structure can thus accommodate the elements Sr, Ba and Pb. Sr in
particular is an important trace element in aragonite. Aragonite crystals forming in
marine environments today contain 5,000–10,000 ppm Sr.

The calcite structure requires cations which are 1 A or smaller. Since Ca^{2+} has
an ionic radius close to 1 A, both crystal structures (aragonite and calcite) are
possible for $CaCO_3$. Analogous with calcite we have magnesite ($MgCO_3$), siderite
($FeCO_3$), rhodocrosite ($MnCO_3$), and smithsonite ($ZnCO_3$) which are all hexagonal. These metals all have an ionic radius of about 0.6–0.7 A, and calcite can
contain considerable concentrations of these cations. Magnesium content in particular may be high, and when the Mg content exceeds 4%, we call it *high-
magnesium calcite* or *Mg-calcite*. When the Mg content is less than 4% we use the
designation low-magnesium calcite or simply calcite.

Dolomite ($Ca.Mg(CO_3)_2$) is a carbonate mineral in which layers of $CaCO_3$ alternate with layers of $MgCO_3$. *Ankerite*, $Ca(Fe,Mg)$ $(CO_3)_2$, has a similar structure, with Fe^{2+} ions substituted for Mg^{2+}.

The magnesium end member, $MgCO_3$, is magnesite, which is also common in sedimentary rocks.

Although sea water contains considerably more magnesium than calcium ($Mg^{2+}/Ca^{2+} = 5$), it is essentially calcium carbonate which is precipitated. This is because Mg^{2+} ions are very strongly hydrated, so that it is kinetically difficult to form magnesium carbonate. The presence of sulphate ions in sea water also seems to inhibit the precipitation of dolomite. At high temperatures ($60–100°C$), however, hydration is much weaker, and magnesium carbonates (e.g. dolomite) are formed far more easily even at low Mg^{2+}/Ca^{2+} ratios (e.g. $Mg/Ca = 0.1$).

Aragonite and high-magnesium calcite are precipitated as metastable phases, and in most cases will go into solution in freshwater and reprecipitate as calcite, which has a lower free energy, during the course of a few thousand years. But in dense shales in particular, aragonite may be preserved, even in Palaeozoic and Mesozoic rocks. Calcite formed under oxidising conditions contains practically no iron because only Fe^{2+} can replace Ca^{2+}. In reducing environments, e.g. during diagenesis, any Fe^{2+} in the porewater can enter the calcite structure and so-called "ferroan calcites" will form. These are calcites which contain a few thousand ppm of iron and can be stained blue with a solution of potassium ferricyanide and alizarin red. In the sulphate-reducing zone, high concentrations of sulphur will cause all available Fe^{2+} to be precipitated as sulphides (FeS_2), so that very little is available to enter the calcite structure. Dolomite normally contains considerable amounts of iron, which substitutes for magnesium. Dolomite formed early in diagenesis is fine-grained and may often have a magnesium deficit in relation to calcium [e.g. $Ca_{55}Mg_{45}(CO_3)_{100}$]. This is called protodolomite.

Classification of Carbonate Rocks

Classifications of carbonate rocks may be analogous with those of sandstones, and the schemes proposed by both Folk (1959) and Dunham (1962) show this tendency (Table 8.1 and Fig. 8.3). They are based on the relative amounts of grains and mud (carbonate mud) and the types of grains (fossils, rock fragments or minerals).

Carbonate rocks consist of grains which are of sand and gravel size, and carbonate mud which corresponds to clay. Carbonate rocks span a wide register of types of different origin. They may be purely clastic sediments deriving from older carbonate rocks on land, biogenic clastic sediments from organisms in the basin (sea) or chemically precipitated carbonate.

Folk's classification calls grains of sand or gravel size *Allochems*. They may be (1) fossils (2) ooids (3) pellets or (4) fragments of carbonate rocks (lithoclasts and intraclasts).

Poorly sorted sediments which contained a considerable amount of carbonate mud at the time of deposition are called micrites. They are thus indicators of a low-energy environment, e.g. deposition below the wave base. Well-sorted car-

Table 8.1. Classification of Carbonate Rocks (Folk 1959)

Percent Allochems	OVER 2/3 LIME MUD MATRIX				SUBEQUAL SPAR AND LIME MUD	OVER 2/3 SPAR CEMENT		
	0-1%	1-10%	10-50%	OVER 50%		SORTING POOR	SORTING GOOD	ROUNDED AND ABRADED
Representative Rock Terms	Micrite and Dismicrite	Fossili Ferous Micrite	Sparse Biomicrite	Packed Biomicrite	Poorly Washed Biosparite	Unsorted Biosparite	Sorted Biosparite	Rounded Biosparite

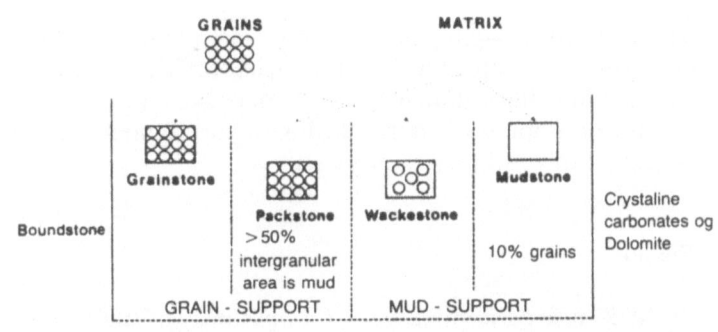

1959 Terminology	Micrite and Dismicrite	Fossiliferous Micrite	Biomicrite			Biosparite		
Terrigenous Analogues	Claystone		Sandy Claystone	Clayey or Immature Sandstone		Submature Sandstone	Mature Sandstone	Supermature Sandstone

■ Lime mud matrix   ▨ Sparry calcite cement

CARBONATE TEXTURAL SPECTRUM

	Limestones and Primary Dolomites					Secondary Dolomites		
	>10% Allochems		<10% Allochems		Bioherm	Allochem Ghosts	No Allochem Ghosts
	Cement> Matrix	Matrix> Cement	1-10% Allochem	<1% Allochem			
>25% Intraclasts	Intrasparrudite Intrasparite	Intramicrudite Intramicrite	Intraclastic Micritic			Intraclastic Dolomite	
>25% Oolites	Oosparrudite Oosparite	Oomicrudite Oomicrite	Oolitic Micritic			Oolitic Dolomite	
Ratio Fossils to pellets >3:1	Biosparrudite Biosparite	Biomicrudite Biomicrite	Fossiliferous Micrite	Micrite	Biolithite	Biogenic Dolomite	Crystalline Dolomite
3:1 to 1:3	Biopelsparite	Biopelmicrite	Pelletiferous Micritic			Pellet Dolomite	
<1:3	Pelsparite	Pelmicrite					

GRAINS   MATRIX

Boundstone | Grainstone | Packstone >50% intergranular area is mud | Wackestone | Mudstone | Crystaline carbonates og Dolomite

10% grains

GRAIN - SUPPORT   MUD - SUPPORT

High energy   Low energy

Fig. 8.3. Dunham's classification of carbonate rocks

bonate sand and gravel without a muddy matrix has primary porosity which may fill with cement during diagenesis. The cement, which is precipitated from aqueous solution, consists of clear, transparent crystals (spar) which are easy to distinguish from micrite, which tends to be brownish because of the organic content. *Sparite* is thus a term for well-sorted carbonate sand, originally with purely primary porosity, which has later been filled with calcareous cement (spar). As such it normally represents a high-energy environment because of good sorting. The terminology in Folk's classification (1959) is based on allochem type and whether the limestone has a micritic or sparitic matrix. *Biosparite*, for example, means lime sand or gravel consisting of fossil fragments without a matrix, but with sparite cement.

Oomicrites and *pelmicrites* are sediments consisting of grains or oolites or pellets deposited in a carbonate mud matrix. *Intrasparite* consists of carbonate fragments eroded inside a basin to form well-sorted carbonate sand. Sediments with a considerable percentage of grains larger than sand size are called *rudites*. Dunham's classification (1962) is based on packing of sediment grains and whether there is a matrix between the grains. *Wackestone* corresponds to the greywackes of clastic sediments, with the grains floating in the matrix. *Packstone* has a grain-supported structure, i.e. the sand grains rest upon one another, and the matrix is interstitial.

Intraclasts

These are fragments of early-cemented carbonate sediment which have been broken down by erosion or biological boring. Intraclasts thus consist of carbonate of the same age as the matrix. Early cementing, e.g. in the beach zone, or freshwater cementation due to regressions, may result in cemented sediments which are later broken down. We may also find early cementation in a marine environment, and tidal channels, for example, will often give rise to characteristic conglomerates of intraclasts (intraformational conglomerates).

Lithoclasts

These are rock fragments transported in from outside the basin, e.g. from lime-stones exposures on land. Such fragments, often called extrabasinal clasts, will thus be older than the sediments they are deposited in.

If we are not sure whether the limestone fragments are intraclasts or lithoclasts, we can simply call them lime clasts.

Peloids

"Peloid" is a descriptive term for particles of microcrystalline calcite without internal structure. Most peloids are the excrement (faeces) of marine organisms, and are 0.1–0.5 mm in diameter, but they may also be larger. Animals which eat mud, like snails, bivalves and crustaceans, deposit large quantities of carbonate

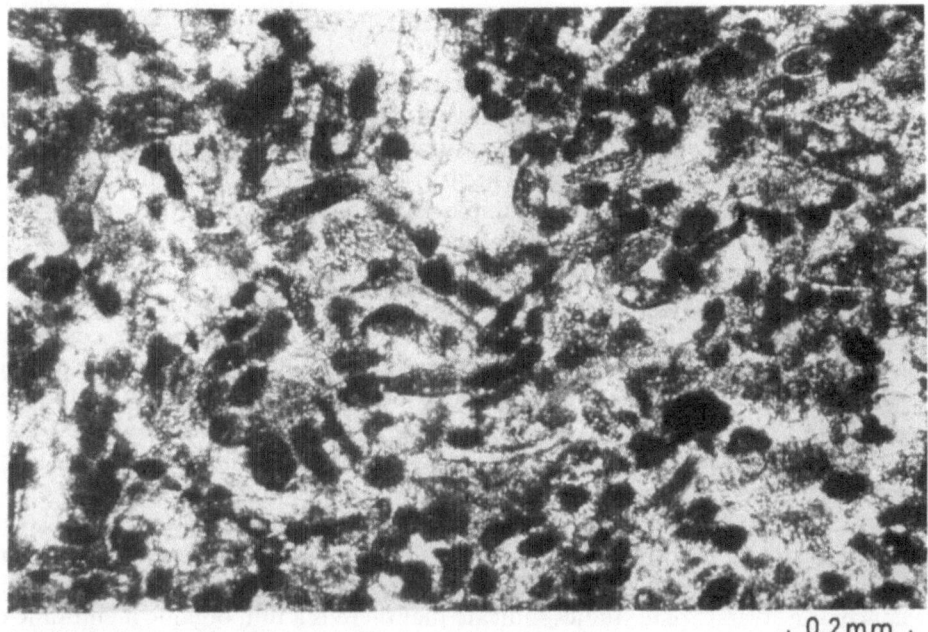

Fig. 8.4a. Pellets with a matrix of sparitic calcite — *pelsparite*. From the Silurian, Oslo field. The pellets were deposited as well-sorted, fine sand-sized grains. The spary cement then filled the primary pores

INCREASING ENERGY IN THE DEPOSITIONAL ENVIRONMENT

 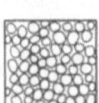 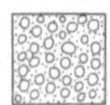 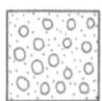 

BOUNDSTONE	GRAINSTONE	PACKSTONE	WACKESTONE	MUDSTONE
REEF OR OTHER TYPE OF ORGANIC BUILDUP	SAND SIZED GRAINS WITHOUT A MUD MATRIX	SAND SIZED GRAINS IN CONTACT FORMING A GRAIN FRAMEWORK WITH A MATRIX OF MORE THAN 50% MUD	SAND SIZED GRAINS "FLOATING" IN A MATRIX OF MUD	<10% GRAINS

Fig. 8.4b. Increasing energy in the depositional environment

mud pellets in carbonate environments. The pellets are somewhat harder than the surrounding mud because slight carbonate cementation takes place inside the organism which produced them. Their organic content gives them a characteristic brown colour. Nevertheless, the pellet composition does not differ so much from ordinary micritic mud, and recrystallisation of mudstone may produce peloids which resemble organic pellets. Pellets may also become well sorted like fine sand and form pelsparite (Fig. 8.4a).

Ooids (ooliths)

Ooids are round grains of carbonate formed through chemical or biological precipitation. A sediment which consists of ooids is called an *oolite,* but the term "oolith" is also often used for single grains, as a synonym for ooids. Ooids are by definition less than 2 mm, but most are 0.2–1 mm in diameter. Similar concentric structures which are larger than 2 mm are called pisolites, and these are usually precipitated in caliche and speleothems.

Ooids have a concentric structure with layers of carbonate around a core which may consist of a small quartz or felspar grain or a carbonate fragment, for example a fossil or a pellet. Modern marine ooids consist of aragonite with a concentric tangential structure. This structure is composed of small aragonite crystals ($< 3 \mu$) with their c-axis parallel to the lamination, so that an extinction cross is produced under the microscope with crossed nicols. There are also less well-oriented laminae, yellowish-brown in colour, which may be due to accumulated organic-rich carbonate mud particles.

We find ooids only in very warm marine environments and in some saline lakes. The water must be saturated with carbonate. Constant wave movement is necessary to roll the ooid around and over the bottom to give even precipitation and form concentric layers. Some studies indicate that there is a thin organic membrane of bacteria on ooids which helps to precipitate aragonite, and which may help to trap small aragonite particles suspended in the water. We would expect that direct chemical precipitation would enable the needles to orientate themselves radially on the surface of the ooid, while snowball-type growth through the accumulation of small aragonite needles would give concentric layers.

Radial ooids have now been found in some modern lakes and in some hypersaline environments, where they contain radial layers of high-Mg calcite. Radial ooids are common in ancient limestones, i.e. of Palaeozoic age (Fig. 8.5), and this suggests that high-Mg calcite was more common in the form of ooids and marine cement at that time than in the Mesozoic and Cenozoic, when aragonite tended to precipitate from sea water.

Because ooids require warm water and constant wave agitation, they are only found in very shallow water, normally less than 2–3 m deep. However, they may be carried out to greater depths and deposited there. Ooids are typical of the West Indies (Bahamas), the Trucial coast (Persian Gulf), and in some areas around the Indian Ocean and islands in the Pacific Ocean. On the east side of the Atlantic Ocean — along the coast of Africa — however, the water is too cold to permit sufficient carbonate saturation. Ooids are therefore an important indicator of depositional environment and climate.

Fossils

Carbonate-Secreting Animals

Foraminifera. These are single-shelled benthic and planktonic organisms with a diameter ranging from 0.1 mm to 3–4 cm. Modern planktonic foraminifera have

Fig. 8.5. Fragments of oolites with large crystals of calcite, probably formed by meteoric water. The radial cracks in the oolite suggest that at the time of formation (the Ordovician) it consisted of Mg-calcite. From the Upper Ordovician in the Oslo field

shells consisting of calcite with 5% $MgCO_3$, and aragonite shells are rare. Foraminifera occur from the Cambrian up to the present day, and are particularly important index fossils in the Tertiary-Quarternary, but also in Mesozoic horizons and in the Upper Palaeozoic (Fusulinidae).

Foraminifera are also good indicators of the temperature of aqueous environments.

Sponges (Porifera). We have three types here, one which secretes a calcareous shell (*Calcisponga*) and two which secrete silica shells (*Hyalosponga* and *Demosponga*). Calcareous sponges were most widespread in ancient, warm seas only a few metres deep. They secrete a calcite shell, and occur in many fossil calcareous reefs. Many sponges are important boring organisms which help to break down the shells of large organisms to finer-grained material. The holes are 0.2–1 mm in diameter and up to 0.5 cm long.

Stromatoporoids. Stromatoporoids were lime-secreting, often reef-forming organisms which occur in nodular or branched colonies. They are important in many Palaeozoic limestones. The structures in stromatoporoids indicate that they secreted an aragonite skeleton.

Corals. Some corals (hermatypic) live in symbiosis with single-celled algae. They do best in very shallow areas (1–20 m deep) and in warm water (25–18°). These corals tolerate no lowering of the salinity, nor turbid water.

The other main type of corals (ahermatypic) do not live in symbiosis with single-celled algae, and often occur below the photic zone. They can live in water with temperatures right down to 0°C. Ahermatypic corals can form reef-like structures at depths of several hundred metres. *Lophelia* reefs or biostromes are found off Lofoten, Norway, for example.

Rugose corals, hexacorals (*Scleractinia*), tabulate corals and octacorals are all important rock-forming corals. Corals have aragonite or calcite skeletons.

Bryozoa. Bryozoa are small colony-forming animals which live in clear, relatively shallow water. They may be rock-forming, and particularly in Palaeozoic rocks one may find bryozoan reefs. Bryozoa secrete an aragonite or calcite skeleton.

Brachiopods. Brachiopods occur from the Cambrian to the present day. Inarticulate brachiopods often have a phosphate shell, whereas articulate species have carbonate shells of calcite or Mg-calcite. Because brachiopod shells were originally calcitic, and do not undergo recrystallisation, as aragonite fossils do, their structure is very well preserved.

Worms (Vermes). Worms are important organisms which disturb sediments, destroying the primary lamination. In many marine environments all sediment will pass through the alimentary canal of a worm one or more times. The sediment becomes partially cemented into small pellets. A number of worms secrete carbonate along the walls of their tunnels. The space may then fill with sparry calcite and calcium-secreting worms can thus contribute to rock formation. *Serpalinomorphous* worms secrete pipes of lime which may form large aggregates.

Molluscs include:

1. *Lamellibranchiata* (pelecypods, mussels, bivalves)
2. *Gastropoda* (snails)
3. *Cephalopoda* (e.g. orthoceros, cuttlefish).

Molluscs live largely in shallow water (< 50 m), since they feed mainly on algae, but they are also found in deep water. They occur in both cold-water and warm-water environments and are important producers of carbonate, particularly in shallow areas. Molluscs which live on a soft substratum often live buried 10–20 cm down in the sediment. The shells become exposed and concentrated through erosion, and may form lag conglomerates. Molluscs have aragonite or calcite shells.

Trilobites. These existed from the Cambrian to the Permian. Cambrian limestones often consist largely of trilobite shells. Although they are now extinct, we assume from preserved structures that they had calcite shells. Trilobites are important index fossils, particularly for the Cambrian and Ordovician.

Ostracoda (mussel crabs). Ostracoda have two shells which are hinged together at the back. They are about 0.4–1.5 mm in diameter and live in shallow water, generally less than 100 m deep. They live in both fresh and salt water, and have calcitic skeletons. Stratigraphically they range from the Ordovician to the present, but they are particularly important index fossils in Mesozoic and Tertiary strata.

Echinoderms. Echinoderms are common fossils in marine limestones, and are easy to recognise through a microscope because most of their skeletal plates consist of one large calcite crystal. Many echinoderms, for example crinoids (sea lilies), are easily broken down to small sand grains and are often an important component of limestone.

Microcrystalline Lime Mud (Micrite)

Lime mud which is deposited in areas with carbonate sedimentation has a grain size of about 1–4 μ. This means that it cannot be studied particularly effectively under an ordinary microscope, and only with an electron microscope can each individual grain really be seen.

Carbonate mud (micrite) was previously assumed to be chemically precipitated carbonate, as opposed to fossils or fossil fragments, which were naturally of organic origin. However, as mentioned above, a very large amount of modern lime mud, for example from the Bahamas, was found to be formed of aragonite needles from the breakdown of calcareous green algae, particularly *Halimeda, Rhipocephalus* and *Penicillus.*

In shallow marine areas near the equator, like the Bahamas, the sea water is often saturated with respect to aragonite. Nevertheless it has not been possible to prove with certainty that any purely chemical precipitation of aragonite proceeds in these areas. In the Persian Gulf, however, chemical precipitation of carbonate crystals does occur, but only when the salinity is very high, about every fifth year. A sudden proliferation of diatoms could also consume so much CO_2 that the pH rises, causing aragonite to be precipitated.

Microcrystalline lime mud may also be formed by mechanical abration. Skeletons which lie exposed to wave and current activity will be abraded mechanically, and a fine-grained lime mud is formed. However, this process will rarely form the well-crystallised aragonite needles that modern lime muds are found largely to consist of. We may therefore conclude that mechanical abrasion of fossils or carbonate fragments is not the main source of lime mud on carbonate banks with green algae. In other areas precipitation by red algae or mechanical abrasion may be important.

Through diagenetic processes, the microcrystalline mud will dissolve and be replaced by somewhat larger crystals, microsparite, which is distinguished from other sparites by its brownish colour, due to the organic matter in the original carbonate mud (Fig. 8.6).

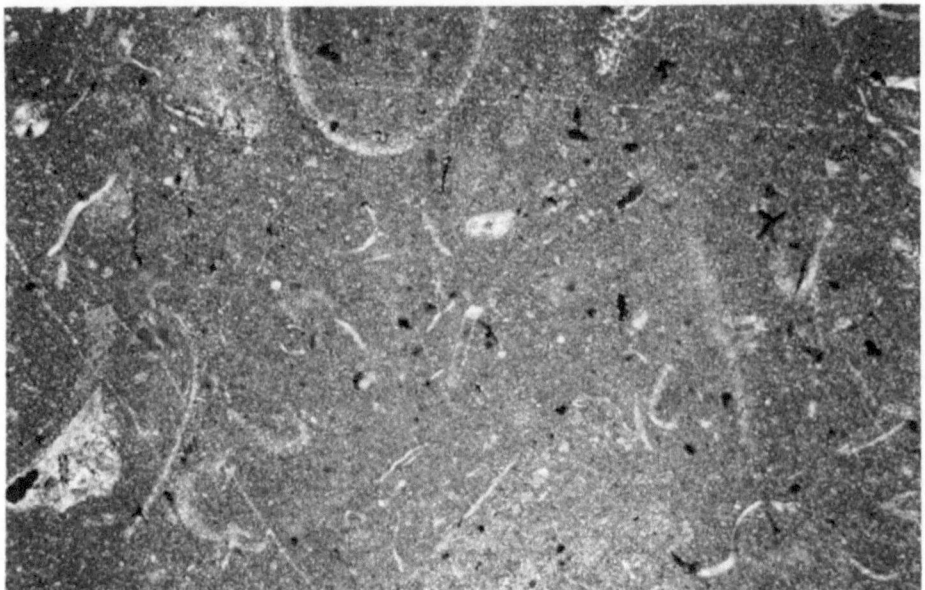

Fig. 8.6. Biomicrite. Brachiopods in a micritic matrix. The micrite is partially recrystallised into somewhat larger crystals (microsparite). Pentamerus limestone from the Oslo field

Micritisation

This process, which was first described by Bathurst (1975), is due to carbonate fragments or fossils being pierced by boring calcareous algae. The holes left by the algae then fill with finely crystalline (2–8 μ) high-magnesium calcite or aragonite. This is called a "micritic envelope". It has been clear for a long time that some bacteria can secrete carbonate, and that if such bacteria are cultivated in sea water, aragonite will be secreted. Crystallisation of aragonite in holes left by boring algae may be due to bacteria which live on the algae.

Modern Carbonate Sedimentation Environments

Modern carbonate sediments are mostly of biogenic origin, except in evaporite basins. The composition of the sediments is therefore largely a function of the type of organisms which produced them. In warm tropical waters, calcareous green algae are important producers of the small carbonate needles which are a major constituent of carbonate mud. Hermatypic corals are also restricted to tropical waters (minimum temperature must exceed approx. 15°C). They are therefore restricted to the photic zone.

The association of hermatypic corals and green algae is called a chlorozoan association. In this environment we also find a large number of other carbonate organisms such as molluscs, forams, red algae and echinoderms.

It is important to remember that the distribution of warm-water carbonate facies is not governed by latitude, but depends to a large extent on ocean circulation patterns. On the western side of the Atlantic Ocean, warm-water carbonate facies are found up to about 30°N from the Equator, while on the eastern side, off West Africa, the ocean is generally too cold, even close to the Equator.

This pattern is due to the east-west equatorial winds producing upwelling on the eastern side of the ocean, and accumulation of warm surface water on the western side.

The Bahamas

The Bahamas Bank is one of the best researched areas of modern carbonate sedimentation. The area is an example of a carbonate platform, where carbonate precipitation has been proceeding since Mesozoic times. As the sea bed has subsided, carbonate sedimentation has built up the area so that the bank has remained in the photic zone. This has led to the bank being surrounded by deeper areas of sea which have acted as sediment traps for clastic sediments from neighbouring continents. Sediments on the bank are therefore pure carbonate sediments, and the carbonate-producing organisms are not subjected to pollution from the clay minerals produced by weathering. East of Andros Island the edge of the carbonate platform gives way to a very steep submarine slope (up to 20-30° in many places). At the Tongue of the Ocean it slopes down to 2500 m, and east of the little Bahamas Bank, which lies to the north, right down to 4500 m in the Atlantic Ocean. Reef structures along the eastern side of the bank have produced a framework which has made the submarine slope more stable. The rocks consti- tuting the slope, particularly the deepest parts, are under lithostatic pressure (the weight of the limestone series) which is very much higher than water pressure alone at the same depth. Diving with diving ships has revealed great blocks of limestone at the foot of the slope. These blocks may have been released by fracturing because of the difference between the lithostatic pressure and the hydrostatic pressure at the same depth.

The Bahamas Bank is a large area, about 700 km N-S and about 200-300 km wide (Fig. 8.7a). The greater part of this area is less than 10 m deep. The Bahamas Bank and most of the other carbonate banks in the world were exposed during the glacial periods when sea level dropped more than 100 m. The bank was exposed to freshwater, which caused the solution of aragonite and precipitation of calcite, and this process converted loose carbonate sediments into hard (lithified) carbonate rock. The Holocene sediments, which are loose carbonate sediments, are not more than 3-4 m thick, and rest unconformably on well-cemented, Quarternary car- bonate rocks. The thickness of the Holocene sediments, mainly from the last part of the Holocene, can be measured by pushing a pole into the sediments until it meets solid limestone.

The sedimentation on the Bahamas Bank reflects the climatic and bathymetric conditions. The temperature of the surface water varies from about 20-22° in winter to 30-32° in summer. Because of the limited circulation between water overlying the

Bahamas platform and the surrounding ocean, the water in the interior of the bank, particularly in summer, has a higher salinity than normal — up to 40‰. In winter the salinity is reduced by increased exchange of water with the surrounding ocean and addition of rain water from Andros Island. Since the prevailing winds are from the east, where we also have deep water, wave power is strongest on the east side of the platform. We therefore find reef facies along the east side of the platform, but not along the west side. Oolite banks, which require somewhat less wave energy, occur on both sides of the bank, however. On the west side of the bank there are also corals and algae (coralalgal facies), but these do not form proper reef structures.

In the middle areas there are very different types of carbonate mud facies. This is because wave energy is damped along the edge of the bank, and the whole of the shallow area within is a low-energy environment. Tidal currents are also damped in the shallow water, and the tidal range drops from about 0.7–0.8 m along the edge to practically zero in the centre. This is because most of the tidal energy is dissipated in overcoming the friction against the bottom. The most important sediment transport mechanism is probably hurricanes. These can cause sediments which are stable under normal conditions to move and form large dunes or sandwaves. Large oolite banks may move along the edge of the platform, and storms also cause some erosion of the reefs. Towards the centre of the platform wind stress can cause changes in sea level of up to 3 m, and create great turbulence which brings large quantities of mud and sand into suspension. Some of this material will be transported down the submarine slopes.

Sedimentary Facies

Practically all carbonate precipitation on the Bahamas Bank — apart from in supratidal environments — is biogenic, and sedimentary facies are therefore very largely a function of a biofacies. Biofacies, in their turn, are governed by the physical environment on the platform, such as depth, currents etc. Ooids, however, are chemically precipitated, probably with some help from microbiological processes.

The reef facies is built up as a wave-breaker structure, with powerful networks of corals which are braced by an encrustation of coral algae. The lagoon behind the reefs is only 2–6 m deep, and in it we find carbonate sand and fragments which have been transported from the reef during storms. At the bottom there are areas with Thallassia grass. The lagoon facies also contains green algae such as *Halimeda* and *Penicillus*. These are low, brush-like forms which cannot resist very high wave power (Fig. 8.4). However, they secrete small (about 1 μ) needle-like crystals of aragonite. When the algae die, the aragonite needles are released, and form carbonate ooze on the bottom. Isotope analyses of these aragonite needles show δ ^{18}O values similar to those found in carbonate mud. Green algae are therefore considered to be the most important sources of carbonate ooze on the Bahamas Bank. Chemical precipitation is of minor importance. Other characteristic fauna in the lagoon are molluscs, gastropods, echinoderms and annelids. Of particular sedimentological importance are crabs and shrimps which dig tunnels and churn up the sediments, destroying the primary structures. The shrimp *Callionassa* produces tube-like structures called *ophiomorpha* which are typical of shallow marine sediments.

Oolite banks form in shallow environments with constant wave agitation. In water which is deeper than the normal wave base, we find large bedforms of oolites and other carbonate sand which move only during major hurricanes. Aerial photographs show that large dunes or sandwaves, with a wavelength of about 50–100 m, have not moved in 20–30 years. In large areas with lower wave power than the oolite banks, the bottom is covered with a mat of blue-green algae. The algal threads of which this mat consists help to protect the sediments against erosion so that there is no transport along the bottom causing current ripples to develop. They also contribute through photosynthesis (consumption of CO_2) to the creation of a chemical environment, with local precipitation of carbonate in this mat. In this manner small particles, e.g. pellets (faeces) and fossil fragments, may become cemented together into larger units. *Grapestone* consists of cemented grains (ooids, pellets etc.) reminiscent of bunches of grapes, that are probably formed in this way. Grapestone facies are found particularly in the northern and southern parts of the bank, while the areas west of Andros Island, which is the shallowest and best-protected against storms from the west, consist mainly of limestone mud. This mud is formed in situ, partly through accumulation of green algae. Pellets from many marine organisms dominate the sedimentary texture over wide areas. The pellets consist of mud which has passed through the alimentary canal of marine organisms e.g. molluscs and annelids, and form small grains (about 0.1–0.5 mm) from clay and silt particles so that lime mud which is transported by currents behaves like fine sand. Crustacea such as shrimps are also important pellet producers.

In the supratidal area which is flooded during storms and springtides, the sea water remains lying and evaporates, and a crust containing dolomite forms. The more soluble salts, like chlorides and in some cases also gypsum, may dissolve again during the next flood or rains.

Further reading: Bathurst 1975.

The Persian Gulf

This gulf is an important example of a carbonate-producing environment which is fundamentally different from the type represented by the Bahamas Bank (Fig. 8.8).

Whereas the Bahamas Bank is surrounded by deeper water, and therefore constitutes a pure carbonate environment, the Persian Gulf lies in a fold zone between the alpine mountain chain of Iran in the north, and the stable Arabic shield in the south and southwest. The gulf is at the most only 80–90 m deep, and in the northwest the rivers Eufrates and Tigris build a delta out into the Gulf. Clastic sediments are also being added from the north, and consist largely of carbonates (marl). Only on the south side (the Trucial Coast) are there purer carbonate sediments, because there is little runoff from the deserts on the Arabian shield. Here too, however, there is a certain supply of clastic material, particularly through eolian transport from the deserts. Sedimentation has been proceeding in a basin in this area since Mesozoic times, and conditions have changed so little that the modern environment is still a good example of depositional environments in the Jurassic-Cretaceous period, when the sediments which contain the world's greatest oil reserves were deposited.

The temperature in the Persian Gulf varies from 20°C in winter to 34°C in summer, and in shallow areas it may be even hotter. The salinity is round 39–42‰, and may increase through evaporation from the lagoons. The tidal range varies from 0.5 to 2 m. The intertidal and supratidal area consists of a 10–15 km broad belt which is covered by blue-green algae (cyanobacteria) and salt deposits. This belt is called *sabkha* in Arabic, and the word has become a geological term. The slope of the sediment surface here is only 0.4 m/km, and during powerful storms and spring tides water flows in and later evaporates. Because of the high temperature and the salinity there are few organisms which can live on the sabkha. Blue-green algae therefore predominate and form a crust of precipitated carbonate.

The algae may grow in layers parallel to bedding or develop dome-shaped, columnar or irregular structures (stromatolites).

In the inner part of the sabkha we also find precipitation of anhydrite, dolomite, magnesite and halite, but the halite is easily dissolved again. Precipitation of gypsum and anhydrite leads to the water developing a higher Mg^{2+}/Ca^{2+} ratio, which will favour the formation of dolomite and magnesite. After an overburden has accumulated, gypsum will dissolve and anhydrite will precipitate and form characteristic "chicken-wire anhydrite".

In the Persian Gulf there are few calcareous green algae, such as *Halimeda* and *Penicillus*, which are important producers of lime mud in the Bahamas and other tropical areas. Chemical precipitation is probably the most important process here, particularly during periods when diatoms proliferate and raise the pH by consuming CO_2. Biogenic carbonate accumulates on the continental shelf in the form of shells of foraminifera, molluscs, gastropods, ostracoda, bryozoa and echinoderms, that become broken down by boring algae. Along the edge of the shallowest part of the shelf are reefs with those corals that are particularly well adapted to the high salinity.

Tidal channels connect the lagoons with the gulf outside. At the mouths of the channels there are ebb-tidal deltas with oolite banks in the shallowest parts (less than 2 m). In addition to oolites there are also grains of bioclastic material. A tidal delta of this sort might form a good oil reservoir because of the high primary porosity, and in the Mesozoic series we find similar reservoir rocks. Shoreface deposits are not as well sorted as we normally expect them to be. They consist of fine sand and lime mud. This is possible because the shoreline is stabilised by plants which bind the sediments. In the lagoons, mud and lime sand, consisting largely of pellets, are deposited. Small amounts of coralline algae and green algae are also found in the lagoons.

Gastropods are particularly important in the lagoons and in the intertidal environment. The sandy area is characterised by crabs which burrow and disturb the sediments.

The lagoons are surrounded by swamps with bushes and mangroves or tidal flats characterised by algal mats.

In the event of progradation, the facies we have just described, from the open marine lagoon to the supratidal environment, will form a characteristic vertical sequence. Periods of transgressions and subsequent regressive development will result in a sequence, or series of sequences, starting with shallow marine (subtidal) sediments and topped by evaporites (anhydrite). These cycles are also found in the Mesozoic, and the anhydrite bed forms an ideal cap rock above oil reservoir rocks because it is impervious and prevents the passage of oil. Carbonate sand facies and in particular oolites are good reservoir rocks. Marine mud, sediments and algal mat facies may contain sufficient organic matter to be a source rock.

Further reading: Purser 1973.

Reefs

Reefs are structures formed by marine organisms which project up from the sea bed. In Precambrian times, organisms which could precipitate carbonate were limited to blue-green algae, but these can also form reef structures. Carbonate precipitated by algae can form small, upward-projecting stromatolite structures on the sea bed or large algal reefs which grow faster than the surrounding sea bed. Such algal reefs or domes may be over 100 m high, and protrude like small mountains when the shale which forms the surrounding sediments is eroded away. Examples of such algal reefs are found in Death Valley, California, and in Morocco in Late Precambrian sediments.

With the development of a rich and varied fauna in the Cambrian and Ordovician, competition was too great for the blue-green algae to be able to build reefs in normal marine environments. A number of organisms, e.g. gastropods (snails) would then feed on the algal mats. There were also other organisms precipitating $CaCO_3$ from ocean water, probably reducing the amount of dissolved carbonate. In the Mid-Ordovician appeared the first corals which could build what we normally think of as reefs, i.e. coral reefs. Because reefs stand on, or form the edge of a continental shelf bounding deep water, they are subject to very great wave stresses. They therefore have to build up as an extremely strong wave-breaker with a solid framework. This framework may consist of corals braced by stromatoporoids and red algae. Bryozoa, sponges and brachiopods may also be important reef-building organisms. In younger (Cretaceous) reefs rudist bivalves and foraminifera may be important.

In modern and Cenozoic sediments we also find reef-like structures produced by marine grasses.

The formation and shape of the reef complexes are very largely a function of the ecological environment of the reef-building organisms.

1. Reef-building corals require warm surface water and their occurrence is therefore limited to the lower latitudes. When making palaeogeographical reconstructions of earlier geological times, however, it is important to remember that surface water temperature is not merely a function of latitude. Along the coast

of West Africa and along the west coast of the American continent cold water wells up, making the water too cold in most cases for reefs to form, even near the equator. Off the coast of East Africa and the east coast of America, on the other hand, the warm surface water is swept against the coast and coral reefs may extend into higher latitudes (30–35°).

2. The hermatypic corals which build reefs live in symbiosis with algae and require sunlight, and are therefore sensitive to changes in sea level. If the sea level rises faster than the coral reef can grow, it may drop below the photic zone and "drown", but this is very rare because the reefs can gain height rapidly.

3. Many organisms in coral reefs live by filtering water to trap organic material. If the water contains siliciclastic mud, for example, the clay minerals will block the filtering organs so that they die. Corals are particularly sensitive to the clay content in the water and can therefore only live in clear water. Addition of clay, for example from a delta, will kill a coral reef. Pollution will have the same effect.

4. Clear water, however, is usually very poor in nutrients, and in order for the organisms in a reef to obtain enough food, there must be good water circulation. Consequently, reefs tend to grow on the edge of ocean basins, or as structures projecting high up from deeper waters. In this way, high temperature is combined with low mud content.

5. Reefs provide a special ecological environment for animals which are not part of the reef structure itself, but which live on other organisms.

6. A number of animals live by breaking down the reef structure, so that many of the primary structures of fossil reefs have been destroyed. The most prominent of these are the boring mussels, which form holes about 1 cm in diameter in corals, algae etc. Boring sponges and algae are also important destructive organisms, forming small holes for the most part (only a few microns). Coral reefs form a very favourable environment for a number of varieties of fish, and the cavities in the structure offer protection against predatory fish. Many types of fish also live by literally eating the reef. They take small bites, and obtain nutrients from the surface. In this manner they help to break down the reef and deposit lime mud on the sea bed.

Reef Geometry

Reefs will continue growing and keep up with basin subsidence or sea level rising as long as the organisms which build reefs grow and build faster than the rate of breakdown through biological and mechanical decomposition. Reefs will grow fastest on the outside, where wave power and the supply of nutrients is greatest. In the case of barrier reefs a lagoon forms on the inside and fills with fine-grained carbonate sediment eroded from the reef, and also from the green algae which live in a lagoon environment (Fig. 8.7). Free-standing reefs will form a ring with a lagoon in the middle, because the reef grows largely along the edge, and sediment accumulates in the middle. The reef will be sensitive to changes in sea level since reef organisms do not tolerate exposure or "drowning" below the photic zone. The Holocene transgression about 10,000 years ago raised the sea level about 100 m in a few thousand years, but most reefs managed to grow quickly enough to keep pace with the rising sea level.

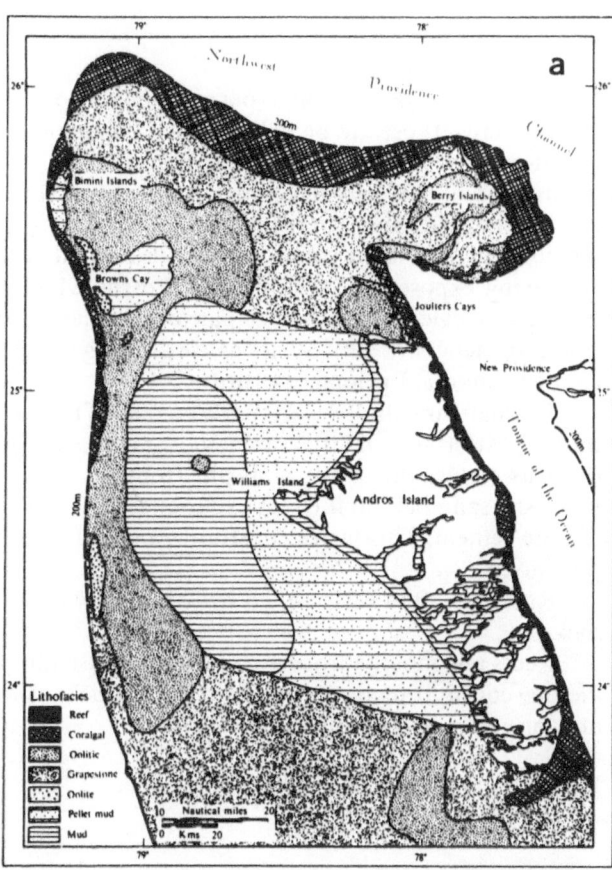

Fig. 8.7. **a** Distribution of types
of carbonate sediments on the
Bahamas Bank. (After Bathurst
1975). Note that the coral and
oolite facies, which require high
energy, are distributed along the
edge of the basin, where wave
power is greatest. Reef facies,
which require the highest wave
energy, only develop on the east
side against the prevailing
winds. **b** The bottom of a lagoon
in the Bahamas with green algae
(*Penicillus*) which look like a
barber's brush. These green algae are responsible for much of
the production of carbonate
mud. (Photo R.G.C. Bathurst)

b

Reefs as Reservoir Rocks

Modern reefs have a very high porosity, made up of everything from small holes to large caverns. It used to be believed that this same primary porosity also occurred in older reefs. Boring and blasting into these reefs, however, revealed that the hollows rapidly fill up with cement, fragments of fossils or lime mud, and that little porosity remains. Even in the framework facies we find that much of the porosity is secondary, due to the solution of fossils or cement. This may happen through the reef being exposed to groundwater. Fossils which consist of aragonite or high magnesium calcite will dissolve particularly easily, and low Mg calcite will form. The diagenetic process may not cause a strong increase in the overall porosity, but a redistribution of primary porosity.

Marine cementation starts early on reefs. The outside in particular, where water flux is greatest, will undergo rapid marine cementation. The coarse carbonate block deposits (talus) down the slope in front of the reef (the fore-reef facies) and bioclastic sand behind the reef experience less water flow and tend to develop less marine cement. This facies constitutes a better reservoir rock in terms of primary porosity.

Reefs are often surrounded by organically rich shale which may form source rocks.

Reefs are good hydrocarbon traps, because they project up from the sea-bed, thereby constituting projecting structures which may become sealed when the reef drowns and is covered by shale.

Further reading: Toomey 1981.

The Role of Algae in Carbonate Sedimentation

Calcareous algae are plants which need sunlight to live. During photosynthesis, algae consume CO_2 and produce oxygen. The absorption of sunlight by water is, however, not equally rapid for the various wavelengths of which light consists. The red section of the spectrum (long-wave light) is more rapidly absorbed than the blue (short-wave). Red algae, which can also use blue short-wave light, can therefore live at greater depths than, for example, green algae, which are dependent on long-wave red light and can therefore live only in somewhat shallower water. Areas with carbonate sedimentation normally have little clastic material in suspension, and the water is therefore very clear. The absorption of light drops off exponentially with depth. Below about 50 m there is very little light left, but some algae can live in somewhat deeper water, down to about 100 m in very clear water. Most calcareous algae require a relatively high temperature in sea water which is close to the carbonate saturation point. They are therefore found mainly in equatorial regions.

Fig. 8.8. General map of the Persian Gulf and detail map of the carbonate facies in the Abu Dhabi area. (After Purser and Evans 1973)

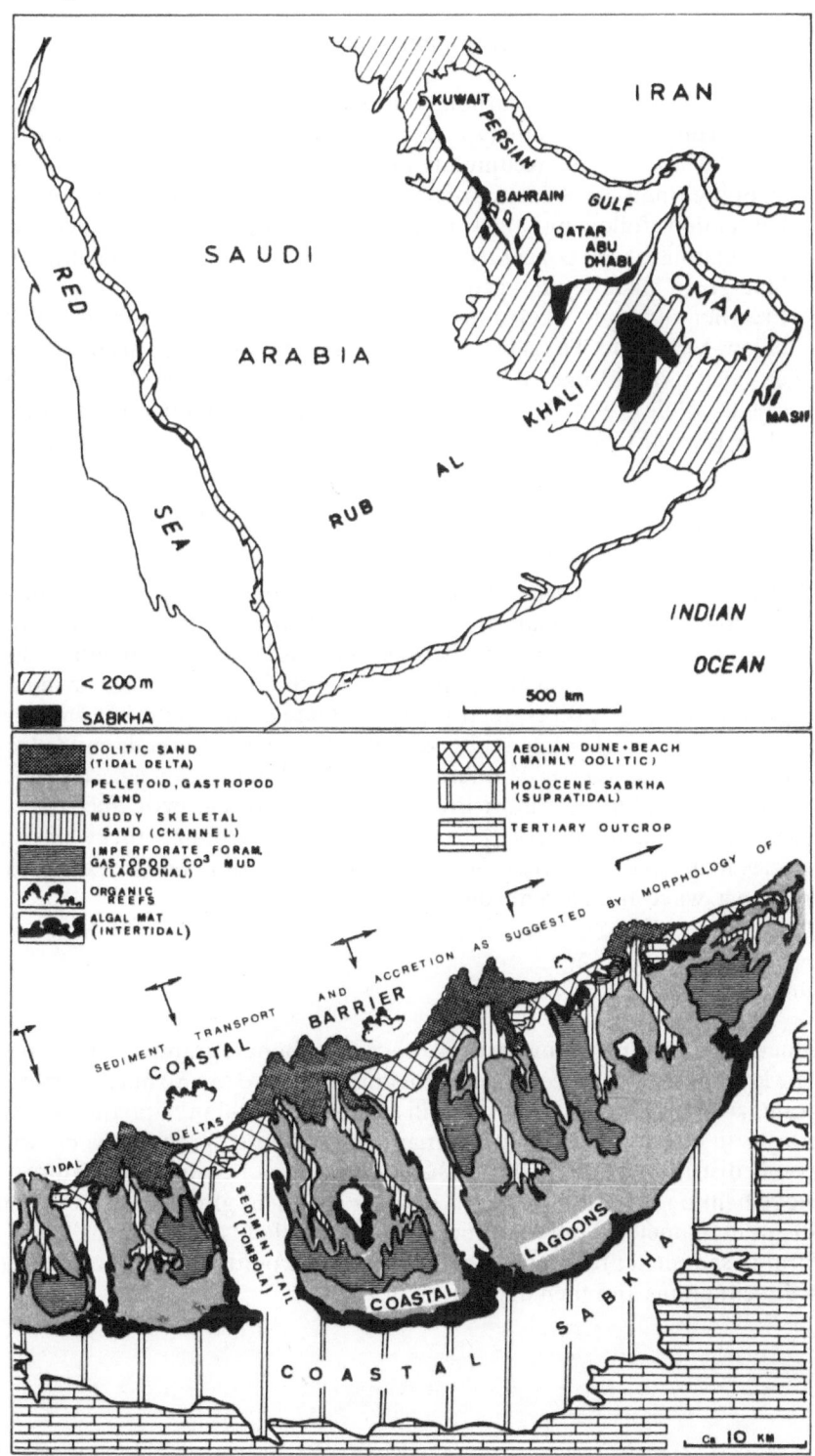

Fig. 8.8

We can distinguish between three types of algae:

1. Those which precipitate carbonate and live on the sea bed. These are mainly red and green algae, plus one type of brown algae.
2. Those which help to accumulate lime mud and to form various structures by trapping small carbonate particles. The end result is the formation of parallel lamination following algal mats, more complex algal growth structures (stromatolites) or a concentric type of structure — ooids and oncolites. These are blue-green algae (cyanobacteria).
3. Free-floating algae. Coccoliths are examples of free-floating algae with skeletons of low-Mg calcite. The skeletons precipitated by algae consist either of aragonite or calcite, often high-Mg calcite. Red algae can secrete both aragonite and calcite, while green algae largely secrete aragonite. Only one species of brown algae (*Padina*) secretes a calcareous shell, and it consists of aragonite.

Stromatolites

"Stromatolite" is the term for lamination in carbonate rocks due to accumulation or precipitation of carbonate as a result of algal growth. Stromatolites have a rather confusing status, intermediate between skeletons and sedimentary structures. There are not always algal remains to be found, and the only evidence is the laminations in rocks. Blue-green algae may secrete carbonate, but they also have a sticky surface which traps suspended fine-grained calcareous mud. In modern deposits one finds a series of different varieties of blue-green algae (cyanobacteria) which form algal mats or more irregular types of algal growth. Species determination is very difficult, however, particularly in older carbonate rocks, since the same species may have a variety of growth forms, depending on the external environment, e.g. wave and current conditions.

Oncolites

Blue-green algae may grow on the surface of carbonate grains, e.g. fossil fragments. The layer of sticky algae on a sediment grain will tend to trap microscopic sediment grains suspended in water. There will also be chemical precipitation which causes a layer to develop around the primary grain. These "growth layers" will not be evenly distributed around the grain, as with oolites, because this requires constant movement on the sea bed. In this manner sediment grains may grow to be up to 10 cm in diameter. Large oncolites are often called "algal biscuits". So oncolites form in sediment after deposition, in contrast to oolites, which are transported as sediment grains and then deposited.

Coccolithophores

Coccolithophores are algae which consist of round coccospheres 2–20 μ in diameter.

Coccospheres in sediments will often disintegrate, so that we only see the separate coccolith plates. These in turn consist of calcite crystals which are 0.25–1 μ in diameter. Because they are so small it has only been possible to study coccolithophores systematically by means of the electron microscope.

These organisms live mainly in the photic zone. In areas of high productivity, for example in the fjords of Norway, the concentration of coccolith cells may be several million per litre, but 50,000 to 500,000 is a more normal level. Although they consist of low-Mg calcite, their size makes them relatively soluble in cold water. In consequence, although production is greatest at high latitudes, it is only at lower latitudes that large quantities of coccolithophores are deposited. Shallow, warm seas with little other carbonate production provide particularly favourable conditions for deposition of high concentrations of coccoliths which may ultimately form chalk deposits. The seas of northwest Europe in Cretaceous times were a good example. The climate in the Mesozoic was undoubtedly considerably warmer than today, and northwest Europe also lay further south. Chalk forms a characteristic rock which is exposed in Denmark, South England and France, continues under the southern and middle sections of the North Sea, but is missing in the north, possibly for climatic reasons. Chalk sediments were probably deposited at depths of a few hundred metres, mostly below the photic zone. Since chalk is a micritic limestone, one would not expect it to form a suitable reservoir rock. The Ekofisk field is in fact the world's only major oilfield in such rocks, and the low permeability of this fine-grained rock creates problems, although production is aided by small fractures.

The fact that the coccoliths consist of low-Mg calcite make them more stable during diagenesis.

Deep-sea drilling production (DSDP) has shown that coccolithophores were very much more extensive during the Cretaceous and Lower Tertiary than they are today, and the colder climate which began in the Miocene can be interpreted from the quantities and composition of coccolithophores in the sediments. Many types have very specific temperature requirements and their occurrence may therefore make an important contribution to the palaeoecology of marine sediments of various ages.

Coccolithophores first appeared in the Mesozoic, and we therefore have no Palaeozoic chalk deposits.

Distribution of Algae in Modern and Older Carbonate Sediments

If one places a profile extending from an ocean basin to a coast with reef development (Fig. 8.11), we see the following distribution of algae: In the ocean basins sedimentation consists largely of planktonic algae. In reef facies we find mainly red algae which build strong, solid structures of carbonate, which in conjunction with the corals can resist waves which break against the reef. The Solenoporaceae family is a typical example of red algae which flourished in the reef

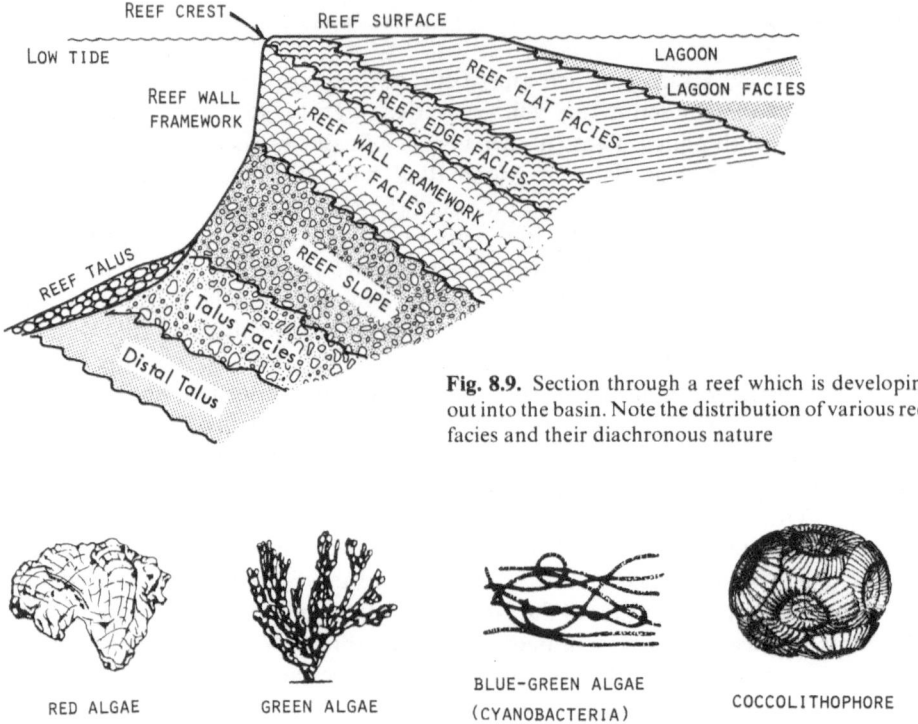

Fig. 8.9. Section through a reef which is developing out into the basin. Note the distribution of various reef facies and their diachronous nature

RED ALGAE   GREEN ALGAE   BLUE-GREEN ALGAE (CYANOBACTERIA)   COCCOLITHOPHORE

Fig. 8.10. Different types of algae

facies. They formed ball-shaped lumps of carbonate which may be eroded and form conglomerates near the reef.

In the lagoons within the reef we find green algae. These have bush-shaped skeletons which would not be able to grow amongst breakers (Fig. 8.10). Among them we find the major producers of small aragonite crystals which form lime mud. The green algae therefore ensure that lime mud is deposited in the lagoon. We find these in protected parts of lagoons, while towards the tidal zone the sea bed is covered with blue-green algae which lie like a gelatinous carpet over the sediments (Fig. 8.10). These algae consist of a network of threads which hold the sediment in place and protect it against moderate currents and wave erosion. However, they may be eaten by animals, e.g. snails. By the shore itself, once more in a high-energy environment, we find red algae again. Blue-green algae and calcareous red algae are also found in temperate environments.

Blue-green algae have existed since early Precambrian times (Fig. 8.12). We find unambiguous stromatolite structures which must be due to blue-green algae in rocks which are approximately 3×10^9 years old. Green algae developed only in the Cambrian, which led to the production of large amounts of lime mud without chemical precipitation. As we have shown, green algae are an important factor in modern carbonate sedimentation. Red algae also developed in the Cambrian. The Solenoporaceae family, which was widely distributed in the Palaeozoic, died out

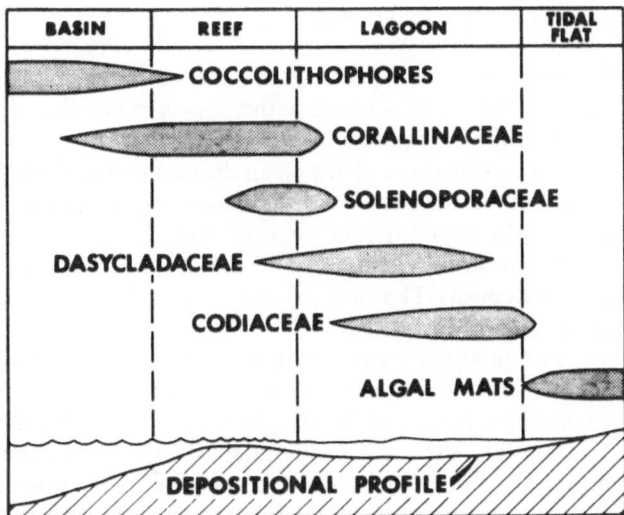

Fig. 8.11. Different types of algae and their typical distribution in a sediment basin. (After Ginsburg et al. 1971)

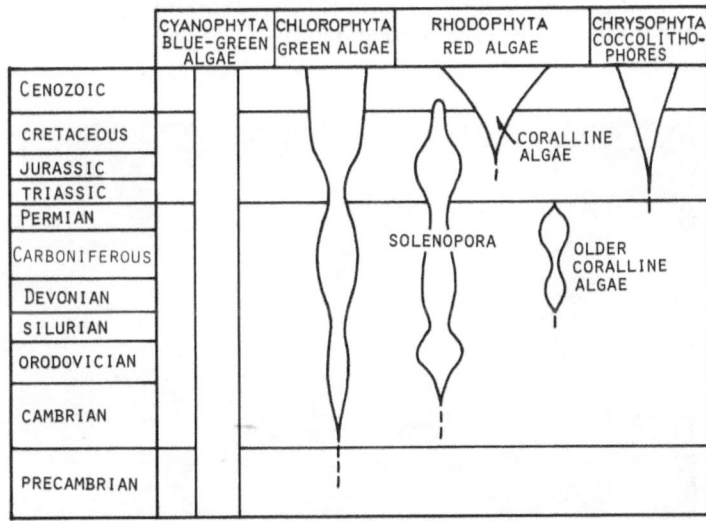

Fig. 8.12. Distribution of some of the important species of algae through geological time. (Ginsburg et al. 1971)

during the Cretaceous. Coralline algae appeared in the Jurassic, and still exist. The arrival of the planktonic algae, mainly coccolithophores, in the Triassic, has been a vital factor in global carbonate sedimentation, and may have resulted in sea water having a somewhat lower carbonate content now than previously.

Further reading: Ginsburg 1971.

Pelagic Carbonate Deposits

Pelagic carbonate deposits consist largely of planktonic organisms which live in the upper water levels, sinking to the bottom when they die. The sedimentation rate is a function of productivity in the upper water layers minus solution as the dead organisms sink down. How clean the carbonate deposits are depends on how much other biological sedimentation there is, e.g. from diatoms and radiolaria, and how rapidly clastic sedimentation takes place.

Foraminifera and coccolithophores form the most important deep-sea carbonate deposits. They are also important as an indication of environment, e.g. water temperature (Fig. 8.13).

Foraminifera are a very important group of one-celled animals. Foraminifera which live on the bottom may have shells of precipitated calcium carbonate, usually calcite, or a shell of small sand grains, cemented together (agglutinated shell). Some foraminifera also have aragonite shells. The calcite shells vary in composition from high-Mg to low-Mg calcite, so that their solubility in sea water varies greatly. Benthic foraminifera with high-Mg calcite are found especially in warm-water environments, where they are less soluble.

Some foraminifera live in symbiosis with an alga which lives in their protoplasm, and planktonic foraminifera live for the most part in the upper 100 m of the sea. At greater depths, foraminifera occur mainly as tests which fall to the bottom, but some can also live at great depths. The biggest tests are found at low latitudes,

Fig. 8.13. Coccoliths from the upper part of the Upper Cretaceous in the Ekofisk field. Petroleum occurs between the small plate-like coccolithophore shells (about 1–5 μ). The picture was taken with a scanning electron microscope. The limestone has 32% porosity and 1 md permeability

where precipitation conditions are best. Foraminifera with low-Mg calcite shells will have the greatest preservation potential at greater depths, since they dissolve more slowly.

The Globigirinidae family is very widely distributed. Their skeletons consist of low-Mg calcite (1% $MgCO_3$), so they are less soluble than skeletons of aragonite and high-Mg calcite. Together with related families, these organisms therefore represent the greater part of the carbonate sediments in the ocean basins. Many species require very specific ecological conditions. Some have left- or right-curved shells, depending on the temperature. As a result, foraminifera provide detailed information about palaeoecological conditions. The amount of planktonic foraminifera relative to benthic declines in shallower water, and can also be used to reconstruct the coastline in marine basins.

Planktonic organisms are normally so small that they settle very slowly in water. It takes a very long time, often many years, for such particles to reach the sea bed, and in consequence they will have plenty of time to dissolve. Most plankton are eaten by higher organisms and are formed into pellets which are aggregates of microfossils. These have a far higher settling velocity and will be less prone to dissolve before they reach the bottom.

Other Carbonate-Secreting Planktonic Organisms

Planktonic gastropods (*Pteropoderma*) also live mainly in the upper 100 m of the sea. However, they have aragonite tests, and are therefore less resistant to solution than other planktonic organisms, but they can form a mud (pteropod ooze) which is limited to a lesser depth than, for example, foraminifera mud. These small gastropod shells will usually dissolve completely in the sediment, and may precipitate out as calcitic cement so that it is not possible to prove that the sediment originally contained gastropods.

Cysts from dinoflagellates are also found in deep-sea sediments.

Calcispheres are round structures with a three-layered cell wall. They are 0.05–0.5 mm in diameter, and usually the hollow later fills with calcite sediment which has grown inwards from the wall (drusy mosaic). Calcispheres have an uncertain systematic position, and appear first in Palaeozoic limestones. They are very common in the chalk sediments of the North Sea.

Further reading: Hsu and Jenkyns 1974; Warme et al. 1981.

Carbonate Diagenesis

Carbonate sediments are such a heterogeneous assembly of rocks that it is difficult to deal with them as a single group. We shall distinguish here between: carbonate mud, which turns into micritic limestone; carbonate sand; and carbonate rocks which form solid rock from the moment of deposition, such as reefs (framestone).

The carbonate mud which is deposited on carbonate banks, e.g. the Bahamas Banks, is only 2–3 m thick, and has been deposited in Holocene times. It overlies

Quarternary sediments which were exposed and lithified by freshwater diagenesis during the last glacial period. Since all shallow marine carbonate platforms were exposed and lithified during the last glaciation, we cannot study how shallow water carbonate muds are compacted. In older carbonate rocks which consist largely of carbonate mud (micritic limestones), we find little sign of compaction, and fossils have in most cases not been deformed due to pressure. However, this raises the question of how the cement necessary to fill the pores was added, since we have 60–70% porosity to begin with.

Attempts to compress carbonate mud experimentally in cylinders have in fact shown that it is possible to reduce the porosity (% water) to 30–40% without deforming fossils. However, we have a very incomplete understanding of the diagenesis of carbonate mud, and in particular the cementing process.

Lithification of Carbonate Sediments

Lithification is the process which turns loose sediment into solid rock. It occurs through new minerals (cements) being precipitated which bind together the primary particles or fragments. To cause precipitation of carbonate cement, we must have porewater which is oversaturated with respect to a carbonate phase. This may happen through the sediment being flushed with oversaturated porewater. Cementing of beach sand takes place because it is flushed by surface water which washes in over the shore. Beach sand which is cemented early in this manner is called *beach rock*. Beach rock may form in the space of only 10–20 years. This can be proved by the bottles, beer cans etc. which are embedded in this early cemented rock. Cement formed in a marine environment is aragonite or high-Mg calcite, which forms needle-shaped crystals (Figs. 8.14 and 8.15).

In zones with meteoric water (groundwater), calcite cement is formed because the low Mg^{2+} content makes it easier to precipitate calcite directly. Freshwater may dissolve the more soluble aragonite from fossils and ooids, and precipitate calcite which grows in large crystals (block-shaped cement) (Fig. 8.16): Above the water table, in the vadose zone, water will flow through and drops will remain suspended below grains, depositing cement there when the area dries out. This is called *pendant* cement (Fig. 8.14).

Early marine aragonite cement may grow as evenly distributed layers of aragonite needles perpendicular to the surface of the grains. This is called isopachous fibrous cement because a layer of uniform thickness is formed. Isopachous calcite cement may also be precipitated in meteoric (phreatic) porewater. Rim cement is cement grown in continuity with the original grain. This is also called syntaxial overgrowth, and is particularly common in crinoid fragments consisting of large single crystals, which offer a preferred substratum for further growth.

When calcite replaces earlier aragonite or high-Mg calcite by neomorphism, we sometimes see "ghosts" of the earlier crystals. Radiaxial fibrous cement mosaics frequently form as a result of replacement of aragonite cement but may form by direct precipitation of calcite. These are mosaics of calcite crystals with optic axes which converge away from the substratum (cavity wall) they grew on. They contain twin laminae which are convex towards the substratum.

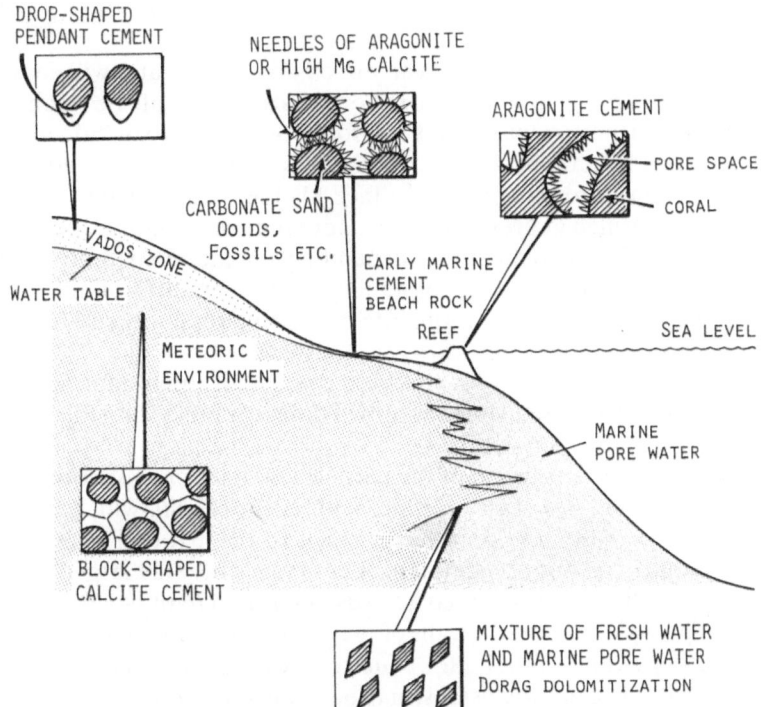

Fig. 8.14. Types of carbonate cement as a function of diagenetic environment. Reefs containing many fossils with high-Mg calcite and aragonite will turn into calcite, particularly if fresh water later percolates through the reef

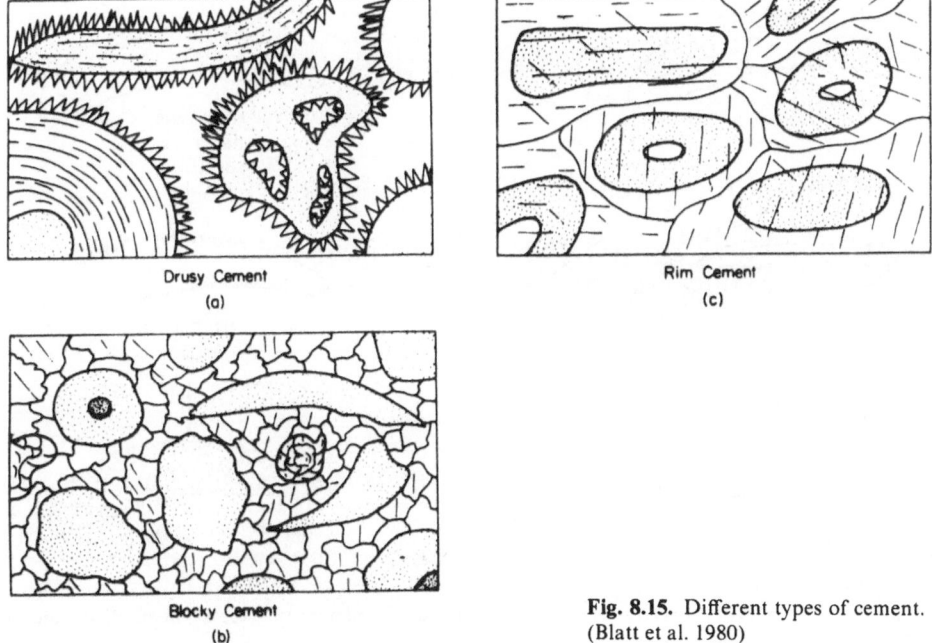

Drusy Cement
(a)

Rim Cement
(c)

Blocky Cement
(b)

Fig. 8.15. Different types of cement. (Blatt et al. 1980)

Calcite cement formed early in oxidised porewater will contain practically no iron, since Fe^{3+} is not soluble in the oxidised state. Only Fe^{2+} can substitute for Ca^{2+} in the calcite structure, and the formation of ferroan calcite therefore requires reducing conditions. Calcite precipitated in the sulphate-reducing zone is free of iron (non-ferroan), however, since all the available Fe^{2+} will form sulphides. Calcite formed at greater depths under reducing conditions will normally contain some iron and manganese, depending on the availability of such ions in the porewater at the time of formation. At temperatures of about 100°C and above iron-rich carbonates like ankerite become increasingly stable and are often found in minor quantities.

A particle, e.g. a fossil which consists of aragonite, may be dissolved and replaced by calcite by means of two different processes (Fig. 8.16):

1. By complete solution of the particle and later precipitation in the empty space. The mould may then be refilled with calcite crystals which grow inwards from the walls towards the centre as a result of competition between the crystals and reduced rate of precipitation. The crystals become larger towards the centre of the hollow. This is called a "drusy mosaic" (Bathurst 1975).
2. Through gradual solution of aragonite and immediate precipitation of calcite along a thin solution film. Much of the original structure of the aragonite fossil, such as organic inclusions, may be preserved, albeit not perfectly, even after they have been replaced by calcite. This is called *neomorphism*.

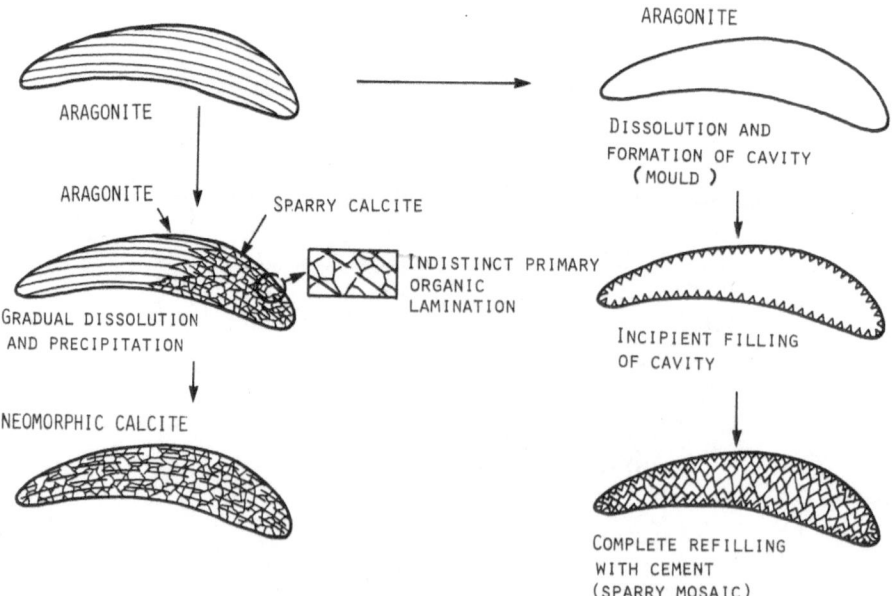

Fig. 8.16. Two types of conversion of aragonite to calcite. *Left*, neomorphic replacement. *Right*, solution and precipitation in the cavity

Grains exposed on the sea bed may be bored by blue-green algae and occasionally also by fungi.

The holes bored fill with small calcite crystals which are preserved when the aragonite shell dissolves. The precipitation of this low-Mg calcite may be aided by bacterial decomposition of organic material in the holes. A micritic envelope is then formed (Bathurst 1975, Fig. 8.17a,b).

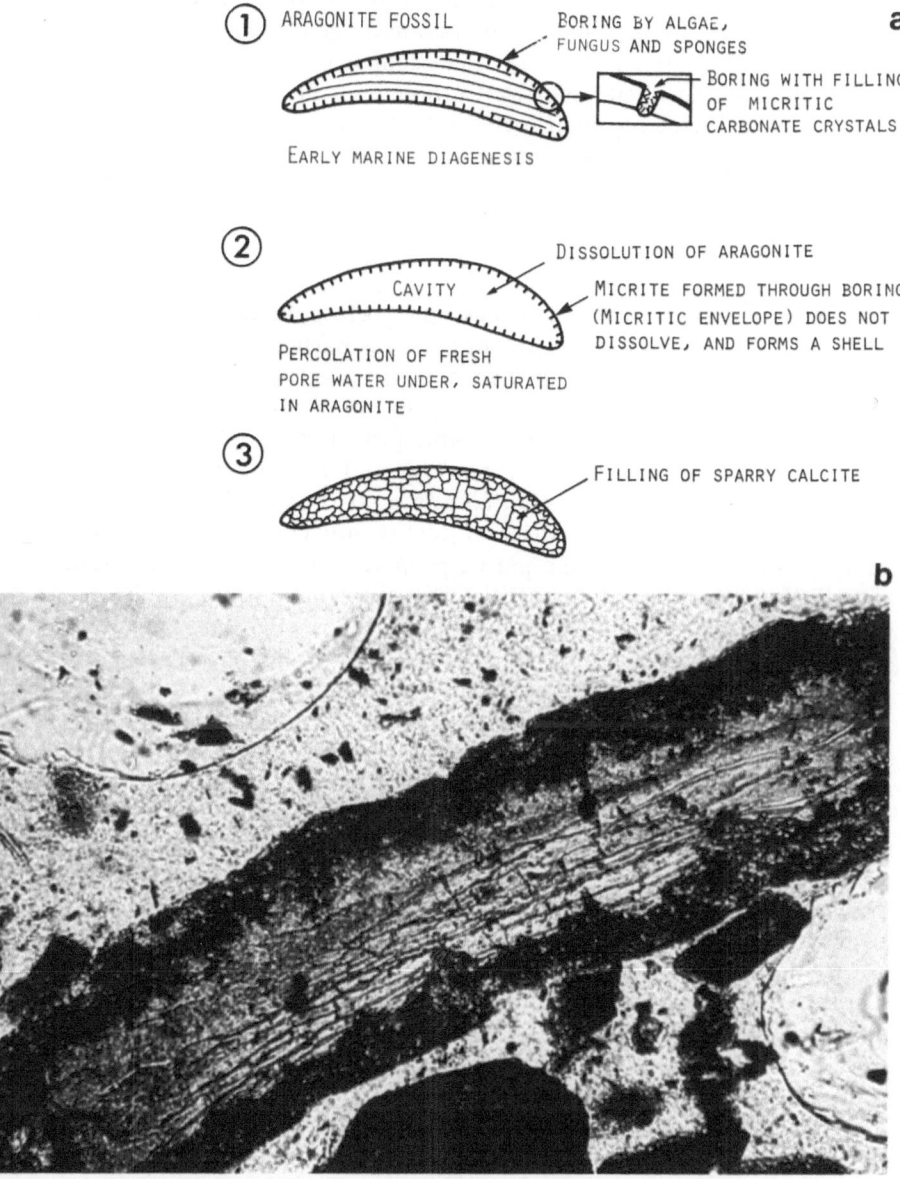

Fig. 8.17. **a** Formation of micritic envelope. (After Bathurst 1975). **b** Micritic envelope formed through algae boring into a mollusc shell from Bimini Lagoon. (After Bathurst 1975)

If the boring process is allowed to continue for a very long time (due to slow sedimentation rates), the whole grain may be eaten away, leaving only a lump without any record of the original grain.

Cementation in environments with marine porewater will form high-magnesium calcite or aragonite. It often occurs in beach rocks or reefs, but also in other places on the sea bed. Cemented sea floor is called "hard ground". Aragonite and high-Mg calcite will in most cases dissolve and reprecipitate as calcite during the course of a few thousand years. At greater burial depths and higher temperatures block-shaped calcite may precipitate as late-diagenetic cement (Fig. 8.14) at the expense of earlier cements. Coarse calcite cement enclosing more than one grain is called poikilotopic calcite cement. Carbonate cement is also common in many sandstones. In clastic reservoir rocks in the North Sea poikilotopic calcite cement is common in the carbonate-cemented intervals.

Diagenesis in Carbonate Sand

The diagenetic processes in carbonate sand correspond in many ways to those we observe in clastic siliceous sandstones. This is particularly evident if we look at Dunham's classification. Mudstone corresponds to clay and wackestone and packstone corresponds to greywackes. Grainstones are well-sorted sandstones with little matrix, and have many of the same properties as well-sorted quartz sandstones, both as regards primary porosity and diagenetic transformation. We distinguish between grain-supported and mud-supported sandstones (Fig. 8.4). Mud-supported sandstones will be subject to compaction of the matrix between the grains, and have many of the plastic properties as fine-grained sediments. The degree of compaction will depend on the proportion of sand grains.

Grain-supported sandstones have a framework of grains which rest upon one another. Compaction cannot take place without the grain framework being deformed. This may take place through the sand grains being crushed mechanically due to the force exerted by the overburden, or being chemically dissolved, particularly at the contacts between grains (pressure solution). The mechanical strength of carbonate fragments from fossils may be relatively low, but that of thick-shelled fossils and ooids is high. The contacts between grains will initially be very small: round ooids will only have very limited areas of contact. As the overburden increases, so will the pressure per unit area at the contact points between grains. Pressure solution will then occur at the contact points, so that the contact area expands and the pressure per unit area decreases. Overpressure will reduce the effective stress caused by the overburden.

If some cement is formed in the early diagenetic stages, before a great deal of overburden has been added, e.g. marine aragonite cement around the grains or calcitic cement in meteoric groundwater (Fig. 8.14), this cement will brace the grain framework and distribute the grain-against-grain stress over a larger area.

Although the early cement reduces the porosity markedly, the remaining porosity may be preserved down to a depth greater than in carbonate sandstones which were not cemented early.

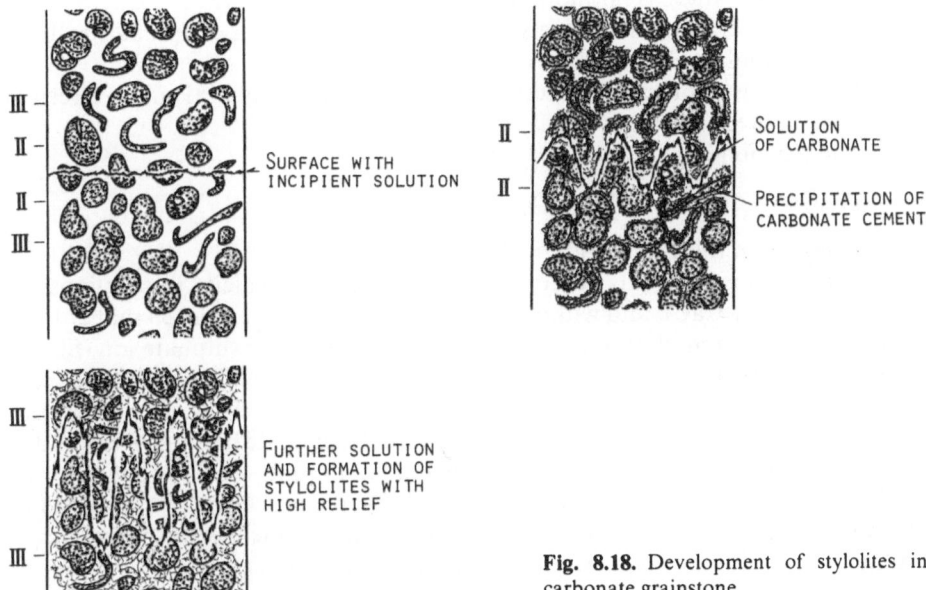

SURFACE WITH
INCIPIENT SOLUTION

SOLUTION
OF CARBONATE

PRECIPITATION OF
CARBONATE CEMENT

FURTHER SOLUTION
AND FORMATION OF
STYLOLITES WITH
HIGH RELIEF

Fig. 8.18. Development of stylolites in carbonate grainstone

Solution of grains due to pressure will often be concentrated in particular horizons. *Stylolites*, which are surfaces where a considerable amount of solution has taken place, will then form. The horizon will be enriched with finely divided silicate minerals and other insoluble material in the limestones (Fig. 8.18). Clay minerals seem to enhance pressure solution and stylolytes may start forming along primary clay laminae. As the rate of solution along the surface varies, a very irregular relief develops which may be taken as a minimum measure of the thickness of the carbonate layer dissolved.

Dolomitisation

The term "dolomite" is used to designate both a mineral and rocks in which this mineral is the main constituent. To avoid confusion, the term "dolostone" has been introduced for the rock, but this is not used very consistently.

The mineral dolomite [$CaMg(CO_3)_2$], consists of layers of CO_3^{2+} groups alternating with layers of Mg^{2+} and Ca^{2+}. This is a highly organised structure (trigonal rhombohedral) and the organisation of more or less pure layers of Mg^{2+} and Ca^{2+} leads to high kinetic energy being required for the crystallisation of dolomite. This is particularly true at low temperatures, and as yet it has not been possible to synthesise dolomite in the laboratory at low temperatures (less than 100°C).

The other obstacle to the formation of dolomite is that Mg^{2+} is very strongly hydrated [$Mg(H_2O)^{2+}$], and in this state cannot easily enter a crystal position. This effect will decrease with increasing temperature.

In sea water there is far more magnesium than calcium ($Mg^{2+}/Ca^{2+} = 5$), and but for this hydration effect, we would expect magnesium carbonates to be the predominant minerals formed.

Dolomite is not formed directly, but as a secondary mineral and as a result of reactions between different forms of $CaCO_3$ and Mg^{2+}.

The reaction

$$2CaCO_3 + Mg^{2+} = CaMg(CO_3)_2 + Ca^{2+}$$

is dependent on the Ca^{2+}/Mg^{2+} ratio.

Experiments show that dolomite forms rapidly on calcite in solutions with $MgCl_2 + NaCl + CaCl_2$ and without sulphate. Another major reason that dolomite is not common in modern marine environments is that the sulphate ion (SO_4^{2+}) is very efficient at preventing dolomitisation. Consequently it takes place more easily when there are few sulphate ions (Baker and Kastner 1981). For this reason dolomite is formed in many lakes and in zones intermediate between fresh and salt water (see the Dorag model). In a sulphate reduction zone, the SO_4^{2+} concentration will be lower, and ammonium (NH_4^+) can be formed through the action of nitrate-reducing bacteria. NH_4^{2+} can replace Mg^{2+} which is adsorbed on clay minerals. In this way magnesium is liberated for dolomitisation.

Models

Dolomitisation means that $CaCO_3$ is dissolved and dolomite precipitated. The conditions for this are:

1. That calcium carbonate is unstable and that the solution is supersaturated with respect to dolomite.
2. That Mg^{2+} is added to the solution so that dolomitisation can continue.

Solution of calcium carbonate takes place most easily if we have aragonite at the outset. Fine-grained carbonate mud has a large specific surface which enables it to react more rapidly than massive carbonate.

Introduction of Mg^{2+} will depend on the rate of percolation of porewater containing magnesium. The dolomitisation process is accelerated if the sediment concerned has a high permeability. If carbonate sediments already contain a good deal of magnesium, dolomitisation will be able to proceed without any addition of Mg^{2+}. Carbonate sediments which are rich in high-Mg calcite, as in reefs, will undergo considerable dolomitisation without any further addition of magnesium being necessary. In shales we often find thin dolomite beds or finely divided dolomite which may be due to addition of Mg^{2+} from clay minerals.

Sea water is a very complex solution. We cannot simply predict the way in which it will react from the concentrations found through chemical analyses. Some of the Mg^{2+} and Ca^{2+} are associated with Cl^- through ion pairing, and cannot be included when carrying out calculations regarding activities involving carbonates.

Theoretical figures indicate that dolomitisation should occur when the Mg^{2+}/Ca^{2+} activity ratio is about 0.6. Although these figures are somewhat uncertain, it is clear that sea water in which the Mg^{2+}/Ca^{2+} ratio is 5.6 is oversaturated with respect to dolomite. The fact that dolomite does not form is ascribable to kinetic reasons, one probably being the high SO_4^{2-} concentration, and only when the ratio is over about 7 will dolomitisation take place in sea water. In freshwater the ion strength is lower, and here dolomite can be formed at lower Mg^{2+}/Ca^{2+} ratios (about 1). However, it is seldom that fresh water has such a high Mg^{2+}/Ca^{2+} ratio.

Freshwater-Salt-Water Mixture Model (Schizohalin-Dorag model)

Although freshwater normally has a relatively low Mg^{2+}/Ca^{2+} ratio, if it is mixed with salt water its magnesium content will rapidly increase because of the high concentration of Mg^{2+} in sea water ($Mg^{2+}/Ca^{2+} = 5.6$).

Mixtures of 5–30% sea water in freshwater result in such a low ionic strength that dolomite becomes stable at a lower Mg^{2+}/Ca^{2+} ratio (down towards 1). The mixture will then be oversaturated with respect to dolomite, but undersaturated with respect to calcite. This model is called the Dorag model, after the Persian word for mixing blood. Even if dolomite is precipitated from a mixture like this, it is clear that we must have extremely thorough mixing of freshwater and salt water in order for the process to be able to continue. Freshwater contains little magnesium, and Mg^{2+} must therefore be supplied by the sea water, and large amounts of sea water must circulate through the limestone. A steady percolation of freshwater will therefore not lead to any great degree of dolomitisation, because freshwater will expel salt water, and the mixing zone will be too small. Even small islands only 1–2 m above sea level will trap freshwater which will form lenses of freshwater surrounded by salt water below sea level. Dolomite could be formed in the mixture zone, and some protodolomite $[Ca_{55}Mg_{45}(CO_3)_{100}]$ has been demonstrated, but no massive dolomitisation, perhaps because of inadequate mixing.

This model explains why dolomite is also found in rocks with no sign of evaporite conditions. How important it is in relation to other models we do not yet know, but recent research regarding the composition of porewater in rocks may well give us new leads with regard to where "dolomitising" porewater is to be found.

Since one prerequisite for this model is percolation of fresh porewater, we can predict in what types of basins and where in the basin a zone of this type, where fresh and salt porewater mix, could occur. Oil geologists observed long ago that dolomitisation was a particularly frequent phenomenon along the edge of sedimentary basins, or in connection with structures which might have formed islands, and this fits well with the model.

Evaporite Model for Dolomitisation

It has been known for a long time that dolomite is often associated with evaporite environments. The first definite example of a dolomite being formed today was found in evaporating sediments in a supratidal environment in the Bahamas in about 1960. It is clear that when sea water evaporates, and aragonite and also gypsum ($CaSO_4 2H_2O$) are precipitated, the composition of the fraction which is still in solution will become increasingly enriched in magnesium, and dolomite will only be precipitated at higher Mg/Ca ratios (about 12). In addition we will find the formation of magnesite, $MgCO_3$. As a result it is easy to explain why dolomite is an important part of most evaporites. Evaporite minerals, particularly chlorides, are very soluble, however, and are often not preserved in a series. Gypsum may also dissolve and be replaced by carbonate. Dolomite might therefore have been deposited in an evaporite environment despite the fact that we do not find any of the original highly soluble salts preserved in the sequence. For this reason it is important to look for indirect evidence of evaporite conditions and solution, and replacement of evaporite minerals.

The most important indicators of evaporite conditions are:

1. Lack of ordinary marine fossils apart from algae, which can tolerate a high salinity. In Palaeozoic and younger horizons stromatolites are typical of evaporites, because under normal marine conditions blue-green algae (cyanobacteria) have too much competition from other organisms.
2. Breccias which may have been formed through solution of salt layers so that beds collapse and form a *collapse breccia*. Such breccias are characterised by angular fragments from an overlying bed, for example of carbonate, which have fallen down into a solution cavity.
3. Pseudomorphosis (replacement) of evaporite minerals, e.g. halite (NaCl) and gypsum ($CaSO_4 \cdot 2H_2O$). Evaporite minerals, which are disseminated through a matrix of less soluble minerals, for example carbonates, are often replaced through pseudomorphosis, so that the crystal form may reveal the original mineral. The cubic halite crystals are typical, and the characteristic swallowtail twins of gypsum crystals are easily recognised even if they have been replaced by other minerals.
4. Sulphate *chickenwire nodules*. Anhydrite often forms very characteristic nodules or continuous layers which look like chicken wire. Even if the anhydrite layers are converted to calcite, these structures may be preserved.
5. Authigenic quartz and felspar. Evaporites are often associated with chert (flint) and chalcedone. A large amount of authigenic felspar may represent a high primary content of zeolites, which are also typical of many evaporites. Under diagenesis or low grade metamorphism (200–300°C) zeolites will dissolve and felspars crystallise out.

Late Diagenetic Dolomite

While early diagenetic dolomite is normally relatively fine-grained, late-diagenetic dolomite usually forms larger crystals, often well-defined dolomite rhombs. The smaller crystals are formed by rapid crystal growth under conditions of high supersaturation. The larger crystals are formed by slow crystal growth from a few nucleation centres at a very low degree of super-saturation combined with deep burial. Dolomite is formed more easily under late than under early diagenesis because the temperature is higher. Less kinetic energy is then required to form dolomite, and porewater with lower Mg/Ca ratios can cause dolomitisation.

As mentioned previously, hydration of Mg^{2+} decreases with increasing temperature, and Mg^{2+} is then more readily available for the dolomite structure. At temperatures up to about 100°C, dolomite may form in porewater with an Mg/Ca ratio between 1.0 and 0.1. Most analyses of porewater in rocks reveal Mg/Ca ratios of between 0.1 and 0.6, declining with depth even if ionic strength varies. Although dolomite is precipitated in solutions with a low Mg/Ca ratio, it is more difficult to explain how magnesium is added so that dolomitisation can take place. The magnesium contained in water from shales is probably not sufficient to dolomitise thick limestone beds, while thin carbonate beds occurring in shales may be dolomitised by Mg^{2+} from the shales. In many cases dolomitisation may be associated with porewater derived from greater depths which has migrated up along faults, and we see that dolomitisation may be intense near faults.

At depths of 2–3 km, where smectite is transformed to illite, Mg^{2+} and Fe^{2+} may be released and react with carbonate to form dolomite.

We often find dolomite enriched along stylolytes, probably because dolomite is less soluble than calcite, and the solution and precipitation round a stylolite will concentrate clay minerals which, in turn, may release some magnesium.

If the composition of the porewater later shifts towards a low Mg/Ca ratio, dolomite may dissolve and calcite reprecipitate. We see evidence of this in a number of cases where distinctly dolomite-type rhombs are found to consist of calcite. One common cause of reversed dolomitisation — often called *dedolomitisation* — is porewater containing dissolved gypsum. This gives the porewater a high Ca^{2+} concentration. However, many people have recommended that the term "dedolomitisation" should be dropped, and the positive term "calcitisation" be used instead.

The Significance of Dolomitisation

For many years there has been intensive research into the processes which lead to dolomitisation. A great deal remains to be learnt, however, before we really understand the precise conditions for dolomitisation, so that we can predict the extent of dolomite in sedimentary basins. The reason for this great interest is that dolomitic carbonate rocks are very important reservoirs for oil and gas. The dolomitisation process creates secondary porosity because calcite or aragonite dissolves and the precipitated dolomite does not fill the entire volume which has been dissolved away. Dolomite has an approximately 12% smaller molar volume than calcite, and this fact may help to explain why dolomitisation often leaves extra pore space. Since dolomitisation involves large-scale percolation of porewater, we may also have net leaching associated with the process. This means that the volume of dolomite precipitated is less than the volume of calcite dissolved. Fine-grained, early diagenetic dolomite has a low permeability, and if it is not fractured it has poor reservoir qualities. Dolomite formed at greater depths tends to have larger crystals (0.1–1.0 cm), and its permeability may consequently be very high if we have intercrystalline porosity.

Micritic limestones are in most cases too dense to be regarded as reservoir rocks. Exploring for oil in these rocks is therefore a question of finding porous dolomite or fractured limestone. Keeping in mind what has been said about dolomitisation models above, which areas of a sedimentary basin rich in limestone are most likely to be dolomitised? It is difficult to generalise, but it is necessary to use all available information to reconstruct the basin, and see whether conditions correspond to the dolomitisation models we know. It was already clear early on, before it was known how dolomite was formed, that it often occurred along the edge of sedimentary basins. This fact can now be related to evaporite conditions here in lagoons or sabkha-like environments. One also expects to find dolomitisation along the edge of sedimentary basins, where freshwater and salt water mix, but perhaps also a little further out, where the edge of the shelf gives way to the deeper part of the ocean basin. Islands in the basin will also be a goal, since fresh water collects in pockets under structures projecting above sea level which have been reefs or tectonically-governed structures.

Further reading: Milliman 1974; Bathurst 1975; Zenger et al. 1980; Toomey 1981; Scholle et al. 1983; Schroeder and Purser 1986; Scoffin 1986.

9 Other Biogenic and Chemical Sediments

Silica (SiO₂) Deposits

Silica is liberated through weathering and goes into solution as silicic acid ($H_4SiO_4 \rightleftharpoons 4H^+ + SiO^{4-}$), much of which is carried out into the sea. The total amount of dissolved silica added to the sea by rivers is estimated to be about 4×10^8 tonnes per year (Heath 1974; Lisitzyn 1972). Silica is largely precipitated biologically from sea water. Radiolaria and diatoms are particularly efficient at removing silica. As a result, even though quartz is only very slightly soluble in sea water (3–6 ppm SiO_2), surface water is usually undersaturated with respect to quartz. This is because diatoms can precipitate silica from sea water even if the concentration of SiO_2 is less than 1 ppm.

The most important silica-producing organisms are:

Phytoplankton; diatoms and silicoflagellates
Zooplankton: radiolaria and sponges.

These organisms are built up of amorphous silica (opal). The total production of organic silica in the oceans has been estimated to be from 2×10^{10} to 10^{11} tonnes per year. The largest contribution comes from diatoms, and a very large percentage (50-70%) of the total primary production of carbon (2×10^{10} tonnes per year) is ascribable to diatoms, which consist of about 60% silica and 40% carbon.

As we have seen above, the addition of silica to the sea only represents about 1% of all organic silica precipitation. Some silica is added to the sea through submarine volcanism along the oceanic spreading ridges and through transformation of basalt, but this is possibly of minor significance compared to what is brought by rivers. Since we may assume that the composition of sea water has always been relatively constant, this means that only about 1% of the overall organic silica production is retained in sedimentary deposits. Most of the silica from plankton redissolves in the undersaturated sea water before it reaches bottom, and some also dissolves in deep-sea sediments and diffuses up into the water. Consequently it is only when the rate of organic precipitation is more rapid than the rate of solution that we find deposition of silica. Diatoms dissolve because sea water is undersaturated with respect to silica, and large quantities of organic material can thereby be released without oxidation taking place. Organic matter produced by the solution of diatoms thus constitutes a large part of the total organic matter accumulated. Phosphates and various trace metals are also released through the disintegration of plankton, and offer a new basis for organic production when water wells up to the photic zone (Fig. 9.1).

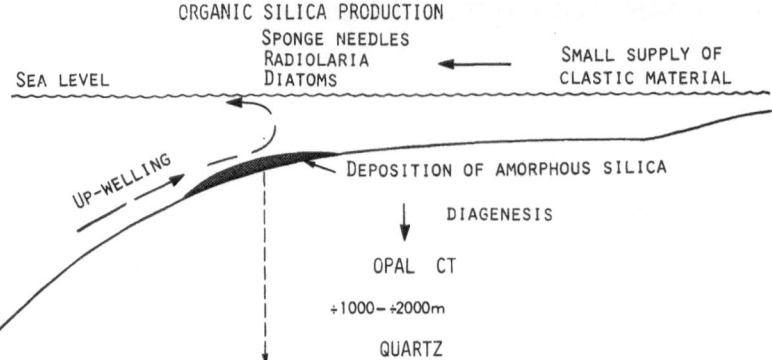

Fig. 9.1. Sedimentation of organic silica in areas with upwelling. With increasing burial biogenic and amorphous silica turn into opal CT and then into quartz

Chert

This is a general term for microcrystalline silicate rocks.

Novaculite is more or less synonymous with light, laminated chert. The term was originally applied to the Lower Palaeozoic rocks of the southern USA.

Amorphous silica is a chemically unstable phase with relatively high solubility (about 150 ppm at 25°C). With time, the amorphous phase will therefore be converted into more stable, less soluble phases. Amorphous silica often occurs in organic ooze formed from radiolaria, diatoms or other organisms with silica skeletons. These will decompose to release bladed crystals called lepispheres, which consist of metastable high-temperature SiO_2 minerals like cristobalite and tridymite. This phase is called Opal CT, sometimes also porcellanite. Opal CT will, when subjected to higher temperatures, slowly dissolve and form quartz.

The transition from amorphous silica to quartz proceeds via cristobalite and tridymite, which are metastable phases. While amorphous silica has a solubility of about 150 ppm, cristobalite and tridymite have a solubility of 6-15 ppm, depending on the degree of order in the crystals. Quartz, which is the most stable phase, has a solubility of 3-6 ppm at > 5°C.

The transition of Opal CT to quartz takes place at about 60-70°C, which corresponds to about 2 km of overburden at average geothermal gradients. The transition will also cause a sudden increase in the seismic velocity of the sediments, and may produce a seismic reflection due to the change in acoustic impedence from sediments with opal CT to chert consisting of quartz.

Laminated chert is very common in Palaeozoic and Mesozoic sequences and was probably deposited on the ocean floor (*ophiolite*) near spreading ridges. It is found overlying basalt with pillow lava in upthrusted sections of the oceanic crust. This chert may be white, grey or dark, depending on the content of organic matter or traces of iron, magnesium etc. If iron is oxidised the chert is red, and called jasper. Chert has also been found overlying the present ocean floor during ocean drilling (Fig.9.2). Although all amorphous silica has been converted to quartz, it is often

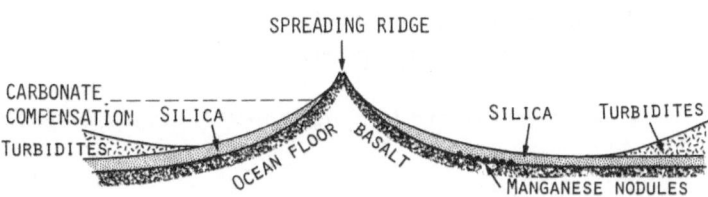

Fig. 9.2. Silica precipitation on the sea bed near the spreading ridge. Chert, which is found above ophiolites, is assumed to form in this way

possible to detect the remains of organic particles, e.g. radiolaria, in chert, and we assume that Palaeozoic and younger chert deposits were biologically precipitated. Chert is not formed only in ocean basins. In areas with upwelling, silica may also be deposited in relatively shallow water. The Monterey Formation in California is a well-known example of chert formed in an upwelling zone. Around the Pacific Ocean similar chert of Upper Miocene age is found. The Monterey Formation is an important oil reservoir rock in California. Subjected to about 2 km of overburden, amorphous silica and opal CT is converted to quartz, and the rock becomes brittle and fractures during the folding and faulting to form good fractured reservoirs.

The Monterey Formation contains organically rich shales containing phosphate. This is a very common association when sediments have been deposited in upwelling areas with a limited clastic supply. Transformation of amorphous silica to quartz appears to depend on the chemical environment (temperature, ionic strength, Mg concentration). Clay minerals reduce the rate of conversion of quartz.

Chert in limestones (flint) occurs typically as nodules, often concentrated along special horizons. The nodules are formed of silica which has been finely distributed in the sediment, largely as sponges and radiolaria. Because small particles with a large specific surface are highly unstable, they will dissolve, and silica will be precipitated in nodules, which are massive structures with a small specific surface, and in consequence more stable.

Precambrian chert deposits typically occur in conjunction with *banded-ironstone formations*. It is uncertain whether there were organisms which could deposit silica in Precambrian times, or whether Precambrian chert was chemically deposited. There are no definite indications of biological precipitation of Precambrian chert, and since silica must also have been added to the oceans, we must assume that the ocean was saturated with respect to silica, and that there may have been inorganic precipitation of silica.

Silica deposits may also be formed in lakes by freshwater diatoms. In dried-up lakes along the rift systems of East Africa, there are thick deposits of diatomite, which is recovered for use in insulating materials. When diatoms proliferate, a great deal of CO_2 is taken from the water by diatoms and algae for photosynthesis, and the water therefore becomes strongly basic (pH 9–10). This increases the solubility of silica, thus increasing the corrosion of silicate minerals. Examples of this are found at Lake Turkana in Kenya. Volcanic rocks and water from hot springs are also an important source of silica.

Chemical precipitation of silica may also take place in evaporite basins and *ephemeral lakes*, which dry up between rainy seasons.

Further reading: Lisitzyn 1972; Heath 1974; van der Lingen 1977.

Evaporites

Evaporites consist of minerals which have crystallised out through evaporation of water. This can happen in many ways.

1. Evaporation of seawater in completely or partly cut-off marine basins.
2. In lakes which have little or no outlet and a high evaporation rate.
3. Through evaporation of seasonal precipitation which collects in topographical depressions without outlets (playas).
4. In soil profiles or sandy sediments, through evaporation of groundwater.
5. In arctic areas freezing of seawater to ice and sublimation of the ice increases the salt concentration of sea water, and evaporite minerals, for example gypsum, may be precipitated.
6. Through solution and precipitation of salts from older evaporite deposits.

Evaporite sediments are formed where there is relatively rapid evaporation and very sluggish water circulation, so that the salt solutions are not diluted by freshwater. The formation of evaporites is therefore no unambiguous indication of a high temperature, and these deposits may also form in Arctic regions. There is, however, a pronounced tendency for evaporites to form in the climatically dry belts about 20–30° from the equator. Large evaporite seas are also found in Siberia. Evaporites contain a number of salts which are too soluble to be precipitated in normal marine or continental environments. The most important salt groups are:

Chlorides
Sulphates
Alkaline carbonates
Ca-Mg carbonates
Borates
Nitrates
Silica deposits.

Marine Evaporite Environments

Although the total salt content of sea water in the world's oceans varies somewhat, the composition of sea water remains relatively constant. The table below shows the percentage composition of dissolved salts in sea water which add up to a salinity of 35‰. To the right are the percentages of the various salts obtained through evaporation.

% by weight of salts in sea water

Ion	% in sea water	Salt	% by weight of common salts after evaporation
Na	30.64	NaCl	77.76
Mg	3.76	$MgCl_2$	10.86
Ca	1.20	$MgSo_4$	4.74
K	1.09	$CaSO_4$	3.60
Cl	55.21	K_2SO_4	2.47
SO_4	7.70	$MgBr_2$	0.22
CO_3	0.21	$CaCO_3$	0.35
Br	0.19		
	100.00		100.00

Experimental evaporation of sea water and chemical calculations make it possible to predict the relative amounts of various salts that will be precipitated. However, absolute figures are only obtained by total evaporation of sea water.

When sea water evaporates calcium carbonate (aragonite) is among the first salts to precipitate, but the amount of carbonate in solution is very small. When the volume of sea water is reduced to $1/3 - 1/5$, $CaSO_4 \cdot 2H_2O$ (gypsum) will precipitate. Only when the volume is down to $1/10$ will NaCl (halite), quantitatively the main constituent, be precipitated, along with magnesium sulphates and chlorides. Polyhalite $(Ca_2K_2Mg(SO_4)_5 \cdot 2H_2O)$ commonly precipitates when the sea water has been reduced to $1/20$. KCl (sylvite) and bromides are among the most soluble salts, and are the last to precipitate when a basin is drying up. In many basins, however, fresh sea water is frequently added, periodically or continuously, so that there is never total evaporation. The resulting evaporites will therefore contain only the least soluble salts, particularly carbonates and sulphates, while chlorides will remain in solution.

Periods of increased evaporation and hence high salt concentration may alternate with influx of sea water with a more normal salt concentration, producing cycles, in each of which the minerals will increase in solubility. These are called evaporite cycles. NaCl and KCl will also dissolve in atmospheric moisture unless the humidity is very low.

The stability of the various salts during evaporation can be determined experimentally, or estimated through physical chemistry calculations. Calcium sulphate may precipitate both as a hydrated mineral, $CaSO_4 \cdot 2H_2O$ (gypsum) and as a non-hydrated mineral $CaSO_4$ (anhydrite). Which of these two phases forms as a result of oversaturation of a solution with respect to calcium sulphate depends on the temperature, salinity and water vapour pressure. In a solution of $CaSO_4$ alone, anhydrite forms only at temperatures of over 60°C, but as the concentration of other salts increases anhydrite may form at temperatures down to 25 -30°C (Fig. 9.3).

Gypsum is, in fact, the mineral which normally forms in marine evaporites, but anhydrite is also observed in modern evaporites in supratidal zones. This is true, for

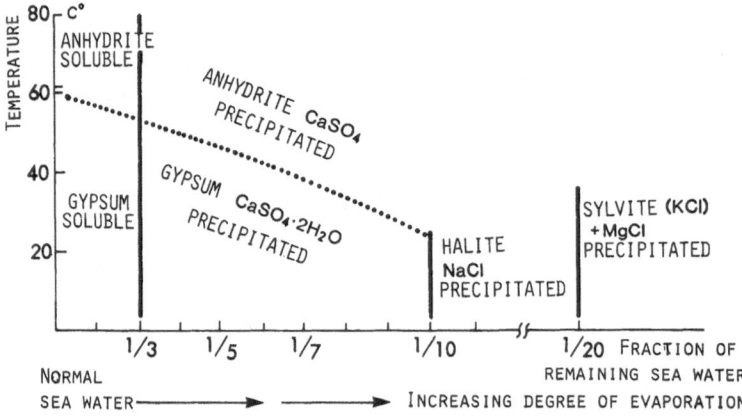

Fig. 9.3. Diagram showing stability areas of some important evaporite minerals

example, of the sabkha deposits in the Persian Gulf (Fig. 8.3). In the black algal mud of the supratidal zone temperatures may be up to 80°C, and anhydrite is deposited in the sediment. In open water, however, where the temperature is lower, gypsum forms.

Today there are only small areas with active evaporite formation. In earlier geological periods there were more extensive evaporite basins.

Evaporite sediments have formed throughout geological history, but appear to have formed more frequently during certain periods, particularly the Permian. In Northern Europe and the North Sea we find thick evaporites from this period in what is called the Zechstein Sea. These salt deposits have been economically important as the raw material for salt production and may also serve as seals for the gas accumulations in Lower Permian sandstones in the southern part of the North Sea. Salt deposits are lightean clastic sediments, and when thick sequences are deposited on top of evaporites, the lighter salt layers will flow upward and gradually form large mushroom or columnar structures. The prerequisite for these domes to start forming, however, is that the salt beds must be at least 100–200 m thick. Salt domes are common in Northern Germany, Denmark and the southern part of the North Sea, where they may form oil or gas traps. Before the opening of the present Atlantic Ocean in Jurassic-Cretaceous times, western Europe and North America were located in the interior of a supercontinent where there would be very little rainfall. New surveys of the bottom of the Mediterranean Sea have shown that there are sizable evaporite deposits there. This indicates that the Straits of Gibraltar were closed for parts of the Tertiary period (the Upper Miocene – 8–12 million years), and that the Mediterranean Sea was a large desert area 2000–3000 m below sea level at that time.

The Sabkha Model

"Sabkha" is the Arabic term for the large flat areas around the Persian Gulf (Fig. 8.3b). The climate is very dry, with only 30–100 mm of precipitation per year, but this is nevertheless sufficient to cause groundwater from surrounding areas to flow out to the sea. Groundwater flows slowly, only a few centimeters a year, but nevertheless causes the diagenetic environment to contain porewater of continental origin. During storms and high tides, marine water floods the sabkha. Some runs off, some evaporates and some filters down into the sediment and mixes with the groundwater. This water is magnesium-rich, and dolomite is formed. Evaporation will increase the salt concentration of the porewater, and if it is intense enough, chlorides are precipitated in addition to gypsum. However, halites will dissolve easily the next time salt water flows in over the sabkha. Moisture in the air will also help to dissolve halite on the surface, so that it is not preserved in the bedding series. Although it is hot, up to 60–80°C on the sabkha, evaporation is limited where there is no open water because of the low permeability of the sediments. When the water table sinks, the rate of evaporation declines considerably. Whereas the rate of evaporation from open water in the Persian Gulf is about 124 cm/year, it is only about 6 cm in the sabkha. We therefore have only moderate "evaporite pumping": chlorides are precipitated above the water table and sulphates below the water table. The chlorides deposited will, however, redissolve as the overburden increases and the water table rises. In some semi-arid regions, e.g. the Coorong region of Australia, dolomite lakes form which evaporate to dryness in the summer (*ephemeral lakes*). Evaporite minerals deposited in summer will redissolve during the winter rains, and will not be preserved in the bedding series.

See McKenzie et al. (1980), Pattersen and Kinsmann (1981) and Muir, Lock and von der Borch (1980).

Marine Evaporites

Evaporation from the surface of the sea will cause the salinity of the water in basin to increase. If there is little wave or current action, the warm surface water will not mix quickly with the underlying, colder water. If the salt concentration of the surface water increases due to rapid evaporation, the density will increase, so that surface water will sink to greater depths and mix with the water there, despite being warmer. In order for the salts in sea water to become concentrated through evaporation, we must have physical barriers around marine areas where evaporation is greater than the total amount of water added to the ocean by precipitation, rivers and ground water (Fig. 9.4).

If the basin is connected by a channel to the open sea, sea water will be able to flow into the basin. Only if fresh marine water is added slowly will we find a build-up of salts in the bay sufficient to form thick marine evaporites. Even the Mediterranean has sufficient circulation by way of the Straits of Gibraltar to prevent the formation of evaporites in the Mediterranean today.

It has been shown that the section through the connecting channel must be less than one millionth of the area of the basin. If a marine basin is totally cut off, the

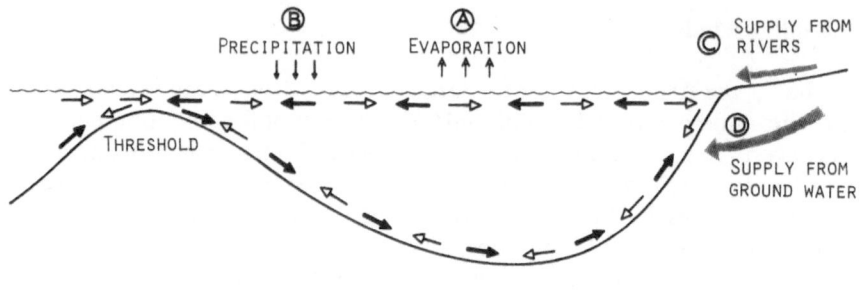

- IF Ⓐ < Ⓑ+Ⓒ+Ⓓ WE GET A LOWER SALINITY THAN NORMAL
  SEA WATER AND A NET FLOW OUT OF THE BASIN.
  EXAMPLE: THE BALTIC SEA, DIRECTION OF CURRENT :

- IF Ⓐ > Ⓑ+Ⓒ+Ⓓ WE GET A HIGHER SALINITY THAN NORMAL
  SEA WATER AND A NET FLOW INTO THE BASIN BUT NOT
  NECESSARILY FORMATION OF EVAPORITES.
  EXAMPLE: THE MEDITERRANEAN, DIRECTION OF CURRENT : ↗

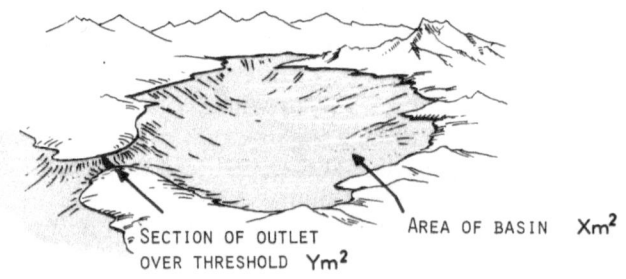

SECTION OF OUTLET
OVER THRESHOLD Ym²

AREA OF BASIN Xm²

TO GET EVAPORITE FORMATION $\frac{Y}{X}$ MUST BE < ABOUT 10^6

Fig. 9.4. Evaporite basins require a marked physical restriction with only a small connection with open ocean basins

water in the basin will evaporate, and it may develop into a freshwater evaporite basin. Episodic influx of marine water into an evaporite basin will produce a characteristic evaporite cycle with increasing solubility.

Tectonic Control of the Formation of Evaporite Basins

Partial or full severing of marine basins may occur as a result of tectonic movements. On a stable continental crust there are limited possibilities for the development of major depressions which can form evaporite basins. Rifting and incipient spreading of the ocean floor provide ideal conditions for the formation of evaporite basins (Fig. 9.4). Rift basins (grabens) subside so that sediment (including evaporites) can accumulate without filling them completely, and the basins may be partly or

entirely cut off by upfaulted blocks (horsts) or volcanoes which block the link
between the basin and the sea (Figs. 9.5, 9.6). In the initial stages of sea-floor
spreading marine evaporites may form and during the Miocene and Pliocene the
Red Sea was an active evaporite basin. Now, however, the opening to the Indian
Ocean is too large for the Red Sea to be a proper evaporite basin. Rifting and
spreading of the ocean floor in connection with the formation of the Atlantic Ocean
led to the formation of thick evaporite series. During the Permian and Triassic
Northern Europe was in the arid zone (about 30°N) and rifting led to the formation

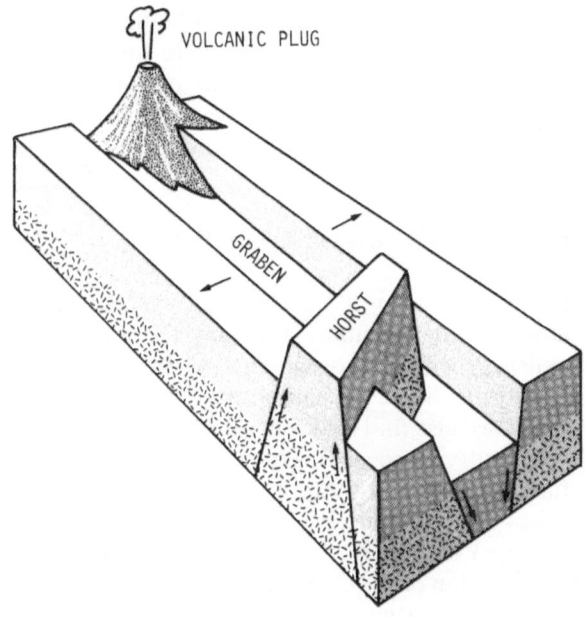

Fig. 9.5. Evaporites in a rift graben
which is partially or fully cut off by
horsts or volcanoes

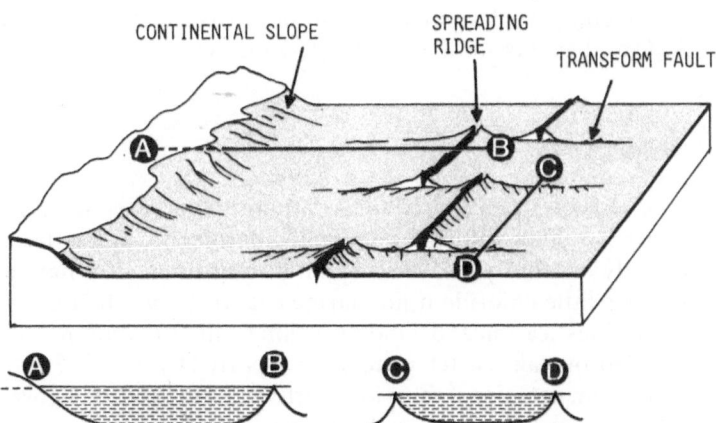

Fig. 9.6. Evaporite basin bounded by spreading ridge, continent and transform faults

of evaporite basins. In Jurassic and Lower Cretaceous times early ocean-floor spreading resulted in the formation of a series of evaporite basins in the area between Africa and South America south of the equator, and in the Gulf of Mexico and North Africa north of the equator. The climate in northwest Europe was then too humid for evaporites to develop. We assume that the basins in these drier zones were separated (blocked) by volcanoes, lava flows or transform ridges. As ocean-floor spreading continued, the Atlantic Ocean widened and the ocean-floor basalts cooled off and subsided more rapidly and after the Mid-Cretaceous no major evaporite basins formed in this area.

As a result we find evaporite deposits of Jurassic-Lower Cretaceous age along the continental shelf in Brazil and Southwest Africa and also in the Gulf of Mexico and North Africa. Where they were sufficiently thick, they formed diapirs which greatly influenced further sedimentation and the structural development of these parts of the continental shelf. Evaporite basins may also develop in basins produced by sagging subsidence on the continental crust without rifting.

Evaporites in Lakes and Inland Seas

Basins with no outlet are formed particularly in tectonically active areas at low latitudes. We find many lakes and inland seas without outlets in connection with rift valleys, such as those of East Africa, and the similar fault-governed basins of California. Since the composition of river water is quite different from that of sea water, we also find different salts in continental evaporites. The composition of the water will vary according to the types of rocks occurring in the drainage area around the lake or sea. If there is volcanism in the area, as is often the case around rift valleys, the weathering of ash and lava and direct addition of volcanic water (springs) rich in dissolved salts will influence lake-water composition and the composition of the evaporites. Lake evaporites normally contain large amounts of carbonate, particularly sodium carbonate and a number of more complex salts, hence the name "soda lakes". The mineralogical composition of these soda deposits is very complex. Examples of important minerals are trona $Na_2CO_3 \cdot NaHCO_3 \cdot 2H_2O$ and gaylussite, $CaCO_3 \cdot Na_2CO_3 \cdot 5H_2O$.

Playas

In playa lakes there may be total evaporation of seasonal rainfall and precipitation of carbonates, including dolomite. If chlorides are precipitated, they will tend to be partially or wholly dissolved again during the next rains. In addition, the water contains little chloride if no marine evaporites are being washed out in the area. Signs of dessication and wind reworking will be common. Salt and clay particles in the dried-out lakes will tend to be transported by the wind and to form small dunes. Water from occasional rains in deserts will often collect between large eolian dunes and form interdune lakes and sabkhas.

Evaporation of Groundwater

Where groundwater evaporation exceeds rainfall there is net upward transport to the surface of the soil, and salts will precipitate out in the soil profile. Gypsum and chlorides have great crystallisation power, and can push aside the sediment matrix so that large euhedral crystals can form. *Caliche* is a hard crust on the soil which forms mainly in arid areas. Here we find cementing of the pore spaces in the soil through precipitation, first and foremost of calcium carbonate, but also of gypsum and silica. Caliche is particularly common in the dry areas of the southwestern USA. We also know of caliche from continental formations from older geological periods.

Solution and Precipitation of Salts from Older Evaporites

Evaporites from older geological periods are easily dissolved by groundwater or surface water. Water draining from areas which contain older evaporites will therefore have an initially high salinity, and it will be far easier to form new salt deposits in lakes or soils.

The Dead Sea is an example of an evaporite basin where water flowing into the basin is already rich in dissolved salts. This is because the Jordan runs through Cretaceous and Tertiary sediments with evaporites. In addition water of volcanic origin enters the rift valley through faults and fractures.

The Stability of Gypsum and Anhydrite During Diagenesis

Rising temperature and pressure will favour the stability of anhydrite. Gypsum formed in evaporite environments will therefore turn into anhydrite when there is sufficient overlying sediment. The transition from gypsum to anhydrite can occur at depths ranging from just under 1000 m to 3000 m (Fig. 9.7). When anhydrite-bearing sediments are uplifted and come into contact with groundwater due to erosion of overlying sediments, gypsum will once more be the stable phase. Anhydrite will then turn into gypsum over a vertical distance of about 50–100 m,

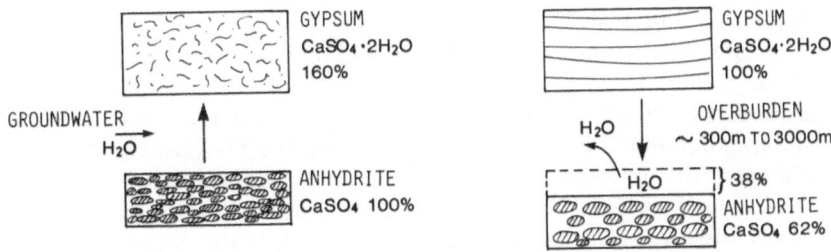

Fig. 9.7. Compaction (dehydration) and expansion (hydration) of the calcium sulphate minerals gypsum and anhydrite. This leads to considerable tectonic activity and geotechnical instability at the surface.

depending on water circulation. The loss of water resulting from the transition from gypsum to anhydrite leads to a decrease in volume of 38%, and the transition from anhydrite to gypsum leads to a corresponding increase. If gypsum layers are folded down to great depths, with transition to anhydrite, great deformations are consequently occasioned by the reduction in volume. In the same way the transition from anhydrite to gypsum will cause great expansion. This creates great geotechnical problems in those areas of Europe, e.g. Switzerland and Germany, where there are many evaporite deposits.

Further reading Kinsman 1975 a, b, 1976; Dean and Schreiber 1978; Eugster and Jones 1979; Taylor 1980.

Iron- and Manganese-Rich Sediments

Iron and manganese are similar in many ways in their geochemical behaviour in sedimentary environments. Both elements are poorly soluble in the oxidised state because they form hydroxides and oxides: $Fe(OH)_3$, $Mn(OH)_4$, Fe_2O_3 and MnO_2. In the reduced state they occur as Fe^{2+} and Mn^{2+} and are then more soluble. In the reduced state, however, both iron and manganese can be precipitated as carbonate ($FeCO_3$, $MnCO_3$), either as separate minerals, or as part of dolomite, calcite or ankerite. They are therefore not very soluble in basic solutions with low redox potentials, but are quite soluble at low pH values and in the reduced state. Consequently we can precipitate iron and manganese in two ways: (1) through oxidation (2) by changing the solution from acid to basic.

The major difference between iron and manganese in sedimentary processes is that while Fe^{2+} forms iron sulphides even at very low sulphur concentrations, manganese sulphides are far more soluble. Manganese will therefore not precipitate in neutral or acid solutions with a low Eh.

Weathering takes place mainly under oxidising conditions, and circulating groundwater will also take oxygen further below the surface, particularly along cracks. As silicates and other minerals dissolve in the oxidising environment during weathering, the concentration of iron and manganese (and also titanium) will increase, along with aluminium, which is an essential constituent of laterite.

When groundwater flows through sediments containing organic material, the oxygen it contains will gradually be used up through oxidation, so that the porewater will become reducing. Iron and manganese may then occur in solution as Fe^{2+} and Mn^{2+}, particularly if the porewater is a little acidic. Where this groundwater flows out of the ground, e.g. at the foot of a slope, we will get precipitation of oxidised iron and manganese ($Fe_2O\rightarrow$, MnO_2) (Figs. 9.8, 9.9). Fe_2 is more easily oxidised than Mn^{2+} in naturally occurring water.

The end products of weathering after quartz and other silicate minerals have slowly dissolved are aluminium and iron hydroxide, which are the main components of laterite. Laterite may contain up to 60% Fe_2O_3.

Bogs, such as we find in the Scandinavian highlands, have conditions ideal for precipitating bog iron. Water flowing through bogs will have a low Eh and a low pH, largely due to humic acids, and iron-containing minerals from underlying rocks will be dissolved by humic acids, transported in the reduced state and precipitated through oxidation where the water flows out of the bog.

Porewater in sediments and sedimentary rocks is normally reducing. This is because most sediments contain reducing agents, particularly organic material. In marine basins and lakes which do not have stagnant bottom water, there will be oxidising conditions only in the upper 0–20 cm of the bottom sediments, and below that they will be reducing. Only where we have rapid downward flow of meteoric oxygen-rich groundwater are oxidising conditions possible at greater depths. Most sediments contain some organic matter and also some minerals that can serve as reducing agents so that oxygen is constantly being consumed. The depth to which the groundwater will be oxidising depends on the supply of oxygen, i.e. the oxygen content and rate of flow of groundwater compared to the consumption of oxygen due to oxidation. Desert sediments contain little organic material which can function as a reducing agent, and the groundwater will therefore remain oxidising longer. Minerals containing Fe^{2+} are thus oxidised and the result is red ferric oxide.

The redox boundary, which normally lies just below the sediment/water boundary on the ocean bottom, represents an important geochemical trap. The concentration of elements on each side of this boundary is very different because of the different solubility of elements in the two different chemical environments.

In the reducing zone there is a higher concentration of elements which are more soluble in the reduced state, such as iron and manganese. In the sulphate-reducing zone however, Fe^{2+} is precipitated as a sulphide. There will therefore be a strong Fe^{2+} concentration gradient and upward transport of Fe^{2+} into the sulphate-reducing zone by diffusion.

If there is insufficient organic matter in the sediments for the sulphate-reducing bacteria, iron may precipitate as $Fe(OH)_3$ above the redox boundary.

Manganese will not be as easily trapped in the sulphate-reducing zone, as Mn^{2+} does not form such stable sulphides as Fe^{2+}, and it will therefore have a greater tendency to be precipitated in the oxidised zone.

During breaks in sedimentation or slow sedimentation, porewater expelled by compaction will cross the redox boundary, and this may cause the precipitation of iron and manganese on the sea floor. *Manganese nodules* are concretions of manganese and iron hydroxides and oxides found on the sea floor, particularly in

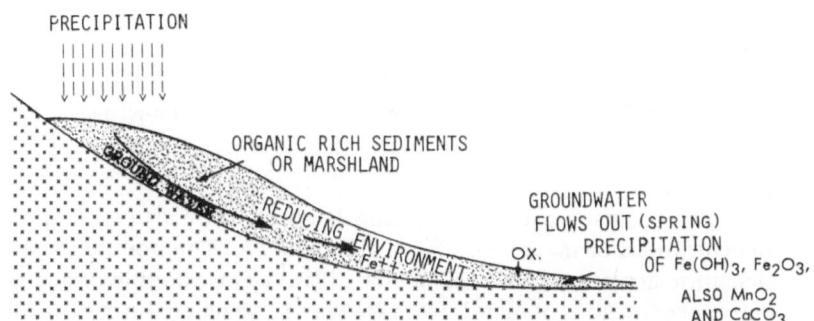

Fig. 9.8. Schematic diagram of the circulation of water through a bog and the precipitation of iron hydroxides and oxides (bog ore). Manganese will be precipitated in the same way

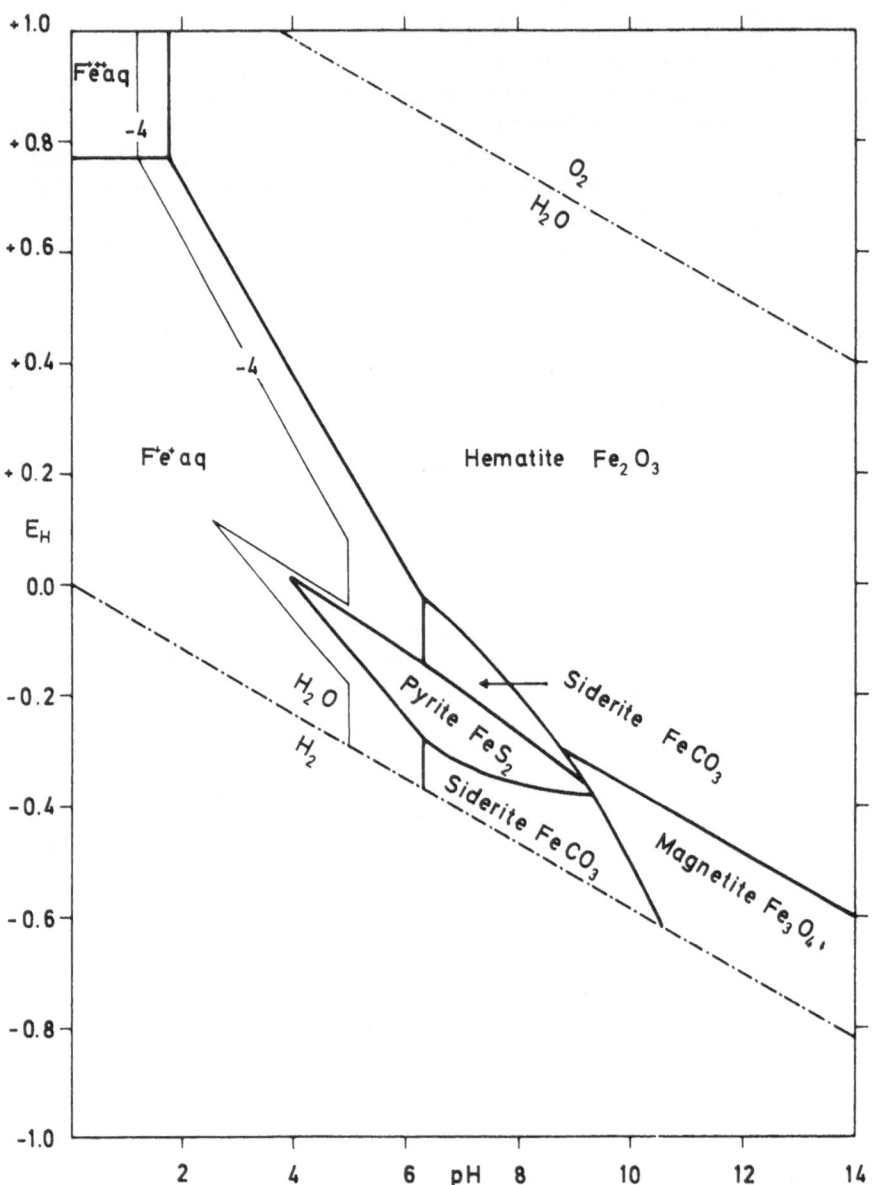

Fig. 9.9. Stability fields of some iron and manganese phases in an Eh-ph diagram (Garrels and Christ 1965)

the ocean basins (South Atlantic and Pacific). They also contain relatively high concentrations of metals like Ni, Cu, Zn and Co. The nodules grow by very slow concentric accretion in a pelagic ooze. Plankton is capable of accumulating very high concentrations of metals from sea water and when the organisms dissolve, the planktonic ooze becomes very rich in these metals, which will be precipitated together with the manganese hydroxides in the concretions.

The sulphate concentration in the porewater, however, declines rapidly from the oxidising zone to the reducing zone over a distance of 1–5 m. Sulphate will therefore diffuse downwards, and be used by sulphate-reducing bacteria to form H_2S, which dissolves to form an acid which can release iron bound in various clastic minerals to form monosulphide, FeS, (machinawite) and disulphide, FeS_2, (pyrite). If the sediments have a high silica content, chamosite may also be formed. Glauconite must be formed right at the redox boundary, since it contains both ferrous and ferric iron. Uranium and vanadium, which are more soluble in the oxidised state will diffuse downwards and be precipitated in the reducing zone below the redox boundary (Figs. 9.10, 9.11).

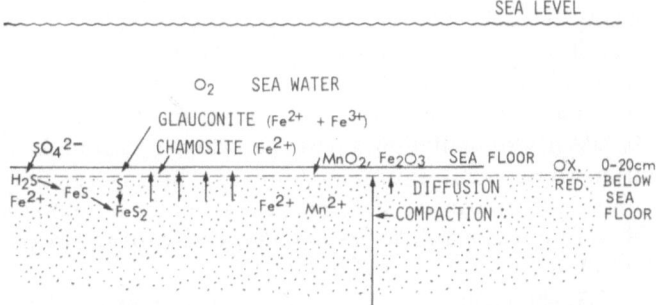

Fig. 9.10. Distribution of early cements in sediments. Some elements are more soluble in the oxidised state (S, U, V), others in the reduced state (Fe, Mn). As a result we will have high concentration gradients and diffusion across the redox boundary. Fe will form sulphides even at low sulphur concentrations and siderite will therefore not form in the sulphide-reducing zone

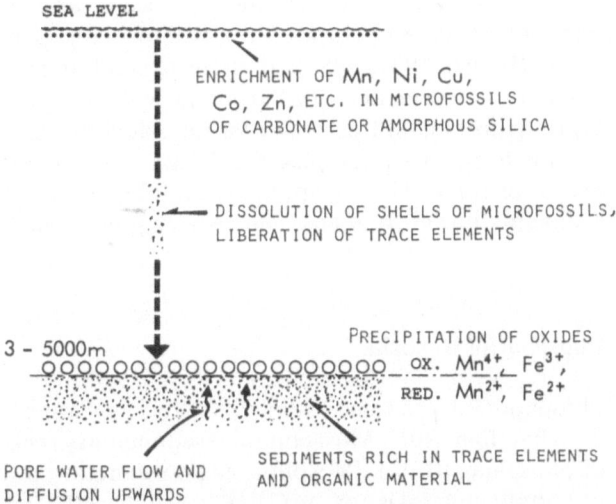

Fig. 9.11. Enrichment of metals in ocean floor sediments takes place firstly through biological precipitation, followed by further solution and precipitation by the redox boundary just below the sea bed. In areas with slow sedimentation rates, manganese nodules will form. Mn deposits also form in some lakes

If there is continuous sedimentation, the redox boundary will keep moving upward, and some elements precipitated above the redox boundary may dissolve again below the redox boundary and be precipitated as sulphides. However, if we have a long period without sedimentation (hiatus), diffusion will take place in the same sediment horizon for a long time. The compaction of the underlying beds will also continue, even if there is no sedimentation, and the porewater flowing upwards will be oxidised in the same horizon. If there is little organic matter in this sediment, the activity of sulphate-reducing bacteria may be reduced. One condition for obtaining concentrated iron- and manganese-rich layers through oxidation by porewater is therefore that we have a marked break in sedimentation.

Iron and manganese are virtually insoluble in oxygenated sea water, and consequently cannot be transported in ordinary solution. However, rivers can carry a good deal of iron and manganese adsorbed onto organic particles in the form of complex organic molecules, or as part of the clastic sediments. Large quantities of iron are carried by rivers like the Amazon each year, but these will be finely distributed in clastic sediments. Iron-rich sediments are often associated with breaks in sedimentation, small hiati when iron from the porewater, in sediments or from iron-containing organic complexes in sea water may have time to concentrate into high-grade iron sediments because of the dilution due to clastic materials. It is therefore reasonable to assume that concentrated iron and manganese beds are a result of diagenetic enrichment.

As shown in Fig. 9.9, iron can be precipitated not only by oxidation, but also in the reduced state in a basic environment. Thin carbonate laminations in shales represent a local high-pH environment, where iron which is normally soluble can be precipitated:

$$Fe^{2+} + CaCO_3 = FeCO_3 + Ca^{2+}.$$

This reaction is a function of the Fe^{2+}/Ca^{2+} ratio and the degree of hydration of Fe^{2+}. At higher temperatures iron will form the carbonate even at low Fe^{2+}/Ca^{2+} ratios (Berner 1971, 1980). If there is sulphur present, iron will first form the sulphide, however, and iron carbonate will not be stable. Siderite is therefore most typically formed in freshwater basins, where the sulphate concentration is low. It cannot form in the sulphate-reducing zone in marine sediments because all available iron will be precipitated as sulphides there. Siderite is common in marine sediments, however, but must have been precipitated below the sulphate-reducing zone.

Phosphorite Deposits

Phosphorite deposits are sedimentary rocks in which the phosphate (P_2O_5) content is higher than usual. Most common sedimentary rocks contain very small amounts of phosphate (0.1%). Phosphorites may contain up to about 37% P_2O_5. The most important mineral is apatite $Ca_5(PO_4)_3(F, Cl, OH)$, which occurs as varieties rich in fluorine, chlorine and hydroxyl groups. Francolite $(Ca,Na)_5(F,OH)(PO_4,CO_3)_3$ is the phosphate mineral most commonly found in marine sedimentary rocks, where some carbonate has substituted for phosphate and where sodium substitutes for calcium. Aluminium and iron apatites such as wavellite $(Al_3(PO_4)_2(OH)_3 \cdot 5H_2O$ and strengite $(FePO_4 \cdot 2H_2O$, are also relatively common minerals in secondary (weathered) phosphorite deposits.

Phosphorite rocks are an important source of fertilizer. Large deposits such as those found in Morocco, Spanish Sahara and Senegal in West Africa, and Florida are of great economic value. Understanding how phosphate deposits form is therefore a matter of considerable economic interest. In most sedimentary rocks phosphorus is a trace element, and very special conditions are necessary for phosphate enrichment to take place. These deposits therefore tell us something important about the environment of deposition.

Phosphorites are practically all marine, apart from rare instances of lacustrine phosphorites and small continental deposits of guano. Guano deposits are formed from bird or bat excrement which gradually becomes enriched in phosphate as the other organic components are washed out. For phosphates to be preserved, the rainfall must not be too high, because then it will dissolve phosphate sediments on land.

The first prerequisite for marine phosphate deposits is that sedimentation must proceed very slowly, i.e. there must be virtually no clastic sedimentation. In consequence we find phosphate beds associated with major or minor breaks in sedimentation, or with periods of very slow sedimentation. It has long been known that phosphate is formed in areas with strong upwelling and high organic productivity. We have good examples of this along the edge of the continental slope off Chile and Peru (Fig.9.12). Water welling up from great depths brings with it nutrients which are liberated when marine organisms disintegrate through oxidation and solution in deeper water. When the water flows up to the surface, the nutrients are used by photosynthesising organisms. They allow high primary

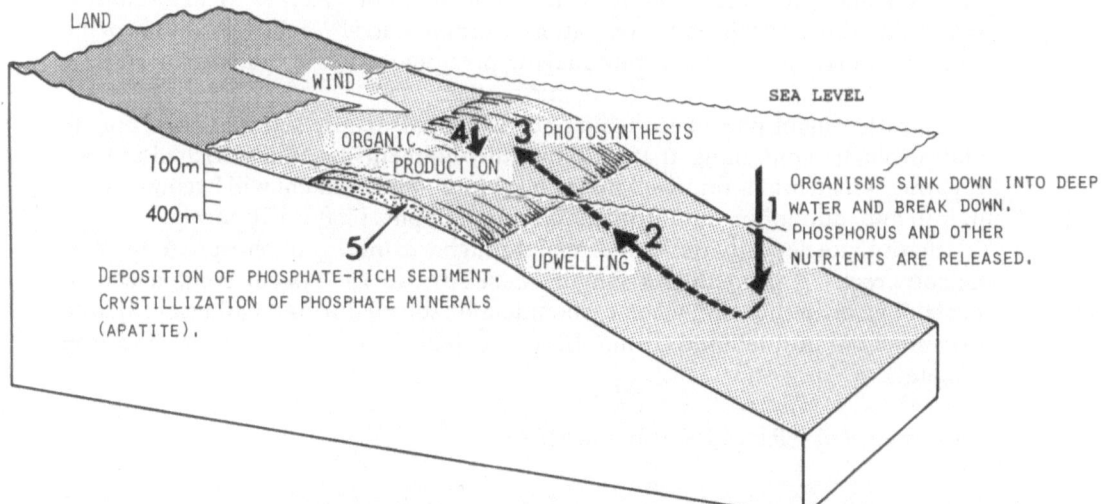

Fig. 9.12. Formation of phosphate deposits on a continental shelf by upwelling. Note that the upwelling will also contribute to reducing the inflow of clastic material. The organic product is rich in phosphorus and forms phosphate minerals on the sea floor, often by replacement of carbonates

production of phytoplankton, zooplankton and higher organisms. The organisms formed contain about 1% P (dry weight), which is an enrichment of 140,000 compared to dissolved phosphorus in ordinary sea water. Organisms with an amorphous silica skeleton (diatoms and radiolaria) will gradually dissolve in contact with sea water, and in some cases carbonate (calcite and particularly aragonite) will also dissolve resulting in further phosphate enrichment. Phosphate minerals then crystallise out of the organic-rich sediments with a high phosphate content and often replace other minerals, e.g. carbonate. Apatite is a heavy mineral (specific weight 3.18) and may also be enriched mechanically by weak traction currents.

Apatite will often crystallise out as concretions in bottom ooze, and erosion by traction currents may concentrate these nodules into a conglomerate. On the continental shelf phosphate forms at depths of between 400 and 100 m, i.e. below the photic zone. However, phosphate nodules and massive beds of fine-grained phosphate mud (phosphate micrite) may also form in lagoons where the water is less clear and the photic zone shallower. These phosphate deposits will be easily eroded even as a result of minor regressions, and form conglomerates of phosphate mudstone (micrite). Phosphate deposits are forming today mainly in restricted areas with strong upwelling, but in previous geological periods we find very extensive phosphate beds, often associated with transgressions. The transgressions will hold clastic sediment back for a while, enabling biogenic matter to concentrate. Phosphate beds are often associated with other authigenic (formed in situ) minerals which take a long time to form, particularly glauconite and manganese deposits.

Marine phosphate deposits may also be formed by vertebrate fossils with a phosphate skeleton, for example fish. This type of phosphate deposit may also be extensive, and is often called "bone beds".

Phosphate minerals such as apatite may contain considerable amounts of uranium and rare-earth metals which substitute for calcium. Weathering of phosphate deposits will lead to oxidation of uranium to U^{6+}, which is soluble in the form of uranyl ions, and uranium may be precipitated again in the reduced state when it comes into contact with organic matter.

Weathering of phosphate sediments will initially cause solution of carbonate and carbonate-containing apatite so that the sediments become enriched in fluorapatite. As calcium is removed through solution, the sediment will become richer in iron and aluminium, and iron and aluminium phosphates will form.

Phosphate deposits are also formed through weathering of phosphate-bearing igneous rocks. Carbonatites from the East African rift system contain a high percentage of apatite, and when carbonate is dissolved during weathering, apatite becomes concentrated and form, phosphate deposits which are also rich in iron minerals.

Further reading: Riggs 1979; Bentor 1980.

10 Stratigraphy

Introduction

Stratigraphy is the study of the succession of rock strata and their properties. Stratification is not limited to sedimentary rocks, but is also found in igneous rocks, particularly volcanic rocks, and in certain plutonic rocks. With all bedded rocks one can set up a stratigraphy, i.e. establish age relations between beds. However, the term "stratigraphy" is used first and foremost for subdividing sedimentary sequences. Stratigraphy involves studying and describing sedimentary successions and on this basis interpreting the geological history represented. In order to reconstruct an environment or special events in our geological history over a large area or on a global basis, it is necessary to correlate sedimentary horizons from different areas. We try to establish which sediment beds were deposited at the same time or by the same or similar sedimentological or biological processes. Correlation is very largely a question of what is possible with the available data. We usually have no possibility of determining absolutely definitely which beds were deposited at the same time, but attempt to use all the information available in the rocks. This information falls into three main groups: (1) Rock composition and structures resulting from sedimentological processes; (2) Fossil content, which is a result of ecological environment and biological evolution throughout geological history; (3) content of radioactive fission products in minerals or rocks which may be used for age dating. These three correlation methods are so different that it has been found useful to work with three forms of stratigraphy which can be used in parallel (see Table 10.1)

1. *Lithostratigraphy:* Classification of rock types on the basis of their composition, appearance and sedimentary structures.
2. *Biostratigraphy:* Classification of rocks according to their fossil content.
3. *Time Stratigraphy:* Classification of rocks on the basis of geological time.

Despite radiometric dating, geological time is not absolute. Age dating gives different ages depending on which half-life is used for radioactive decay, and is encumbered with many other uncertainty factors. Time-rock stratigraphy chronostratigraphy is therefore a theoretical concept which describes an absolute time scale which we cannot quite measure. Rock-time stratigraphy, on the other hand, takes its point of departure in rocks and is based on limits which could function as limits of convenience in greater or lesser areas. For rules for stratigraphic nomenclature, see Hedberg 1972, 1976, Holland 1978.

Table 10.1. Different types of stratigraphic divisions and units

Lithostratigraphy	Biostratigraphy	Chronostratigraphy	Geochronology
(subdivision of sequences based on rock composition)	(subdivision of sequences based on fossil content)	(subdivision of sequences based on time. Time-rock stratigraphy)	(subdivision of geological time)
Lithostratigraphic Units	Biostratigraphic Units	Chronostratigraphic Units	Geochronological Units
Super group	Assemblage zone	Eonothem	Eon
Group	Range zone	Erathem	Era
Formation	Acme zone	System	Period
Member	Internal zone	Series	Epoch
Bed	Biozone	Stage	Age
		Chronozone	Chrone

Lithostratigraphy

Formations are the fundamental units in lithostratigraphy. A formation is part of a bedding series which can easily be recognised in the field or a borehole due to its composition (lithology). A formation is thus a mappable unit which is distinguished by a separate colour or symbol on an ordinary geological map (e.g. 1:50,000) or in the description of the series. There is in principle no limit to the thickness of a formation, but they usually vary from a few tens of metres to several hundred metres. A 100–200 m-thick sandstone will constitute a natural formation. However, a formation will seldom be homogeneous, and for more detailed mapping it will be useful to divide the formation into smaller units. Parts of a sandstone formation which contain, for example, shale or conglomerate beds can be distinguished as *members* of the formation.

The smallest unit in the lithostratigraphic classification is a *bed,* which one assumes to have been deposited by a single depositional process, and is distinguishable from the rocks above and below. It is often called a sedimentary unit. *Beds* are sedimentary layers from a few centimetres to a few metres thick.

For some purposes it is expedient to group a number of formations together into a larger unit called a *group.* A group normally consists of three to six formations. Where this is natural, a group can be divided into two or more *sub-groups.* The largest lithostratigraphic unit is a *super-group* or *suite*, which consists of two or more groups, and this unit is used when one wants a common name for thick sedimentary series deposited over a long geological period.

In sedimentary basins which are not exposed, for example in the North Sea, formations and groups are defined on the basis of records from wells, such as well logs. In offshore areas it is difficult to find enough geographical names to give the stratigraphic units, and other names are then also used, including historical names, names of animals etc.

Lithostratigraphic Terminology

Internationally accepted rules have been established to facilitate comparison and correlation. Each lithostratigraphic unit ought thus to be named according to the following rules:

1. Stratigraphic units should preferably be defined with reference to a type section where the beds are well exposed or occur in a well.
2. Each stratigraphic unit ought to be named after a geographical site, preferably near the type section if exposed on land.
3. The same name ought not to be used for more than one stratigraphic unit. The unit which is first defined has priority.
4. Where geographical names are not available or where other, older stratigraphic names are well established, other types of names may be used.
5. Stratigraphic names may consist of a place name and a stratigraphic unit, e.g. the Brøttum Formation, or may also contain a rock name, e.g. the Ekre Shale Formation. If one wants to shorten the last name, it becomes the Ekre Formation (not the Ekre Shale).

Biostratigraphy

Biostratigraphy is based on the fossils in sedimentary rocks. A biostratigraphic unit (*biozone*) is defined as part of a series of strata characterised by one or more fossils. The beds between the lowermost and uppermost occurrence of a fossils form the *range zone*. The base of a biozone is defined by the first appearance of one or more fossils, and the top as the base of the next zone. If a zone is defined by means of several fossils, it is called an *assemblage zone*.

Biostratigraphic correlation has made it possible to correlate sedimentary rocks on a global basis, and forms the foundation for global stratigraphic classification of sedimentary rocks.

Biostratigraphic correlation is based on the fact that during the course of geological time, there is biological evolution whereby some species die out and new ones appear. Certain species will therefore be represented only in sediments deposited over a limited time span. However, this method has many limitations. Some fossil groups develop rapidly, others slowly, and some are found only in rocks deposited in particular environments. It may also be difficult to give an unambiguous definition of species which appear to evolve into other species vertically in a stratigraphic sequence. The ideal index fossil for defining a biozone ought to belong to a fossil group which has undergone rapid biological evolution, had global distribution, and not had very special environmental requirements. The graptolites constitute such a fossil group. Because of their planktonic lifestyle they were not very dependent on the local environment. However, even this type of plankton may show a distribution pattern which reflects ocean currents and temperatures etc. In recent times microfossils have been used more and more for biostratigraphic correlation. Both animals and plants (e.g. diatoms, radiolaria and the spores of land plants) have wide distribution and have the advantage that even in small samples, e.g. from boreholes, they provide adequate material for a statistical analysis.

The distribution of all fossils will be governed to some extent by the environment, and new species will not necessarily spread simultaneously to all parts of the world. We see this clearly if we study Quarternary and Recent fauna and flora. The first appearance of a fossil in a sedimentary series is therefore not quite simultaneous over large areas. This may have major consequences for more detailed subdivisions of bedding series within relatively short geological time periods, while for large time units environment it is of less importance (Table 10.1).

Although fossils are always helpful in stratigraphic correlations, we now avoid defining important geological boundaries by means of the first occurrence of a fossil. Such a boundary would have to be moved if one found a new occurrence of the fossil. The geological periods are now defined by international committees which select a type section with a continuous and preferably fossiliferous facies, and the boundary is marked as an absolutely concrete point on the section. The boundary is then unambigously defined, and all available means can then be used, including fossils, to correlate the boundary with other areas. This is the principle of the arbitrary boundary. In reality it is not entirely arbitrary. We try to put it on a section which gives optimal possibilities for correlation with other areas. Geologists used to define a stratigraphic boundary at a break (hiatus) in a sequence. This led to sediments which were deposited in other areas in the period included by the break not being included in any unit. For example, the boundary between the Tertiary and Cretaceous was defined in England where there is a break between these units. As a result it was difficult to reach agreement on whether sediments which were deposited, for example in Denmark, during this period, should belong to the Cretaceous or the Tertiary. As a result we now try to define boundaries in sections where we have had continuous sedimentation.

Time Stratigraphy

We distinguish between two types of time stratigraphy:

1. *Chronostratigraphy* is subdivision of sequences and their correlation on the basis of time. Chronostratigraphic units are by definition synchronous.
2. *Geochronology* involves division of the history of the Earth into time units. A geochronological unit is a specific interval of geological time.

A geochronological unit is a geological time unit defining a specific interval of time on the geological time scale. For example, the Ordovician period is such a geochronological unit, and represents a particular period of geological time. A geochronological unit may define the time between two specific geological events.

We can say, for example, that some of the rocks in the Oslo region were deposited during the Ordovician period. The corresponding chronostratigraphic unit (system) signifies the rocks which were formed during the same period. We therefore say that some of the rocks in the Oslo area belong to the Ordovician system.

Chronostratigraphic units are subdivisions of rock sequences made on the basis of geological time. Chronostratigraphic units are defined in type sections, and then attempts are made to correlate the rocks that were deposited at the same time over wide areas. The precision one can attain with chronostratigraphic correlation depends on the rocks containing evidence of well-defined geological events of short duration which were approximately simultaneous over wide areas. These events might be biological (spreading of new species), sedimentological (deposition of ash layers, rapid sediment transgression) or geophysical (changes in the magnetic field of the earth). Correlation of simultaneous events in geological history has a long tradition in geological research, but is dependent on an absolute time scale. Only after establishing radiological dating methods has it been possible to set up something approaching an absolute time scale.

Chronozones

The basic chronostratigraphic unit is a chronozone. A chronozone includes all rocks formed during a particular time interval which are distinguished by a geological phenomenon or a particular part of a sequence. In most cases a chronozone will be defined as the period between the first appearance and the last occurrence of a particular fossil. Whereas a biozone can only be defined where the fossil is found, a chronozone represents all rocks formed during this period, regardless of whether they contain fossils. To follow such a zone into rocks without fossils we would have to rely on other types of correlation. A *chrone* is the geochronological equivalent of a chronozone, and is the time which the rock in a chronozone represents. A chronozone may be named after a biostratigraphic unit, e.g. a *Didymograptus extensus* chronozone, or it may be named after a lithostratigraphic unit, for example the Bjerkåsholmen chronozone, which can be recognised over large areas and has the same age. Such characteristic strata are often called *marker beds, key beds, datum beds*, or *levels*. Chronostratigraphic horizons are very important for correlation within large sedimentary basins and form the basis for all facies reconstructions (Table 10.1). If we have two chronohorizons or datum beds which can be correlated across the basin, we can measure the variation in thickness and composition of the sediments which were deposited in a particular period of time delimited by the lower and upper bed. This may make it possible to map variations in sedimentation rate and the ratio of sandstone to shale in datum beds. Maps showing the sediment

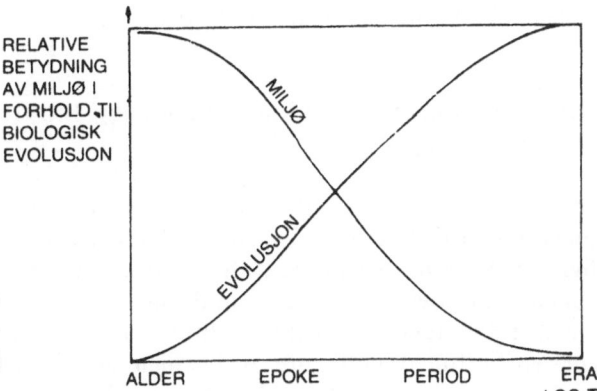

Fig. 10.1. The figure shows the relative importance of biological evolution in relation to local environment as a function of geological time

thickness between two marker beds are called isopach maps. They are very useful for reconstructing sedimentary facies on the basis of borehole data, a standard method in connection with oil prospecting.

The best chronostratigraphic horizons (marker beds) are bentonite (ash) layers, phosphate beds, and thin limestone or sandstone beds or particular fossil horizons. When analysing stratigraphic records from wells (logs) one tends to use beds which produce a distinctive pattern. Seismic reflectors can also in certain instances be used as datum beds.

Reversals of the earth's magnetic field have provided a basis for a series of chronostratigraphic horizons.

A *stage* is a chronostratigraphic unit which includes one or more chronozones, but which nevertheless covers a limited period of time, usually 1–10 million years. A stage or sub-stage is the smallest unit in chronostratigraphic hierarchy which is used for correlation all over the world. A stage is defined in a type section and usually named after a geographical name near the type profile. For example, the Kimmeridge stage is well exposed along the coast at Kimmeridge, near Dorset in England. The correlation of a stage is usually based on biostratigraphy.

An *age* is the period of time (geochronological unit) which corresponds to a stage.

A *series* is a larger chronostratigraphic unit than a stage. For example, the Malm series constitutes the upper part of the Jurassic system. The geochronological unit which corresponds to a series is an *epoch*. We can, for example, say that these limestones were deposited during the Malm epoch.

A geochronological *period* varies in duration from about 30 million years (Silurian) to about 90 million years (Cambrian). The Pleistocene period, however, is much shorter, only about 2 million years. The rocks formed during a period constitute a *system*. An *era* includes two or more periods. The Palaeozoic era had a duration of about 340 million years, but the Cenozoic era did not last longer than the longest Palaeozoic periods (65 million years).

The largest units in the chronostratigraphic scale, *erathem* and *enothem* are not used much, since it is seldom relevant to group rocks which were deposited over such long periods of time. However, when we discuss geological history in relation to the biological development of the earth, for example, it can be useful to speak about the Phanerozoic *eon* which covers the Palaeozoic, Mesozoic and Cenozoic eras. The Precambrian is divided into two eons: the Proterozoic (590–2500 million years ago) and the Archean (2500–4000 million years ago).

The Relationship Between Lithostratigraphy, Biostratigraphy and Chronostratigraphy

These three types of stratigraphy are based on completely different criteria, and long geological experience has shown that it is expedient to maintain this tripartite division. Within a small area, these three ways of classifying rocks may nearly coincide, but over larger areas and under special geological conditions, we get considerable deviations between litho-, bio- and time stratigraphic boundaries. This is illustrated in Figure 10.2 which depicts schematic representation of a sandstone formation with shale above and below.

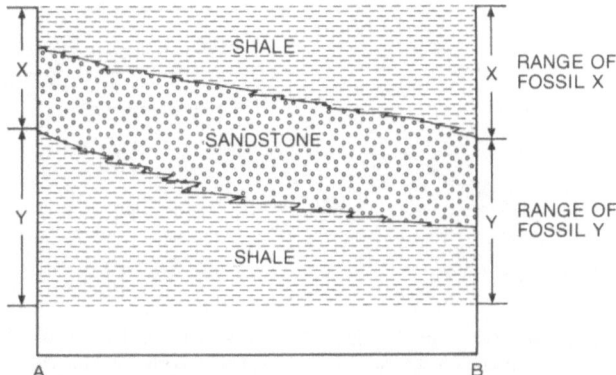

Fig. 10.2. A time-transgressive sandstone formation. X and Y are good index fossils which are little affected by facies. The sandstone is thus younger in area *A* than in *B* and has probably been deposited by progradation from *B* to *A*

Lithostratigraphically, this sandstone formation is unambiguously defined by the boundary between shale and sandstone. However, if we find good index fossils, X and Y, in both the shale and the sandstone (Fig. 10.2), it may turn out that the top of the sandstone in area B corresponds chronostratigraphically to the bottom of the sandstone in area A. Sandstone sedimentation has thus moved from B to A during the course of a measurable period of geological time. Sandstone sedimentation along coasts and on deltas will tend to shift over long periods of time, and the sandstones deposited will therefore have an upper and a lower boundary which are not parallel to a theoretical time plane.

If fossils X and Y had been animals which lived on the sea bed, the picture would have been different. If fossil X preferred a sandy bottom and a Y clayey bottom, the biozones would coincide with the lithostratigraphic division. We would not be able to decide then whether the sandstone was deposited at the same time in areas A and B, unless we found an ash layer, some other marker bed or fossils to give us an indication of the time plane or *chronohorizon*.

In practical geological work one is primarily interested in mapping rock types, in other words lithostratigraphy. Good index fossils or marker beds will provide important information about how forces have acted over the ages. For example, we can tell whether deposition took place during regressive outbuilding or transgressive movements, depending on whether the base of a sandstone suddenly becomes younger or older towards the basin or towards land.

If we study a modern coastal area, we find different fauna in different environments (Fig. 10.3), so we can easily see that the distribution of fauna within a limited geological time period is not controlled by stratigraphic time, but by facies. Many animal groups provide good indices for sedimentary facies.

Palaeomagnetism and Magnetic Stratigraphy

Palaeomagnetism

When clastic sediments are deposited or when volcanic rocks solidify, minerals orientate themselves in the currently existing magnetism in field. Decisive for magnetic rocks is when the temperature falls below the Curie point. Magnetic minerals

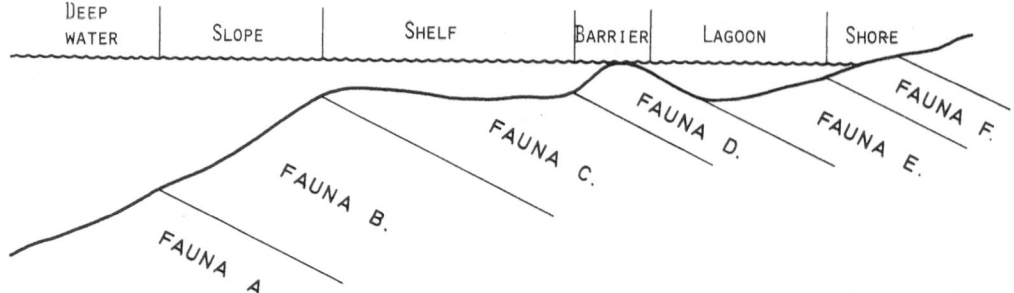

Fig. 10.3. Diagram showing what types of fauna live on the bottom at the same time in different environments. We see that fauna zones here largely coincide with facies. This fauna will not define a time stratigraphy

in clastic sediments can orient themselves in relation to the magnetic field during deposition. Diagenetic minerals will be able to orient themselves during diagenesis. The magnetic minerals in rocks thus constitute a magnetic vector which indicates the direction and strength of the magnetic field during formation. By compensating for later effects, the orientation of this *remanent magnetism* can be measured and can be used as an indication of geographical latitude and longitude. We assume here that the magnetic pole has always circled round the geographical poles, so that an average of a number of measurements will point at the geographical north. Palaeomagnetic measurements are a great help in reconstructing the positions of the continents in early geological periods, and important for plate tectonic and palaeoclimatic reconstructions. From the Palaeozoic onwards there is quite good agreement between palaeomagnetic determinations of latitude and palaeoclimatic indications like glacial deposits, evaporites, fauna zoning etc., but this does not apply to Precambrian deposits. There we find a large number of glacial deposits in the areas which according to palaeomagnetic determinations should lie on the equator. Most geologists have assumed up to now that the inclination of the earth's axis has been constant, and that the continents have moved in relation to this axis. Some theories, however, postulate that it, too, may have changed its orientation over the ages.

Magnetic Stratigraphy

In recent years it has also become evident that the polarity of the earth's magnetic field has been reversed for long geological periods. A number of measurements of magnetism in rocks of known age have also provided us with a time scale with periods of normal and reversed magnetic fields. We thus find that we can divide geological time periods into long epochs with mainly normal or reversed polarisation. Within these we also find shorter periods when the magnetic fields had reversed polarity. Since we must assume that the switch from normal or reversed magnetism has taken place simultaneously all over the world, and that it has taken a relatively short time, such physical changes offer an ideal basis for correlation.

However, there are often major practical problems, since the change has taken place rapidly in many periods of the earth's history. Where we have many measurements and continuous profiles, e.g. from deep-sea samples, we will be able to correlate with relative certainty on the basis of long periods of normal or reversed magnetic fields. Volcanic rocks and sediments deposited in fluvial or shallow-water environments, however, will have many breaks between beds, so that one cannot register continuous variations in the residual magnetism. During the last 700,000 years it looks as though we have had normal polarisation, possibly with the exception of a short period about 200,000–300,000 years ago. If we find sediments or volcanic rocks with reversed magnetism, we know that they are very probably more than 700,000 years old.

Magnetic stratigraphy has already been used a good deal as a method, but is not fully developed. In performing more analyses, we keep finding more reversals in the magnetic field in periods we had believed to be stable.

Radiometric Age-Dating Methods

Our geological time scale is founded on age dating based on radioactive processes. There are a number of other indirect methods of measuring absolute geological time, but only radiometric dating methods give satisfactory quantitative results. The measurements are based on the fact that radioactive nuclei undergo fission at a certain rate. The rate of fission can be determined with some accuracy, and there is no reason to believe that it has varied through the ages, although this is difficult to prove. Radioactive decay proceeds at a certain rate which is proportional to the assumed number of fissionable atoms present. The number of atoms which can undergo fission falls off according to the following formula, however:

$$dN/dt = -\lambda N$$
$$N = N_o e^{-\lambda t},$$

where

N_o = number of atoms at time T_o
N = number of atoms at time T
λ = disintegration constant.

A convenient measure of the rate of fission is the half-life, $T_{\frac{1}{2}}$, i.e. the time it takes for half of the original atoms to decay. Then

$$\frac{N}{N_o} = \frac{1}{2} = e^{-\lambda T_{\frac{1}{2}}}$$
$$\log \frac{1}{2} = -\lambda T_{\frac{1}{2}}.$$

The half-life $T_{\frac{1}{2}} = \dfrac{0.69315}{\lambda}$, where λ is the decay constant, often also referred to as the disintegration constant, which is the fraction of the total number of atoms which will decay in a given time (n/N).

The radioactive decay processes result in new nuclei called daughter nuclei. In order to date geological material we measure the ratio between the fissionable nucleus (mother nucleus) and the daughter nucleus.

One of the most commonly used dating methods is the potassium-argon method. ^{40}K is an unstable nucleus which decays mainly to ^{40}Ca with the emission of beta particles. However about 12% turns into ^{40}Ar through capture of electrons and emission of X-rays. A mineral which contains potassium, for example biotite, will also contain some ^{40}K which goes over to ^{40}Ar. The amount of argon in the mineral is thus an expression of the mineral's age, which can be calculated if we know the half-life ($T^{1/2}$) or the rate of decay (half-life 1.31×10^9 years). The amount of argon gas in the minerals is analysed with a sensitive mass spectrometer. Argon is a very volatile gas, however, and when minerals containing argon are heated up or deformed, for example during metamorphism or folding, the argon will escape, and the radioactive "clock" will indicate an age which corresponds more to the time of metamorphism than to the formation of the mineral. So radioactive age determination does not necessarily give a figure for the age of the rock, but says something about the geological processes which have affected the rock, and thus helps us to reconstruct the geological history.

Another commonly used dating method is the rubidium-strontium method. Rubidium has a radioactive isotope, ^{87}Rb, which decays to ^{87}Sr with a half life of 4.89×10^{10} years. The amount of ^{87}Sr is then a function of the period that ^{87}Sr has been accumulating from ^{87}Rb since the formation of the mineral or rock. Instead of measuring absolute amounts of Sr^{87}, one can simply measure the Sr^{87}/Sr^{86} ratio on the mass spectrometer. We then plot the ratio between Rb^{87}/Sr^{86} and Sr^{87}/Sr^{86} on a graph.

Analyses of minerals or rocks which have the same age will then form points on a straight line (isochrone). The slope of this line will be an expression of the age of the rock.

Using this method we can analyse both minerals and whole rocks, since we assume that the ^{87}Sr formed does not escape from the rock as easily as argon (in the potassium-argon method). When dating sediments one should choose the finest-grained clay sediments with low permeability. It can then be assumed that after deposition the sediment was homogeneous with respect to strontium isotopes and an isochrone will be obtained which dates any diagenesis occurring a relatively short time after deposition. Larger clastic fragments, however, will contain a strontium isotope ratio which corresponds to the age of the source rock, and one may thus obtain a date intermediate between the age of the source rock and the time of deposition of the sediments.

Authigenic minerals forming in sandstones can be separated out and dated by radioactive dating methods. Authigenic illite can be dated by the potassium-argon method, and this will provide information about the time when the authigenic minerals formed in the sandstone by diagenetic processes. In oil and gas reservoirs the formation of illite is assumed to stop when hydrocarbons are introduced and saturate the pore space. The age of the illite may therefore provide an indication of the maximum age of oil emplacement.

There are several other dating methods, of which the uranium-lead method and the so-called lead-lead method, i.e. the relationship between different lead isotopes, are the most important.

The methods mentioned above are based on fission processes with long half-lives, 10^9–10^{10} years, and for young sediments the quantity of decay products to be analysed will be small, and accuracy poor. The $^{40}K/^{40}Ar$ method can be used under favourable conditions on rocks formed as recently as 150,000 years ago, but with great uncertainty. The protactinium method ($^{231}Pa/^{230}Th$) has given age determinations which agree well with the ^{14}C method for younger sediments.

The fission-track method is also well suited to younger rocks. Glass, or minerals which contain a sufficient amount of uranium 238, will show tracks from fission products which we observe as deformations. By counting the number (or frequency) of such tracks, we get an expression of the age (see divisions of the Phanerozoic Eon, Fig. 10.4).

The Carbon 14 Method

The carbon 14 method is the most commonly used dating method for the youngest sediments, from about 50,000 years old up to the present. ^{14}C is formed in the atmosphere when an ^{14}N atom absorbs a neutron and gives off a proton. ^{14}C is unstable, and decays to ^{14}N. ^{14}C is produced only in the atmosphere at over 10,000 m. Thereafter ^{14}C mixes with the lower air layers and the sea. The ^{14}C method is based on the assumption that the ^{14}C content of the atmosphere has been constant for a long time, due to an equilibrium between the ^{14}C added from the atmosphere and the ^{14}C which decays to ^{14}N. The half-life of ^{14}C is 5730 years. This means that after 5730 years half of the ^{14}C atoms will have changed to ^{14}N. ^{14}C enters the carbon dioxide in the air (CO_2), and enters plants through photosynthesis. If we assume that the carbon dioxide in the air in the past had as much ^{14}C as now, the ^{14}C content of older plants or plant remains and charcoal is an expression of their age. This age can be determined analytically with relatively great accuracy, for example 10,200 ± 100 years, depending on the nature of the sample. When the plant material has aged 40–50,000 years after sediment formation, the ^{14}C content will be so small that we will be approaching the limits for detection. This is then the upper limit to the age of material we can analyse.

As already mentioned, the ^{14}C method is based on the assumption that the ^{14}C concentration of the CO_2 in the air has been constant throughout the past few thousand years. However, it is impossible to test this without an absolute time scale. Analyses of rings in old trees (particularly pines) have given us an absolute time scale going back some thousands of years. By comparing these two methods it has been found that on samples more than about 2,000–3,000 years old there is systematic deviation which indicates that the ^{14}C in the atmosphere has varied by up to 10% and perhaps more. As long as we compare ^{14}C datings with each other, they will show relatively good agreement. For archeological ^{14}C dating, however, it is more important to know the deviation from the absolute time scale.

^{14}C also enters sea water and freshwater from the atmosphere. Surface water, which is in contact and undergoes exchange with the air, will have a ^{14}C content which corresponds to that of the air. Groundwater with little circulation and water in ocean basins with slow exchange with the surface has a lower ^{14}C content, however, and is thus old water. Carbon-secreting organisms which live in the sea and in lakes will take up CO_2 from the water, and the amount of ^{14}C in $CaCO_3$ can

EON	ERA	SUB-ERA	PERIOD / SUB PERIOD	EPOCH	AGE	Abbrev	Ma	AGE DURATION	
Phanerozoic 590 (Ph)	Cenozoic 65 (Cz)	Quaternary (Q) or Pleistogene (Ptg) 2.0		Holocene		Hol	0.01	0.01	2.0
				Pleistocene		Ple	2.0	1.99	
		Tertiary 63 (TT)	Neogene 22.6 (Ng)	Pliocene (Pli) 3.1 — 2	Piacenzian	Pia		3.1	22.6
				Pliocene — 1	Zanclian	Zan	5.1		
				Miocene (Mio) 19.5 — 3	Messinian	Mes		6.2	
					Tortonian	Tor	11.3		
				Miocene — 2	Serravallian	Srv		3.1	
					Langhian Late	Lan2	14.4		
				Miocene — 1	Langhian Early	Lan1		10.2	
					Burdigalian	Bur			
					Aquitanian	Aqt	24.6		
			Paleogene 40.4 (Pg)	Oligocene (Oli) 13.4 — 2	Chattian	Cht	32.8	8.2	40.4
				Oligocene — 1	Rupelian	Rup	38.0	5.2	
				Eocene (Eoc) 16.9 — 3	Priabonian	Prb	42.0	4.0	
				Eocene — 2	Bartonian	Brt		8.5	
					Lutetian	Lut	50.5		
				Eocene — 1	Ypresian	Ypr	54.9	4.4	
				Paleocene (Pal) 10.1 — 2	Thanetian	Tha	60.2	5.3	
				Paleocene — 1	Danian	Dan	65.0	4.8	
	Mesozoic 183 (Mz)		Cretaceous 79 (K)	K_2 Late 32.5 — Senonian	Maastrichtian	Maa	73.0	8.0	79
					Campanian	Cmp	83.0	10.0	
					Santonian	San	87.5	4.5	
					Coniacian	Con	88.5	1.0	
					Turonian	Tur	91.0	2.5	
					Cenomanian	Cen	97.5	6.5	
				K_1 Early 46.5	Albian	Alb	113	15.5	
					Aptian	Apt	119	6.0	
					Barremian	Brm	125	6.0	
				Neocomian	Hauterivian	Hau	131	6.0	
					Valanginian	Vlg	138	7.0	
					Berriasian	Ber	144	6.0	
			Jurassic 69 (J)	J_3 19 — Malm	Tithonian	Tth	150	6.0	69
					Kimmeridgian	Kim	156	6.0	
					Oxfordian	Oxf	163	7.0	
				J_2 25 — Dogger	Callovian	Clv	169	6.0	
					Bathonian	Bth	175	6.0	
					Bajocian	Baj	181	6.0	
					Aalenian	Aal	188	7.0	
				J_1 25 — Lias	Toarcian	Toa	194	6.0	
					Pliensbachian	Plb	200	6.0	
					Sinemurian	Sin	206	6.0	
					Hettangian	Het	213	7.0	
			Triassic 35 (Tr)	Tr_3 Late 18	Rhaetian	Rht	219	6.0	35
					Norian	Nor	225	6.0	
					Carnian	Crn	231	6.0	
				Tr_2 Middle 12	Ladinian	Lad	238	7.0	
					Anisian	Ans	243	5.0	
				Tr_1 (Early) Scythian 5 (Scy)	Olenekian — Spathian	Spa		1.25	
					Olenekian — Smithian	Smi		1.25	
					Induan — Dienerian	Die		1.25	
					Induan — Griesbachian	Gri	248	1.25	
	Paleozoic 342 (Pz)		Permian 38 (P)	P_2 Late 10	Tatarian	Tat	253	5.0	38
					Kazanian	Kaz		2.5	
					Ufimian	Ufi	258	2.5	
				P_1 Early 28	Kungurian	Kun	263	5.0	
					Artinskian	Art	268	5.0	
					Sakmarian	Sak		9.0	
					Asselian	Ass	286	9.0	
		Carboniferous 74 (C)	Pennsylvanian 34 (Pen)	Gzelian	Stephanian 10	Gze			34
				Kasimovian		Kas	296		
				Moscovian	Westphalian 19	Mos	315		
				Bashkirian	Namurian 18	Bsh	320		
			Mississippian 40 (Mis)	Serpukhovian		Spk	333	13.0	40
				Visean 19		Vis	352	19.0	
				Tournaisian 8		Tou	360	8.0	
			Devonian 48 (D)	D_3 Late 14	Famennian	Fam	367	7.0	48
					Frasnian	Frs	374	7.0	
				D_2 Middle 13	Givetian	Giv	380	6.0	
					Eifelian	Eif	387	7.0	
				D_1 Early 21	Emsian	Ems	394	7.0	
					Siegenian	Sig	401	7.0	
					Gedinnian	Ged	408	7.0	
			Silurian 30 (S)	Pridoli		Prd	414	6.0	30
				Ludlow		Lud	421	7.0	
				Wenlock		Wen	428	7.0	
				Llandovery		Llv	438	10.0	
			Ordovician 67 (O)	Ashgill		Ash	448	10.0	67
				Caradoc		Crd	458	10.0	
				Llandeilo		Llo	468	10.0	
				Llanvirn		Lln	478	10.0	
				Arenig		Arg	488	10.0	
				Tremadoc		Tre	505	17.0	
			Cambrian 85 (€)	Merioneth 20 (Mer)	Dolgellian	Dol		10.0	85
					Maentwrogian	Mnt	525	10.0	
				St David's 15 (StD)	Menevian	Men		8.0	
					Solvan	Sol	540	7.0	
				Caerfai 50 (Crf)	Lenian	Len		15.0	
					Atdabanian	Atb	570	15.0	
					Tommotian	Tom	590	20.0	

Fig. 10.4

then be measured. Mollusc shells are also suitable for age dating, and chemically precipitated carbonates can be dated in this way. Organisms have a tendency to fractionate the lightest isotopes from the heaviest. The ratio between two stable carbon isotope types, ^{12}C and ^{13}C can therefore be used to correct ^{14}C determinations.

^{14}C analyses of extant (recent) marine organisms will often give apparent ages (low ^{14}C content) and these will be an expression of slow exchange of atmospheric CO_2 with that in sea water. Fossil fuels such as oil, coal etc. have a ^{14}C content which is practically zero. In burning oil and coal we create pollution which is recognisable by the low ^{14}C content of the CO_2 formed through combustion.

Correlation

Stratigraphy is the science of bedding series and their correlation. We attempt to determine which geological events were simultaneous and the time intervals between well-defined geological episodes. This is not so much a goal in itself as a precondition for a detailed reconstruction of processes which have taken place through the geological ages, e.g. spreading of the seabed or formation of mountain chains.

The geological time scale is the result of laborious classification of rocks (chronostratigraphy) and geological time (geochronology) (see Table 10.1). Classification of rocks from the Phanerozoic Eon has been based on fossils. Although biostratigraphic correlation is still the most important method, magnetostratigraphy and radiological dating methods are gaining importance. Figure 10.4 shows the first attempt to establish an absolute means of age-dating geochronological units. Within a sediment basin, however, certain types of lithostratigraphic correlation can turn out to be most accurate. Logging in wells offers good opportunities for lithostratigraphic correlation. Seismic profiles have made it possible to perhaps an even greater degree to make lithostratigraphic correlations over great distances (see next chapter).

Dynamic Stratigraphy

Stratigraphy is often presented as a descriptive science, the object of which is to make a precise and well-defined classification of sedimentary bedding series. However, it is important to remember that a stratigraphic classification is only an aid for facilitating description of rocks and clarifying their mode of formation, and that stratigraphy is not a goal in itself. The stratigraphic relations we find in sedimentary series are a result of geological processes, first and foremost movements of the earth's crust. To stress this connection between the descriptive stratigraphic divisions and the geological processes they reflect, we speak of *dynamic* stratigraphy.

Fig. 10.4. The Phanerozoic time scale. (After Harland et al. 1982)

The extension of marine sediments is an important stratigraphic factor. When the sea, i.e. the coastline, moves in over previous land areas, this is called *transgression*, and when areas which were previously sea become land, we have *regression*. It is a common misconception that transgressions necessarily imply a rise in sea level, and regressions a fall in sea level. Transgressions and regressions depend on the following factors:

1. Uplift or subsidence of the land area in relation to sea level.
2. Raising or lowering of the level of the oceans (eustatic sea level changes).
3. Rate of sedimentation.

Eustatic changes in sea level will cause transgressions or regressions all over the world at the same time. Variations in sea level due to melting or accumulation of ice during glacial periods are good examples of this. The relationship between erosion/deposition in the context of transgression/regression is shown in Fig. 10.5. There are also slower eustatic changes in sea level which are due to plate tectonic movements. However, if we assume sea level to be constant, transgressions and regressions are a function of the movement of the earth's crust and rate of sedimentation. We have *transgressions* if the land subsides more rapidly than the sedimentation rate, or if the coastline is eroded faster than the land is elevated (Fig. 10.5). *Regressions* are caused either by the rate of sedimentation exceeding the rate of subsidence, or by uplift of the land (Fig. 10.6). A regression which is due to uplift of land areas or sinking of the ocean will give an increased rate of sedimentation because rivers develop a higher gradient (lower erosion base). Beach and delta sediments will grow out into the sea, forming an *offlap* sequence. With transgressions the sediment supply to a shelf area will tend to be reduced, since the slope of the river is reduced. Sediments deposited during a transgression onto the land are called an *onlap* sequence. If sedimentation keeps pace with subsidence, the coastline and the sedimentary facies will remain in the same position (Fig. 10.7). A transgression will often lead to a period without any real clastic sedimentation on a delta because the delta moves backwards and the transport capacity of the rivers

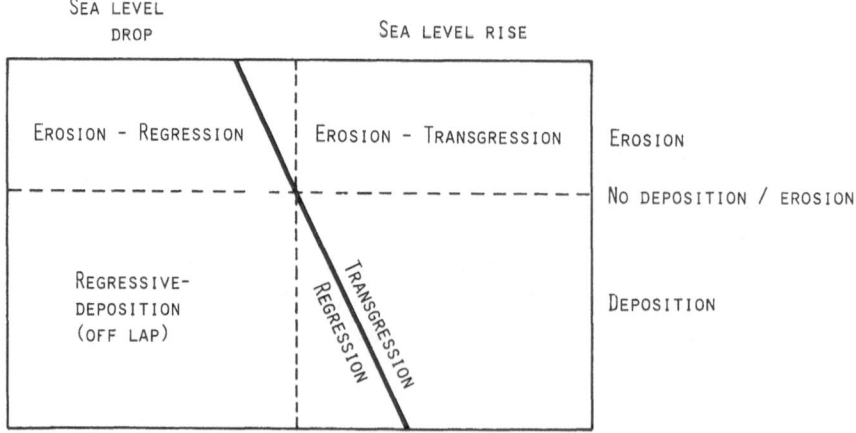

Fig. 10.5. Diagram showing connection between rise and fall of sea level, erosion and transgressions and regressions. (After Currey 1964)

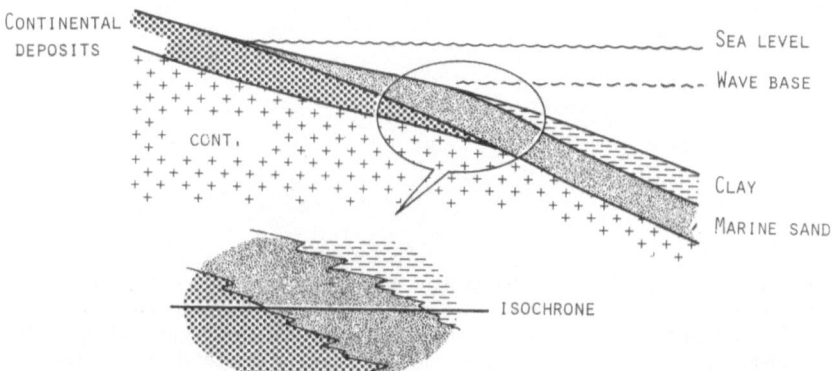

Fig. 10.6a. Schematic representation of transgression along a coastline. The strata become younger towards the land in relation to a time line (isochrone)

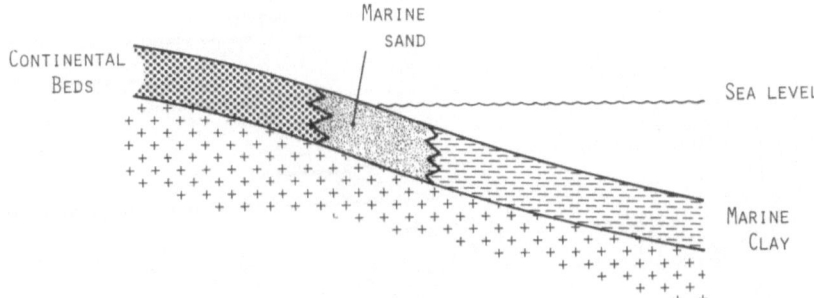

Fig. 10.6b. Schematic representation of a regressional sequence. The shallow marine sandstone or fluvial facies grow younger out into the basin in relation to the isochrone

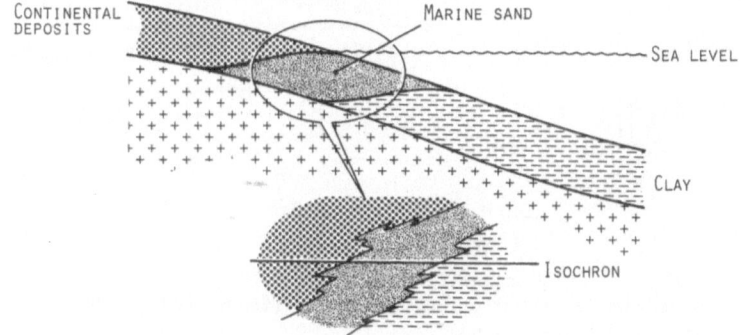

Fig. 10.6c. Coastline where there is equilibrium between sedimentation and subsidence

diminishes. Such a transgressive episode will often be marked by a carbonate bed or a thin sandstone bed which is deposited simultaneously over large areas. Transgressive beds of this sort provide very good marker beds for making correlations. During regressive periods with progradation of deltas or coastlines out into the sea, sediments will grow younger as we go seawards (see Fig. 10.8). This means that a delta front sand or a shoreface sand which we can map in the field

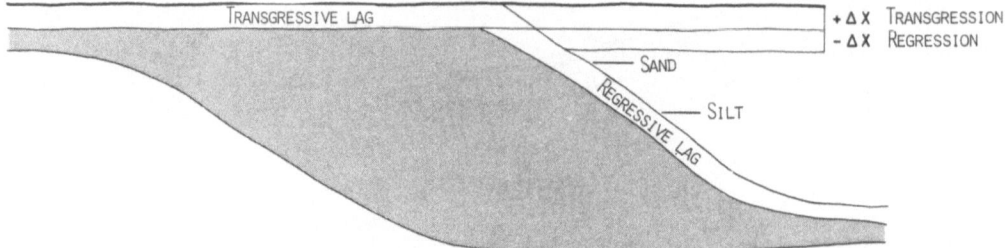

Fig. 10.7. Diagram showing schematically why transgressive beds in a delta are easy to correlate. Carbonate beds or thin sandstones are deposited during relatively rapid transgressions when there is little clastic supply due to the sea level rising. A regressive bed is part of a prograding bedding series which is not always easy to follow in terms of time stratigraphy because of the rapid facies changes

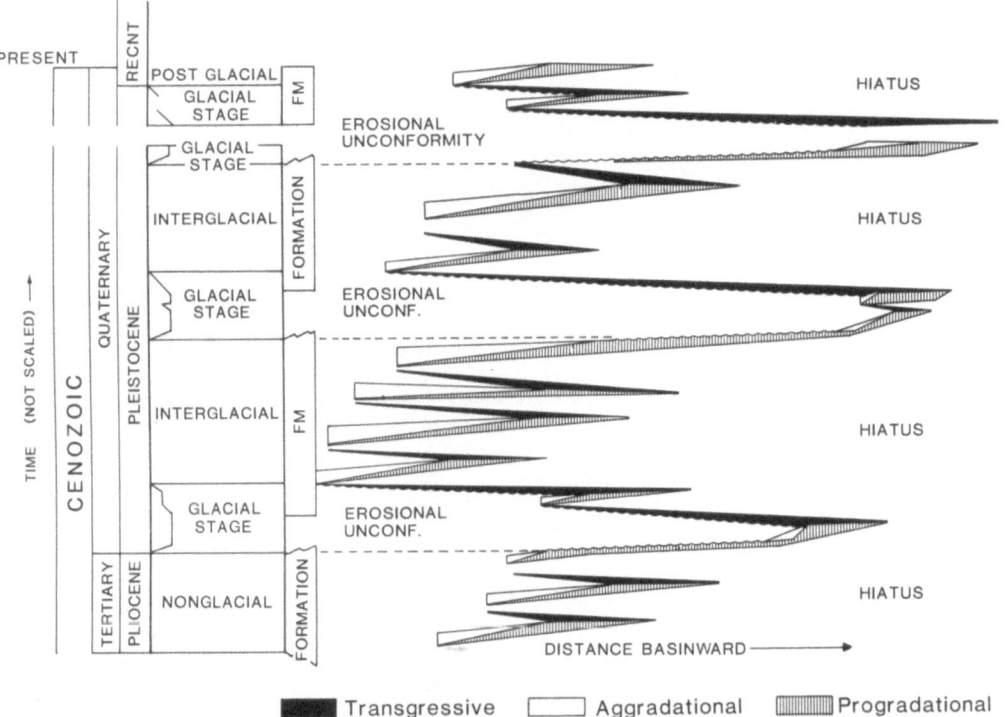

Fig. 10.8. Transgressions and regressions resulting from glacial sea-level changes. Note that in each area sedimentation occurs only during short episodes (depositional episodes). Most of the time there is non deposition or erosion (Hiatus)

becomes younger as we follow it into more distal parts of the basin, even if we give the sandstones the same lithostratographic formation name. Delta lobe shifting along the coast will produce sandstones that vary in age along the coast too. We say the beds are *time-transgressive* or *diachronous*.

Glaciations produce very rapid sea-level fluctuations up to 100 to 150 m and these have a profound influence on shelf and coastal sedimentation resulting in very characteristic depositional sequences (Fig. 10.8).

11 Seismic Stratigraphy and Basin Analysis

Seismic Stratigraphy

Seismic records are based on measurements of the time sound waves (seismic waves) take to travel through rock. The sound or signal is produced by explosives or compressed air (air guns). Rock is an elastic medium and the velocity of sound conveys a lot of information about the properties of the rock. The most important parameters influencing the velocity of sound are: pressure, temperature, porosity, mineral composition, permeability and density. Velocity is proportional to the square of the compressibility of the solid phase. The velocity of sound waves in water is about 1500 m/s, but depends on temperature and salt concentration. Sound passes through unconsolidated sediments at velocities which are only slightly higher than the velocity in water (1500–2000 m/s, and sometimes even lower) because they have a high water content and because the framework on which the sediment grains are based does not offer any real strength as a medium for waves. Cementation, for example of sand with carbonate or siliceous cement, will bind the grains together in a framework which will increase the velocity considerably. Compaction due to overlying sediments which cause water to be expelled will also cause higher velocities, not only because the water content decreases, but because more contacts are formed between the clastic grains. Velocities in moderately consolidated sediments lying at shallow depths, such as the Tertiary sediments of the North Sea, are 2–3 km/s. In more consolidated sedimentary rocks which have not been subjected to metamorphosis, waves will have velocities of about 3–4 km/s. This is the case for many of the Mesozoic sediments in the North Sea. Metamorphic and eruptive rocks buried at depths of 3–5 km will cause velocities of about 5–6 km/s. Limestones will often cause higher velocities than sandstones at the same depth because carbonate cement provides the structure with a high degree of rigidity and low compressibility. Sandstone in turn provides a more rigid medium for sound waves than shale at the same depth because of its grain-supported structure.

If rocks do not contain oil and gas we can assume that their porosity is identical with the water content in the rock. Velocity will then be a function of porosity, and if we know the velocity of sound in the rock matrix, we can calculate the porosity using Wyllie's equation:

$$\frac{1}{V_r} = \frac{(1 - \varphi)}{V_m} + \frac{\varphi}{V_f},$$

where

V_r = velocity in rock when saturated with liquid, i.e. the measured velocity
V_f = velocity in the fluid
V_m = velocity in the rock matrix.

We can then find the porosity, φ.

If the pores are filled with gas instead of liquid (water or oil), the reduction in velocity will be even greater since the velocity of sound in gas is considerably lower than in a liquid.

V_r thus approaches V_m at zero porosity, and for sandstone this is about 5.5–6 m/s.

When sound waves move between sedimentary beds with different velocities, they will be refracted according to Snell's Law:

$$\frac{\sin x_1}{\sin x_2} = \frac{V_1}{V_2}.$$

Here x_1 is the angle the incident sound waves form with the normal to the boundary between the strata, and x_2 the angle the emergent waves form with the same normal (Fig. 11.1). V_1 and V_2 are the velocities in the rock strata.

If the two beds have different velocities, they will as a rule also have different densities, and part of the energy will not be refracted, but reflected. How much of the energy is reflected depends on the difference in the *acoustic impedance*, which is the product of velocity and density (Fig. 11.2).

The coefficient of reflectivity is then:

$$R = \frac{\rho_2 \cdot V_2 - \rho_1 \cdot V_1}{\rho_2 \cdot V_2 + \rho_1 \cdot V_1}.$$

where ρ_1 and ρ_2 are the density of the two rocks, and V_1 and V_2 the velocities in the same two rocks. We see that the greater the difference in density and velocity, the

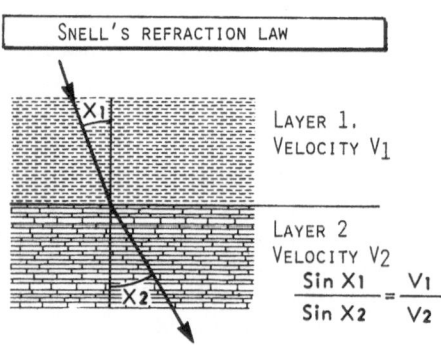

Fig. 11.1. Snell's law for the refraction of sound waves

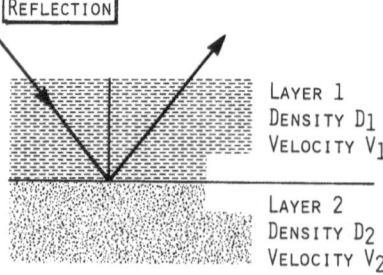

Fig. 11.2. Diagram showing the principles for reflection of waves in a sedimentary bedding series. The amount of energy reflected is a function of acoustic impedance which is the product of the density (ρ) of the beds and the velocity of the sound waves (V)

greater the amount of energy which will be reflected. Sandstone will often have a significantly different acoustic impedance from shale, and a considerable amount of sound energy will be reflected from the boundary between a sand bed and a shale bed. Limestones will tend to have both high velocities and high densities. The result will be even greater contrast in acoustic impedance between limestones and, for example, shales. However, this contrast will always depend on the porosity of the limestone in question.

On a seismic section, beds which have greatly contrasting acoustic impedances stand out as strong reflectors. This makes it possible to map characteristic rock boundaries, e.g. the top of a limestone or the boundary between a shale and a sandstone, using seismic sections (Fig. 11.3).

As we have seen, the critical parameters determining the reflection coefficient are the velocities and densities of the different lithological units. Using a well we may produce a velocity log (sonic log) and a density (ρ) log which record how these properties change through the sequence (see Fig. 11.3a). The product of velocity and density may then be computed and presented as a $\rho \cdot v$ log. Note that at the boundary between limestones and shales there is a very significant drop in both velocity and density, resulting in a marked change on the $\rho \cdot v$ log. Sandstones

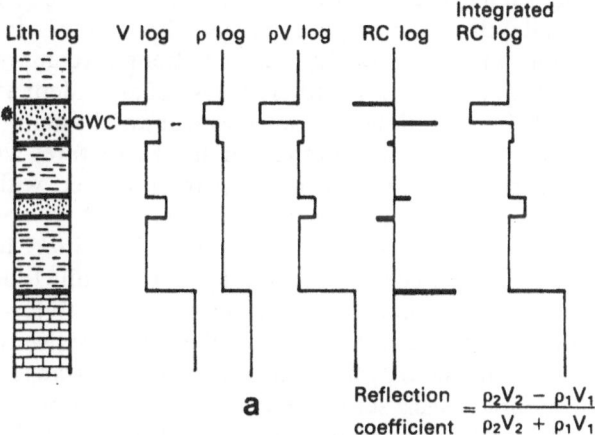

$$\text{Reflection coefficient} = \frac{\rho_2 V_2 - \rho_1 V_1}{\rho_2 V_2 + \rho_1 V_1}$$

Fig. 11.3. a Formation of a reflection coefficient based on velocity (V) and density (ρ). (Anstey 1982). **b** Synthetic seismogram of the different reflection coefficients in a sequence (RC log). The resolution of the seismogram varies with the width of the seismic pulse (Anstey 1982). Normally, sediments have to exceed 30–40 m in thickness to be distinctly recorded on a seismic section

usually have higher velocities than shales, but they may not differ much in density, so the difference in acoustic impedance will be small. The reflection coefficient, which is an expression of differences in acoustic impedance, is a synthetic seismic trace such as we would have seen on a seismic cross-section through the sequence.

If we have gas instead of water in parts of a rock, the velocity will be considerably reduced. The velocity of sound in gas is considerably lower than it is in liquid, depending on composition, temperature and pressure. The boundary between gas-bearing and water-bearing rocks will produce a strong reflection because there is a large difference in impedance between the two layers. For this reason the boundary between gas and oil is often revealed as a strong reflection because it is horizontal and does not always follow the other rocks. It is called a "flat spot" and exemplifies direct indication of hydrocarbons through seismic methods.

Reflections which are multiples of sea-bottom reflections are also near-horizontal, and may be confused with "flat spots". Temperature-dependent diagenetic reactions may also produce horizontal reflections, e.g. the transformation of opal CT to quartz.

Seismic velocities will normally increase as a function of increased overburden. Deviation from a steady increase in seismic velocity as a function of depth may be due to overpressure, i.e. a pressure which is higher than hydrostatic pressure. Shales subjected to overpressure, for example, will have a higher porosity than is usual for that depth. The high pore pressure also results in lower effective stresses between grains and consequently reduced strength (compressibility). This may occur in sandstone too, but the difference is not as evident as it is in shales.

Seismic sections have long been one of the major means of analysing structures in sedimentary basin in the search for structures which might serve as a trap for hydrocarbons. Only in recent years, however, with improved seismic data and processing techniques, have they been used for detailed interpretation of stratigraphic relations and environments of deposition. It also took some time before it was fully realised that seismic reflections usually follow time lines in a sedimentary sequence. In other words, seismic reflections follows surfaces which constituted a surface (the ocean floor) when the sediments were deposited. This means they follow a boundary which tends not to be time-transgressive. Wells have been drilled to double-check seismic sections and it has been seen that reflections can be followed from a sandy facies into a siltstone shale facies. We can, for example, follow seismic reflections from the fluvial part of a delta out into pro-delta clay. This may seem strange, since it is sandstones which should form the greatest contrast in acoustic impedance in relation to the shale. However, there is a gradual transition between the fluvial and delta front sediments and the delta slope sediments, so that the base of the prograding sandstone does not constitute a good reflector. On the delta slope sand will give way to clay which has been deposited simultaneously. Here the seismic reflections will follow the surface from delta front sand to delta slope, where we have an alternation of sand and shale running parallel with the slope. We can often follow the reflection a little further out into the basin, but it will be less marked here because there is less difference in acoustic impedance between fine sand, siltstone and shale.

The shifting of sedimentation from one part of the delta to another as the fluvial supplies switch course (distributary abandonment as part of delta-lobe shifting), probably also contributes to the formation of lithological contacts on the delta slope. Prograding of new delta lobes results in deposition of sheets of sand over mud near time-stratigraphic boundaries. The inactive delta lobes will be compacted and often develop a thin carbonate or transgressive sandstone layer, while sedimentation takes place in the active lobe. The small unconformities produced in this way also tend to produce lithological contrasts which may be recorded on the seismic record. When prograding deltas follow coastlines out into the basin, we will find sloping reflections which are thus a sort of large-scale cross-bedding.

Different Types of Seismic Signatures

A stratigraphic unit which is composed of a conformable bedding series, genetically linked together at the top and bottom by unconformities, is called a *depositional sequence* (see Fig. 11.4). As we shall see, unconformities are due to a break in sedimentation due to changes in sea level or other causes. A depositional sequence is thus a package of sediments deposited during a definite period of time, defined by unconformities above and below.

There may be various relations between depositional series and the lower unconformity:

A *baselap* is formed by gradual progradation of the lower boundary of a depositional series to produce a small unconformity. (Fig. 11.4).

If the beds in a depositional series progress outwards over unconformities, so that the beds above the unconformities grow younger as we go from land out into the basin, we call the result a *downlap*. Building of beds out into the basin is called an *offlap*. We are thus dealing here with a bed which has a primary slope in relation to the unconformity surface.

Onlap is the term for primarily approximately horizontal beds which form a *baselap* with a sloping unconformable surface. It occurs most commonly as a result of sediments gradually covering an unconformity in connection with sedimentation

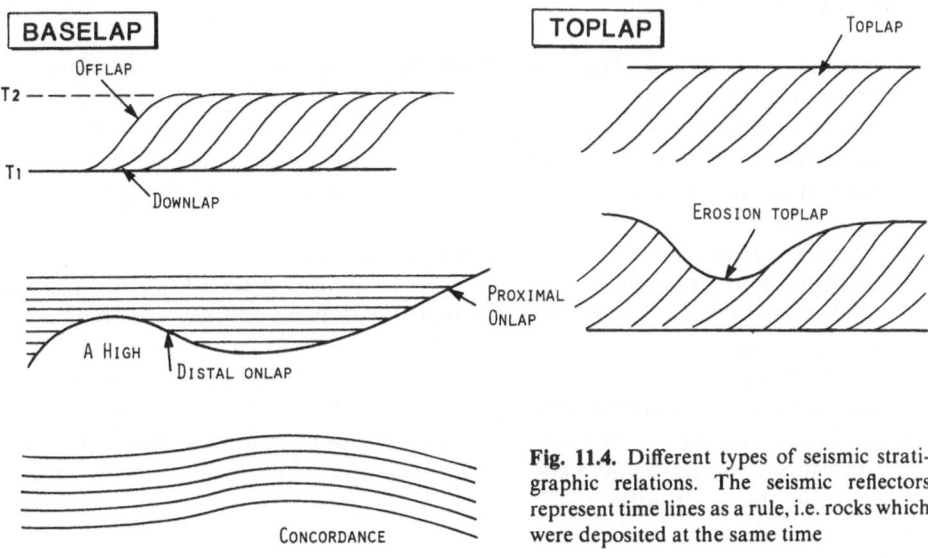

Fig. 11.4. Different types of seismic stratigraphic relations. The seismic reflectors represent time lines as a rule, i.e. rocks which were deposited at the same time

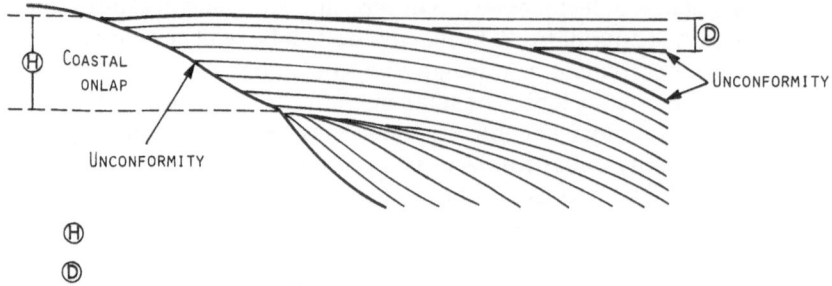

Fig. 11.5. Transgression which forms coastal onlap over an unconformity. Later the relative sea level sank, and a prograding downlap (or offlap) sequence formed. Finally the sea level rose again, and an onlap sequence formed. *H* and *D* are not to be regarded as absolute values for sea level changes. They must be adjusted for isostatic responses to loading and unloading and to tectonic uplift or subsidence

and transgression onto the land, resulting in an unconformable surface (Fig. 11.5). We call this *proximal onlap* or *coastal onlap*. If sedimentation covers a structure which projects up out of the basin, for example a horst or a salt dome, we get *distal onlap* (Fig. 11.4).

A *toplap* is the contact between the seismic reflections and an upper unconformity. An erosion surface will truncate the reflections sharply, forming an *erosional truncation (erosional toplap)*.

Interpretation of Lithology and Sedimentary Facies by Means of Seismic Profiles

In addition to structural data, seismic profile gives us information about the properties of sedimentary rocks. Because seismic reflections mainly represent time lines, i.e. sedimentary beds which were deposited simultaneously, it is also possible to a certain extent to interpret the environment of deposition (see Fig. 11.6). The most important parameters used are:

1. *Reflection amplitude.* The strength of the reflections. As we saw above, the fraction of the energy which is reflected at the boundary between beds is a function of the difference in acoustic impedances, i.e. the velocity multiplied by the density. If we have an alternation at different beds, the distance between the beds in relation to the wave length of the seismic waves will play a major part (Fig. 11.3).
2. *Reflection frequency.* The distance between the reflections will provide information about the thickness of the bed, but there will be a lower limit to the thickness which it is possible to see, which will depend on the wavelength of the seismic waves.
3. *Internal velocity in the beds.* The internal velocity in the bed can provide information about lithology and porosity.
4. *Reflection continuity.* The continuity of reflections will be a function of how continuous the sediment beds are, information which is essential for reconstructing the environment.

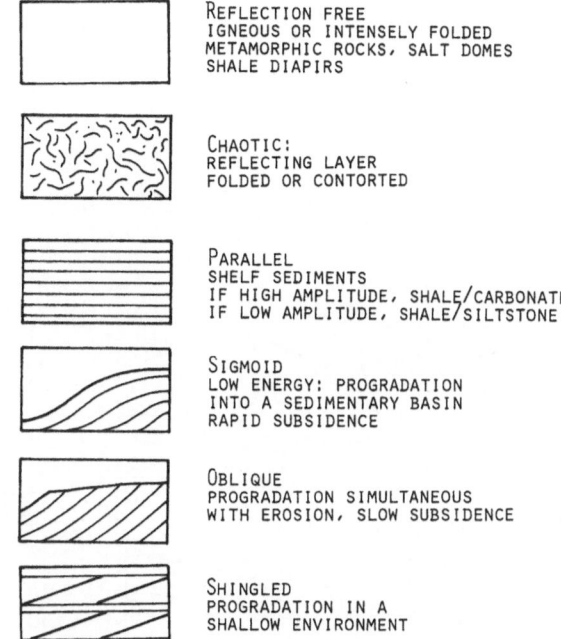

REFLECTION FREE
IGNEOUS OR INTENSELY FOLDED
METAMORPHIC ROCKS, SALT DOMES
SHALE DIAPIRS

CHAOTIC:
REFLECTING LAYER
FOLDED OR CONTORTED

PARALLEL
SHELF SEDIMENTS
IF HIGH AMPLITUDE, SHALE/CARBONATE
IF LOW AMPLITUDE, SHALE/SILTSTONE

SIGMOID
LOW ENERGY: PROGRADATION
INTO A SEDIMENTARY BASIN
RAPID SUBSIDENCE

OBLIQUE
PROGRADATION SIMULTANEOUS
WITH EROSION, SLOW SUBSIDENCE

SHINGLED
PROGRADATION IN A
SHALLOW ENVIRONMENT

Fig. 11.6. Descriptive classification of seismic signatures and interpretation of the types of sediments which give these signatures

5. *Reflection configuration.* If we take the compaction effect into account, the shape of the reflecting beds gives us a picture of the sedimentation surface as it was during deposition. The slope of the reflectors, for example, represents the slope of prograding beds in a delta sequence corrected for later differential compaction and tilting. Erosion boundaries with unconformities will in the same way show the palaeotopography during erosion.

When interpreting lithology and environment of deposition through seismic profiles, it is important to use sedimentological models as aids. If there are well data on lithologies, these must also be integrated. The information we gain from seismic profiles is often not sufficient for an unambiguous interpretation. There may be several lithological compositions and environments which could give similar seismic signatures. Only by looking at the whole basin in a sedimentological context do we have a good point of departure for selecting the interpretation which seems most reasonable.

Filling of Sedimentary Basins

Seismic profiles present a picture of the way the basin has been filled in. This is a result of an interaction between rate of subsidence, rate of deposition and the energy of the depositional environment (see Fig. 11.7). A detailed interpretation of depositional environments and rock types based on seismic profiles off Brazil has been published by Brown and Fisher (1977) (see Fig. 11.8).

ONLAP
HIGH ENERGY
PROGRADING
ONLAP
DIVERGENT
INFILL
LOW ENERGY
UPLAP
INFILL DURING
FAULTING OR
DIAPIRISM

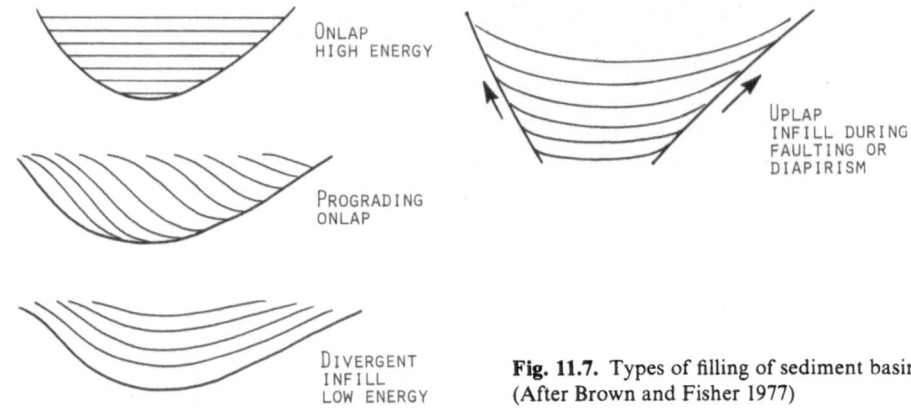

Fig. 11.7. Types of filling of sediment basins. (After Brown and Fisher 1977)

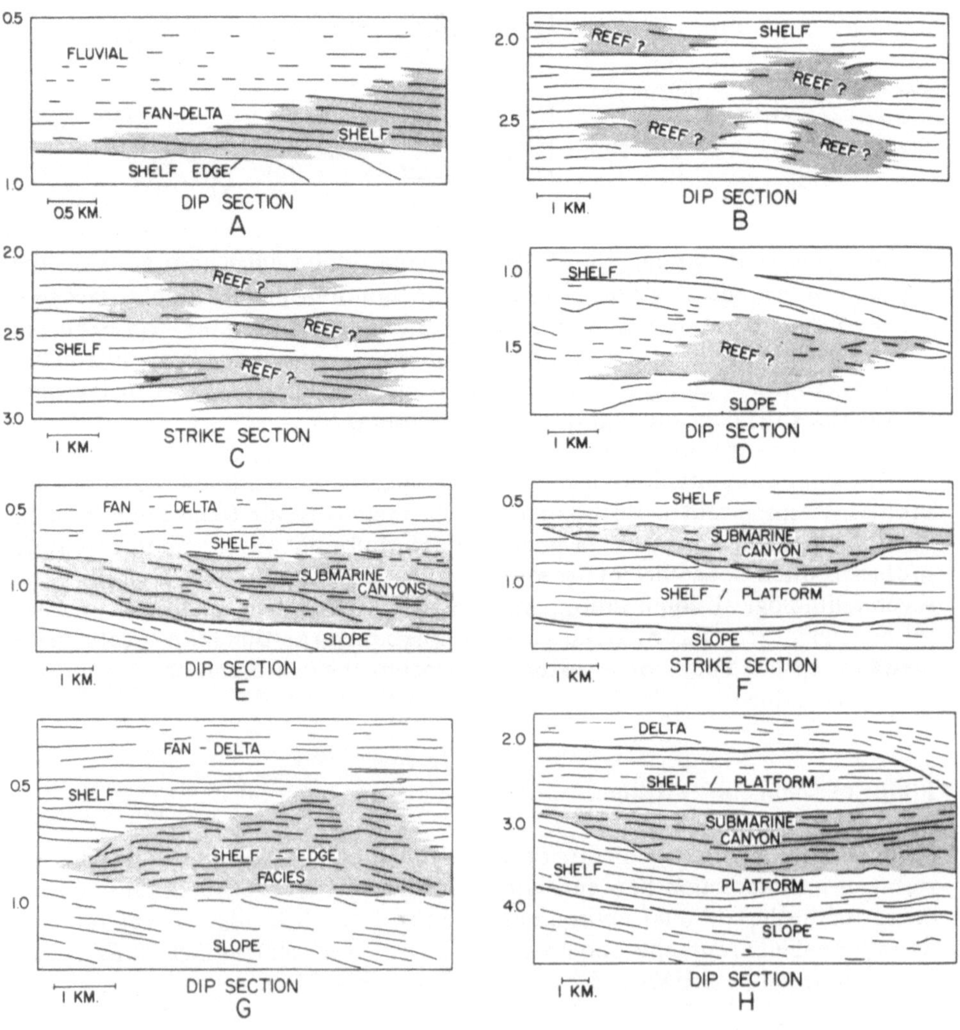

0.5
FLUVIAL
FAN-DELTA
SHELF
SHELF EDGE
1.0
0.5 KM.
DIP SECTION
A
2.0
SHELF
REEF ?
REEF ?
2.5
REEF ?
REEF ?
1 KM.
DIP SECTION
B
2.0
REEF ?
REEF ?
2.5
SHELF
REEF ?
3.0
1 KM.
STRIKE SECTION
C
1.0
SHELF
1.5
REEF ?
SLOPE
1 KM.
DIP SECTION
D
0.5
FAN
DELTA
SHELF
SUBMARINE
CANYONS
1.0
SLOPE
1 KM.
DIP SECTION
E
0.5
SHELF
SUBMARINE
CANYON
1.0
SHELF / PLATFORM
SLOPE
1 KM.
STRIKE SECTION
F
FAN - DELTA
0.5
SHELF
SHELF EDGE
FACIES
1.0
SLOPE
1 KM.
DIP SECTION
G
DELTA
2.0
SHELF / PLATFORM
3.0
SUBMARINE
CANYON
SHELF
PLATFORM
4.0
SLOPE
1 KM.
DIP SECTION
H

Tectonic Boundaries

Primary seismic reflections will be deformed through tectonic deformation so that they become tilted or folded. We can distinguish faults where good reflections suddenly stop, suggesting an abrupt lateral change in lithology. Faults are generally too steep to reflect the sound wave straight back again, and the fault plane itself will not appear as a reflector on the seismic profile. Because of the special "edge" effects near faults, the ends of the reflecting layers which should define faults will not be quite correctly located on the seismic profile, and it may therefore be difficult to trace the fault entirely accurately. The termination of beds against faults may produce diffraction from a point source, giving a curved alignment. Special treatment of seismic data (migration) will correct a fair number of these errors and give a more correct picture.

In recent years seismic lines have been shot with smaller and smaller grid spacings. Now seismic grids with spacings down to 50 m are often shot over reservoirs to obtain a better map of the reservoir structure. A three-dimensional seismic data set is then produced and seismic sections can be constructed at any angle relative to the grid. This method also allows us to construct horizontal time-slice through the structure. This is almost like a topographic or geological map which is a horizontal projection of the geology. It is also a very powerful method of delineating faults and other important structural elements.

Another relatively new development is *borehole seismics,* particularly vertical seismic profiles (VSPs). This method involves firing shots near the sea floor close to a well and recording signals at regular depth intervals in the well. The main advantage of VSPs is that they produce a very good profile of the seismic velocity as a function of depth, better than a synthetic seismic log.

Changes in Sea Level

It has been clear for a long time that there are unconformities in sedimentary sequences which can be correlated over long distances, and that there were periods in geological history with a high sea level and others when it was low.

However, only when seismic stratigraphy was developed and seismic reflectors were correlated with well data where it was possible to double-check against stratigraphy, was a detailed picture obtained of a pattern of onlaps and downlaps which could be interpreted as variations in sea level (Fig. 11.9).

Proximal onlaps are due to sedimentation moving landwards over an unconformity surface. If we are dealing with a coastal deposit, a proximal onlap will mean that the sea level has risen in relation to the land surface which forms the top of the unconformity. On seismic profiles we can see onlaps onto the land, measure the height range between the lowest and uppermost onlaps, and calculate the difference in seismic time, and convert this into approximate thickness.

Fig. 11.8. Typical seismic patterns from shelf and slope. (Brown and Fisher 1977). Strike sections are parallel to the coastline and dip sections are perpendicular running from the coast into deeper water

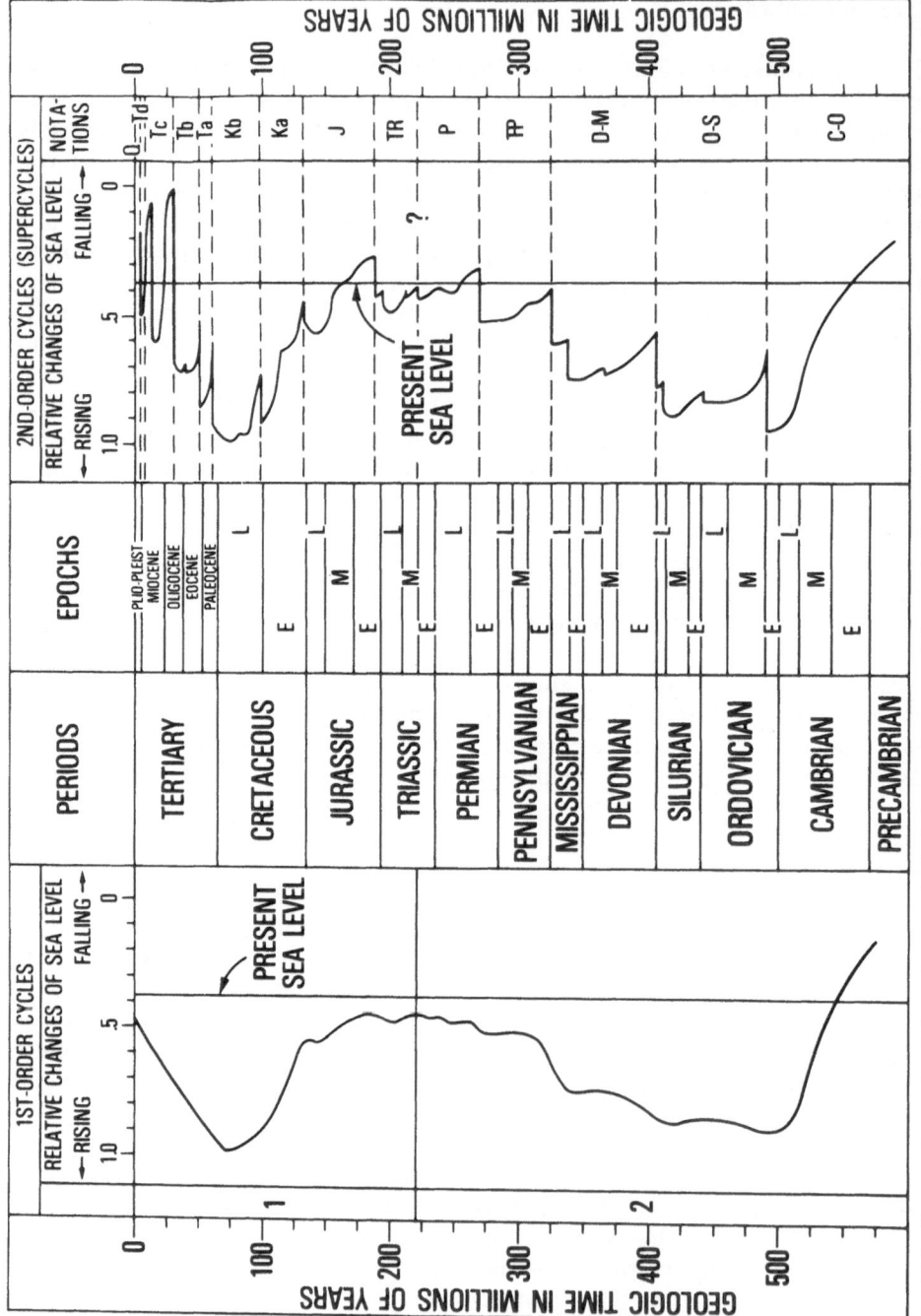

Fig. 11.9. Changes in sea level throughout geological time as interpreted from seismic profiles and well data. The rapid regressions and slow transgressions shown by the diagram are a result of the "method". They are probably not real and have been modified in later publications. (Vail et al. 1977)

However, we must remember that the thickness of the sediments deposited is due not only to a rise in eustatic sea level, but also to local subsidence of this part of the basin. The weight due to increased water depth will cause further subsidence, and sedimentation will increase the load, resulting in further subsidence to attain isostatic equilibrium. Local tectonic subsidence may produce a relative change in coastal onlap in a seismic profile. Regressions are defined as the boundary between land and sea being displaced out into the basin. They may be caused by a fall in sea level which will shift the coastline to further out on the shelf or to the edge of the continental slope. Here unloading of some of the water plus erosion of sediments leads to isostatic uplift of the area landward of the coast line, so that the measured regression is greater than the real lowering of the sea level. Here, too, local tectonic movements will play a part.

Eustatic Regressions and Transgressions

These are changes in sea level which seem to have been simultaneous all over the world. Although one might perhaps expect that the sedimentation in a sedimentary basin would first and foremost be characterised by local tectonic conditions, drainage, depositional conditions etc., studies of thousands of seismic profiles and wells from many sedimentary basins indicate that there have been simultaneous transgressions and regressions in completely different parts of the world: seismic profiles have been correlated with oil wells where the age of seismic unconformities and depositional sequences can be dated by means of biostratigraphy. It turns out that characteristic seismic reflectors which represent falls in sea level are of the same age, e.g. in the North Sea, West Africa, the Mexican Gulf, Alaska etc. This can be explained as being a result of eustatic changes in sea level. The effective change in sea level may vary from one area to the next, however. This shows that the falls or rises in sea level which we measure on seismic profiles are also affected by other tectonic factors which may be relatively widespread, or local. It has become clear that many areas, especially continental margins, have a rather similar tectonic history which is ascribable to global sea-floor spreading.

As a result of systematic work within the oil companies (particularly Esso-Exxon, under the leadership of P. Vail), it has been possible to establish a curve representing changes in sea level right from the Palaeozoic to the Quarternary (Fig. 11.9). This curve is constantly being corrected and improved. It must not be taken absolutely literally, although it is based on a large amount of data. One significant feature of the curves first published is that all regressions appear as very rapid episodes, while the transgressions appear to have been slow. This feature had an effect on the way the curves were arrived at, and they are now referred to as curves of relative change in coastal onlap which form the basis for an eustatic sea-level curve. The time period for transgressions is defined by the age difference involved in the onlapping bedding series. Regression periods are estimated as the age difference between the younger part of the onlapping sequence and the oldest part of the prograding (downlapping) bedding series further out on the shelf or slope. It is quite possible that sedimentation on the slope starts when the sea level is still high as a result of erosion and redeposition on the slope. The difference in age between the transgressive deposits and the base of the prograding units would then be nil.

For a detailed description, see articles by P. Vail et al. 1977 and Brown and Fisher 1979.

The principles of seismic stratigraphy and sea-level curves have been the subject of considerable debate. Vail et al. have produced several more updated versions (Vail et al. 1984). A revised sea-level curve was produced by Exxon research in 1987. Many workers now argue that simultaneous tectonic events may produce very widespread sea level fluctuations, which have interpreted as being eustatic.

The prerequisite for the oldest beds over the unconformity representing the beginning of the sedimentation shift resulting from regression (low stand deposits) is that there is no erosion or sedimentation on the continental slope while it is at high stand. We now know that in Holocene times, with relatively high stand, there is erosion and sedimentation in submarine canyons. If we now dropped the sea level, these sediments would be included in the bedding sequence over the unconformity, and their age would therefore not date the regression, but be somewhat older.

Brown and Fisher (1977) have made a model of erosion and deposition of an onlap sequence at high stand. This model consequently differs somewhat from the one which forms the basis for Vail's sea-level estimations (Fig. 11.10). See also Watts 1982.

Further reading: Payton 1977; Brown and Fisher 1979; Anstey 1982; Berg and Wolverton 1985.

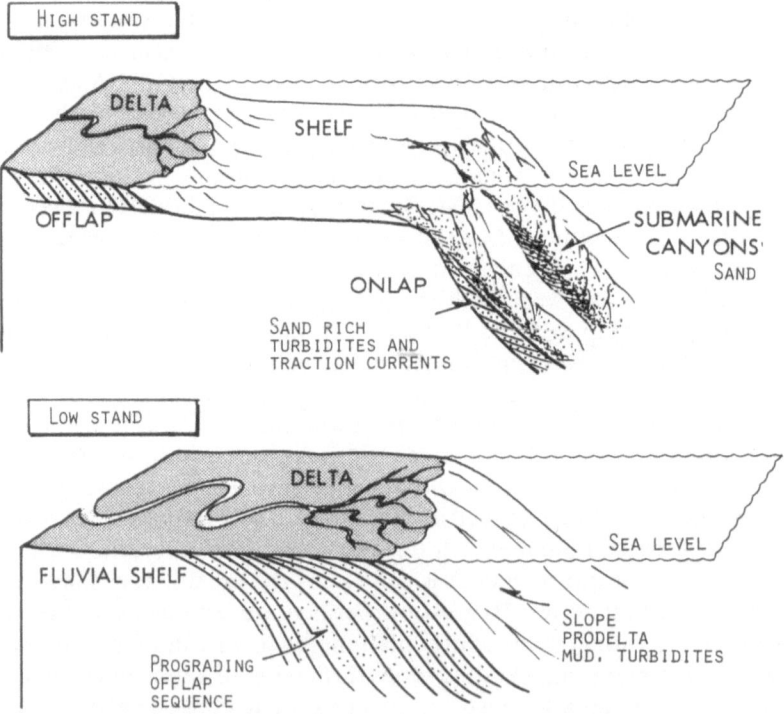

Fig. 11.10. A model for coastal progradation and offlap in the event of low sea level and erosion, and coastal onlap with a high sea level

Reasons for Variations in Sea Level

Sea level is the height of the surface of the oceans relative to the continents. It is useful to attempt to distinguish between *eustatic* changes in sea level, which occur simultaneously all over the world, and local variations which are associated with a particular area. Even when we study recent geological history, for example the Holocene, this is none too easy if we want to be absolutely accurate: what fixed point do we have? We take as our departure point stable parts of the continents where there is little active tectonic movement (neotectonics), and attempt to establish a more or less absolute eustatic sea-level curve. For older geological periods this is even more difficult, but by correlating profiles with good biostratigraphic control or age determination, we can establish that we have radiological periods with transgressions and regressions which appear to be simultaneous over the greater part of the world. However, we must keep in mind that the margin of accuracy of dating from the Palaeozoic, for example, will be some 2–5 million years, and that the sea level changes indicated for each place represent the net effect of many different factors. The cause of eustatic changes in sea level may be:

1. Changes in the volume of sea water. Variations in the amount of water in the atmosphere do not play much part, but if large amounts of ice accumulate, the volume of sea water will be reduced so much that it will result in a demonstrable regression. The last ice age caused a regression of about 100 m. However, it is important to remember that although the last ice age resulted in a regression followed by a transgression in Holocene times in most parts of the world, it was mainly the opposite in the ice-bound regions because of the ice loading on the crust of the Earth. Isostatic subsidence due to the ice load was greater than eustatic regressions. Later, in post-glacial times, the elevation due to unloading of the ice was greater than the eustatic transgression due to the melting of most of the continental ice sheets.

2. Changes in the volume of the ocean basins. As we shall see, sea-bed topography is a function of the age of the sea bed, i.e. how long it has had to cool down since it was formed, and the sediment thickness. Periods with rapid sea-bed spreading will result in relatively broad spreading ridges which cause the volume in the ocean basins to decrease, and sea water will push further in onto the continents (Fig. 11.11). If all sea-bed spreading stopped now, the spreading ridge would slowly sink and have disappeared almost completely within about 100 million years. The volume of the oceans would then be greater, and the result would be a regression. This mechanism may explain the great fluctuations in sea level over the geological ages. Note too that periods with major transgressions can be correlated with periods of rapid sea-floor spreading, for example in the Cretaceous and Carboniferous periods. The deep channels formed in connection with subduction are, in fact, small in relation to the width of the spreading ridge. In Permian and Triassic times we had one big super-continent and little sea-floor spreading. This was a regressive period with a large land area.

   Drying out of cut-off ocean basins may also lead to eustatic changes in sea level. There is much to indicate that the Mediterranean Sea was cut off and dried up in Upper Miocene (Messinian) times. This reduced the world's total volume of ocean

ATLANTIC OCEAN
Cross section

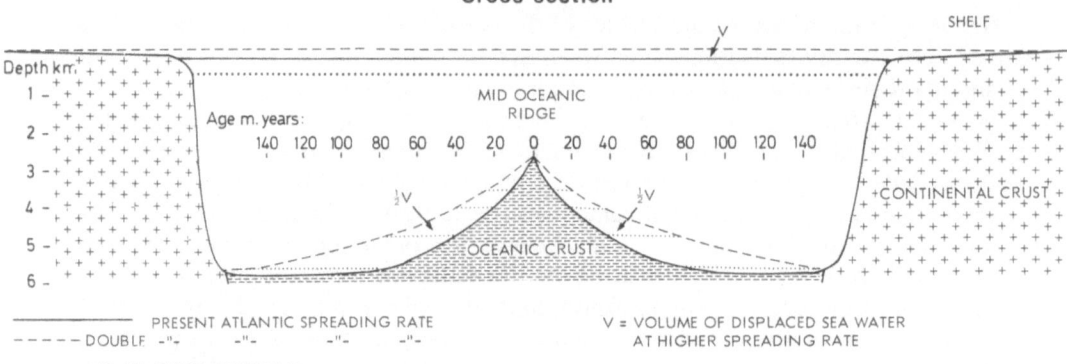

Fig. 11.11. Section through an ocean basin with active sea-bed spreading (the Atlantic Ocean). An increase in the rate of spread would give a broader spreading ridge and transgression. A slow rate of spreading means a narrower ridge, which will displace less water and may cause regression

basins, causing a transgression. It has been estimated that the Mediterranean is not large enough to cause a transgression of more than a few metres (about 5 m).

3. Local changes in sea level due to loading and unloading.

   Changes in the isostatic equilibrium.

A) Ice-loading — glaciation. A 1000-m-thick ice cap will cause about 300 m of isostatic subsidence.
B) Water loading during transgressions and formation of new basins.
C) Loading due to sedimentation — sediment loading.
D) Unloading due to erosion.

These points have been discussed earlier.

4. Causes relating to plate tectonics.

A) Elevation of land (regression or inversion of basins) due to tectonic compression and depression of the continental crust. An increase in the depth of the *Moho* (seismic discontinuity separating the earth's crust and mantle) will lead to elevation of the land. The greatest land elevation is a result of continental collision, when the thickness of the continental crust may be doubled (70–80 km) as in the Himalayas.
B) Transgression movements lead to the continental crust stretching and thinning, and we have subsidence. We see this at the transition between the continental crust and the oceanic crust, and where we have rift formation the continental crust under the rift becomes thin, causing graben formation.
C) Variations in the temperature gradient affect the density of the rocks and thereby the isostatic equilibrium. Rifting causes elevation of the areas along the margin of the rift where the crust is not thinned (e.g. East Africa) and subsidence of the whole area when the rifting ceases (e.g. the North Sea).

How Shall We Explain the Observed Variations in Sea Level?

The greatest variations in sea level, such as transgressions in the Cambrian, Ordovician and Cretaceous, can be explained fairly satisfactorily by means of plate tectonic models. A relatively low sea level at the end of the Palaeozoic and the beginning of the Mesozoic (Triassic) can be explained as being due to limited sea-bed spreading and rifting, for example along the Atlantic Ocean. This increased the geothermal gradient and elevated above sea level areas which were previously covered by shallow epicontinental seas.

The short-term variations in sea level are more difficult to explain, however. For geological periods which experienced complete continental glaciations (Quaternary-Upper Miocene, Permian-Carboniferous, Mid-Upper Ordovician and late Precambrian) we can resort to glacioeustatic changes in sea level (changes in sea level due to glaciation), which are very rapid in geological terms. We find similar abrupt regressions in the Lower Tertiary and Mesozoic as well, and we have no indications of continental glaciation from these geological periods. At present this appears to be an unresolved problem. If the changes in sea level we record on seismic profiles and in other ways are real, we have two possibilities: (1) There were more or less continuous glaciations in Mesozoic times too, but these have not as yet been demonstrated. (2) There are other processes which can lead to rapid variations in sea level which we do not yet know of. On the basis of palaeoclimatic studies it now seems difficult to imagine major continental glaciations in the Mesozoic. Nor is it easy to find other processes which can explain the observed sea level curve. The formation of large evaporite basins like the Mediterranean in Messinian times is one possibility, as mentioned, but very large basins would have to have been involved.

Geothermal Gradients in Sedimentary Basins

What Factors Control the Flow of Heat in Sedimentary Basins?

It has been known for a long time that the temperature in the crust of the earth increases with depth.

How rapidly the temperature increases as a function of depth below the surface of the earth or the sea bed varies considerably, however, and depends on the heat conductivity of the rock and on heat flow.

The *geothermal gradient*, defined as temperature variation as a function of depth,

$$\left(\frac{\Delta T}{\Delta Z} \right)$$

expresses the increase in temperature ΔT over a depth interval ΔZ. This is usually expressed as number of degrees C per km or 100 m, for example 30°/km. In the oil industry it is commonly expressed in °F/1000 ft.

Heat flow (Q_z) is the flow of heat through a particular cross-section per time unit. Heat flow is proportional to the geothermal gradient and the thermal conductivity, k, of the rock.

$$Q_z = k \cdot \frac{\Delta T}{\Delta Z}.$$

Heat flow has traditionally been expressed as μ cal/cm^2/S (1 μ cal $= 10^{-6}$ cal). In the Si system mW/m^2 (1 μ cal/cm^2/s $= 40$ mW/m^2). Thermal conductivity (k) is defined as W/m°C $- \mu$ cal/cm/s°C. Heat flow through a rock is thus a function of both the geothermal gradient and the heat conductivity of the rock. If we want to postulate variations in geothermal gradients and heat flow in sedimentary basins, we must look at how heat flow takes place in sedimentary rocks.

1. By means of conduction through rock and pore water in the rock.
2. Through flow of pore water through the rock. Rather considerable flux of porewater is required and this mechanism is most important locally.
3. Transport of heat through radiation is of little importance below 800°C, and we can therefore disregard it in the context of sedimentary basins.

Movement of rocks in relation to the surface affects the geothermal gradients. Erosion removes the uppermost, colder strata so that warmer strata come closer to the surface, and the geothermal gradient increases. As a result of subsidence of a sediment basin and sedimentation, heat flow upward will be partly offset by rock subsidence. This will result in lower geothermal gradients, than one would otherwise have expected. Sedimentary basins with high rates of sedimentation are therefore often characterised as "cold basins". Tectonic elevation and erosion will increase geothermal gradients in the same way.

The geothermal gradient in sea-floor rocks is consistently greater than it is over the continents, and on the continents it is highest in areas of volcanic activity.

Areas with a high geothermal gradient due to volcanism will slowly cool down to normal gradients when volcanic activity ceases. It may take about 100 million years before a normal geothermal gradient is established (Fig. 11.12). Approximate isostatic equilibrium predicts that cooling and contraction of the crust of the earth will cause subsidence due to increased density. Sea-floor basalt is hot and flows at relatively shallow depths in the earth's crust, and the spreading oceanic ridges are only about 2.2 km below the surface of the sea. After about 180 million years of cooling, the depth at isostatic equilibrium is about 5.7 km without sediment loading (Fig. 11.12a). With sediment loading the theoretical limit to the thickness of a sedimentary sequence overlying oceanic crust is about 17 km before isostatic equilibrium is achieved. Sediment basins on the continental crust will have a sedimentary thickness which is a function of the thickness of the crust and the density of the sediments (Fig. 11.12b). The thinner the continental crust beneath a sedimentary basin, the more sediments can accumulate before the basin is filled and in isostatic equilibrium.

Further reading: Gretener 1982.

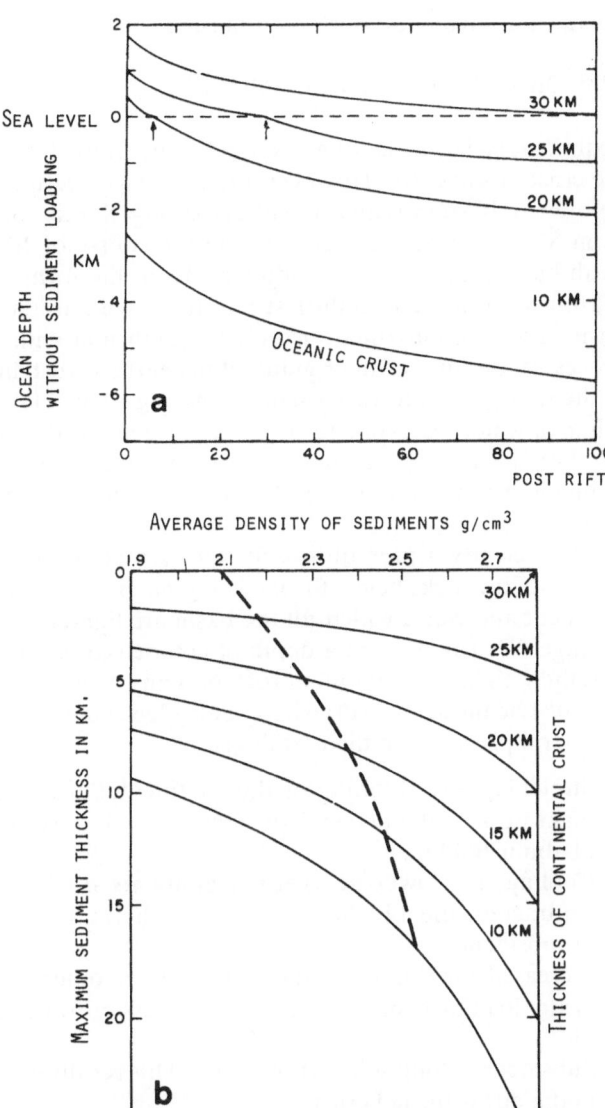

Fig. 11.12. a Subsidence of the sea-bed and continental crust of various thicknesses as a function of post-rifting age. (After Kinsman 1975). **b** Potential loading capacity to isostatic equilibrium on oceanic crust of various thicknesses of continental crust as a function of sediment density. (After Kinsman 1975a)

Sedimentation and Isostatic Equilibrium

Why Do Sedimentary Basins Subside?

Sedimentary basins can be assumed to be in isostatic equilibrium in relation to the crust of the earth. However, the crust has a certain rigidity, and it takes time before equilibrium is attained after loading. Studies of uplift curves, for example from Scandinavia, show that during the course of 10,000 years the crust of the earth has undergone major adjustments in the form of uplift to compensate for the unloading of ice after the last glaciation. We can assume that major sedimentary basins are in approximate isostatic equilibrium with regard to most geological processes. Because of the rigidity of the earth's crust, however, this will not apply to the same extent to very small basins (e.g. glacial lakes). Nevertheless, even the filling of water reservoirs for hydroelectric plants does cause local subsidence.

For the most part, then, we can use the classical Airy isostasis model for sedimentary basins, and we can draw many interesting conclusions by applying this model.

It is clearly a prerequisite for the formation of sedimentary basins that the density of the rocks below the basin is greater than that of the rocks around it. The sediment and water which fill the basin are lighter and ensure that the density is average. We can assume a depth of compensation of about 100 km. This means that the weight of a column of rock plus any water present down to a depth of 100 km must be the same everywhere. Subsidence of a sedimentary basin may be due to several processes in the earth's crust.

1. Stretching and thinning of the continental crust. Heavier mantle rocks then make up a greater percentage of the rock column down to a compensation depth of about 100 km.
2. Cooling, i.e. lower geothermal gradients in the crust of the earth lead to contraction and a higher rock density (thermal contraction), and this will result in subsidence.
3. Increased loading of water, sediments or other rocks will cause subsidence. Water loading could be due to a transgression and sediment loading as the basin fills.
4. Subsidence along subduction zones. This results in lower geothermal gradients in the down-turned crust.
5. Tectonic loading. Thrusting of tectonic plates leads to increased loading on the part of the earth's crust in question and we have subsidence, especially in front of nappes.

How Do Changes in Sea Level and Sedimentation Lead to Isostatic Compensation?

Variations in sea level due to eustatic transgressions or tectonic subsidence will lead to the crust of the earth losing its isostatic equilibrium. We will find isostatic compensation in the form of further elevation or subsidence of the sea floor, which will reinforce the primary change in sea level. If the sea rises 100 m, for example due to ice melting, this will increase the isostatic loading on the sea floor. We can

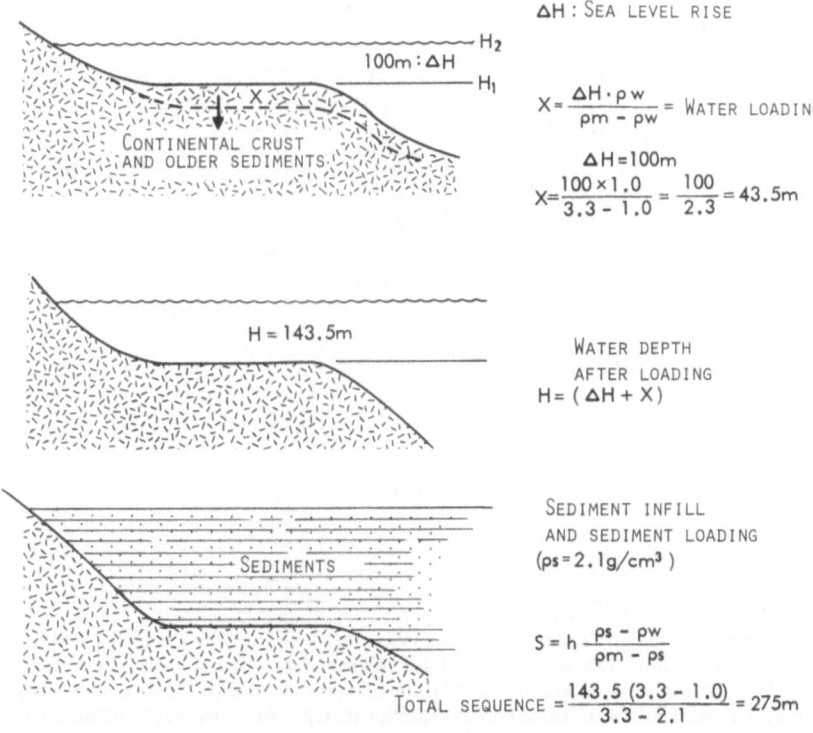

$\Delta H :$ SEA LEVEL RISE

$X = \dfrac{\Delta H \cdot \rho w}{\rho m - \rho w} =$ WATER LOADING

$\Delta H = 100m$

$X = \dfrac{100 \times 1.0}{3.3 - 1.0} = \dfrac{100}{2.3} = 43.5m$

WATER DEPTH
AFTER LOADING
$H = (\Delta H + X)$

SEDIMENT INFILL
AND SEDIMENT LOADING
$(\rho s = 2.1 g/cm^3)$

$S = h \dfrac{\rho s - \rho w}{\rho m - \rho s}$

TOTAL SEQUENCE $= \dfrac{143.5 \, (3.3 - 1.0)}{3.3 - 2.1} = 275m$

Fig. 11.13. Schematic representation of isostatic equilibrium. The weight of a hypothetical column down to a reference level of 100 km should be the same. Where the average rock density is high, the water will be deeper

calculate that there will be a further 43.5 m of subsidence, so that the total depth of water at equilibrium will be 143.5 m (Fig. 11.13)

If a sedimentary basin with this depth of water fills with sediment, we will find further isostatic subsidence because of the weight of sediments, and we could get a total deposition of 250–300 m depending on the density of the sediments. A 100-m rise in sea level will thus lead to deposition of almost 300 m of sediment. In the same way, primary tectonic subsidence due to cooling of the ocean floor will lead to further subsidence due to increased water and sediment loading.

We can thus take our point of departure in stratigraphic data in the form of measured profiles or oil wells and calculate backwards to the primary tectonic or eustatic changes in sea level. This method is called "backstripping".

$$Z = F\left[Y \frac{\rho_m - \rho_s}{\rho_m - \rho_w} \quad -\Delta H \frac{\rho_w}{\rho_m - \rho_w} \right] + (H - \Delta H)$$

sediment loading   water loading   changes in water depth, (Watts 1983)

where Z is the primary tectonic subsidence, F is a factor which is a function of the rigidity of the earth's crust, ΔH is the change in sea level and H is the depth, Y is the sediment thickness with compensation for compaction, i.e. sediment thickness, and

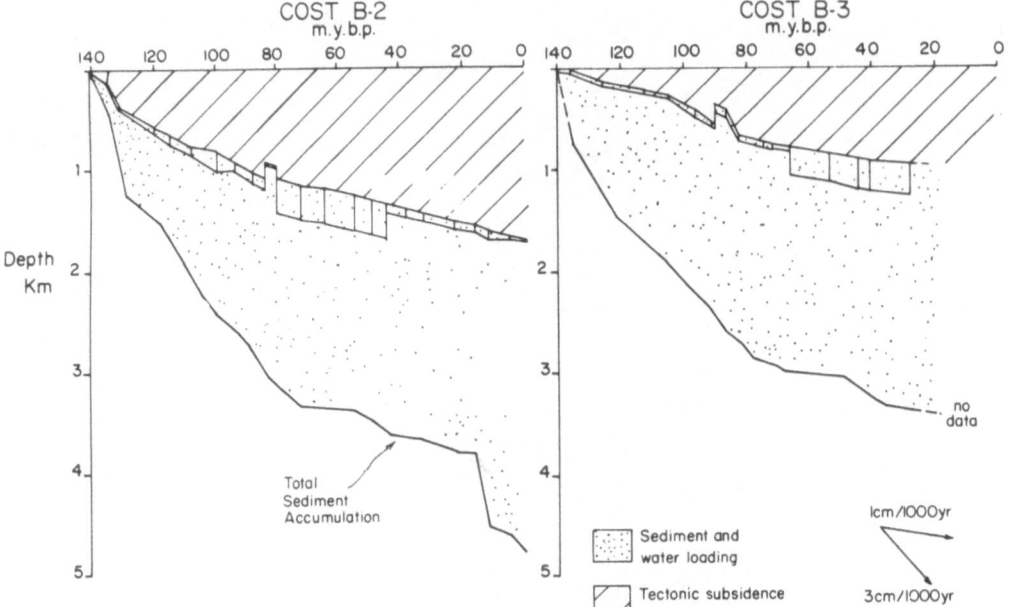

Fig. 11.14. Tectonic subsidence and sediment accumulation at the COST B-2 and B-3 wells (Eastern USA) through time. The tectonic subsidence has been computed assuming an average response of the basement to sedimentary loads (Curve *b*). Note that the tectonic subsidence is generally smoother than the total sediment accumulation curve suggesting that the "backstripping" method adequately accounts for sedimentary processes at the margin. (Watts 1981)

ρ the density, shortly after deposition (Stickler and Watts 1978b; Bally et al. 1981). If we know the variations in sea level and the depth of the water from environmental interpretations, for example, these can be inserted into the equation so that the primary tectonic movement can be worked out (Fig. 11.14.)

The backstripping technique which is used to reconstruct the subsidence history of different parts of a sedimentary basin lends itself very well to computer modelling. The data obtained on the depth and temperature history of the source rock in particular has allowed much better assessments of kerogen maturity and the times of oil expulsion and migration. One parameter which is crucial but difficult to estimate is the variation of the geothermal gradient as a function of geological time. It is also often difficult to estimate the palaeodepth during the deposition of different sedimentary formations. Whether a formation was deposited at a depth of 200 m or 1000 m will make a very significant difference, and it is often difficult to make accurate palaeoenvironmental estimates of the palaeobathymetry.

Fracturing and Subsidence of the Continental Crust (Rifting)

Tension tectonic processes in the form of stretching and thinning of the continental crust take place in connection with rifting, and may be a result of tension forces and uplift due to the high geothermal gradients associated with rifting. The resulting

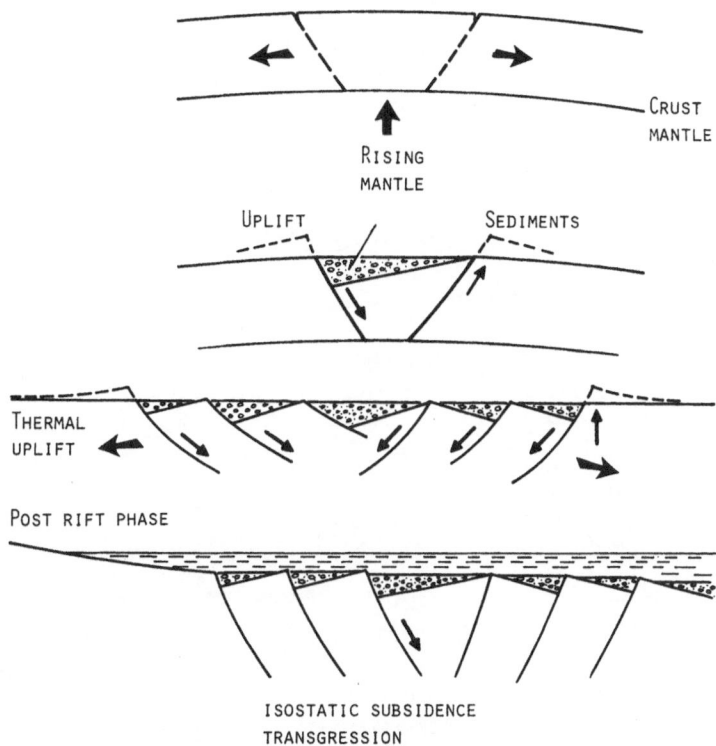

Fig. 11.15. Development of a continental rift system by mantle doming and tension and thinning producing rotated listric fault blocks

thin continental crust is subject to isostatic subsidence. Fracture zones in the continental crust will tend to subside, creating rift valleys, which often form in a sedimentary basin because the continental crust is thin and because heavy rocks from the mantle push their way up. Those parts of the rift valley system which have a lot of volcanism will develop a more pronounced topography and less sedimentary basin than areas with little volcanism (see the East African rift system). Along the edge of the rift system, where the continental crust is not stretched, uplift occurs due to the geothermal gradient. This causes the basement rock at the surface to slope away from the rift valley. This will to some extent be compensated for by the formation of erosional valleys on the side of the rift valley, which cut backwards into the raised shoulders on the side.

Horsts which are associated with rifts are pieces of continental crust that are not stretched, and being in a region with a high geothermal gradient, they will "float" isostatically and form high mountains like the Ruwenzori between Uganda and Zaire.

The subsidence due to rifting causes the rocks on both sides to be unstable, and gravitational sliding of blocks in the crust in towards the rift structure may occur. There is a tendency for *listric* faults to form, i.e. parallel, curved fault planes which start as normal faults and curve round with depth until they are almost horizontal (Fig. 11.15). The blocks then become rotated so that they turn over and slope away

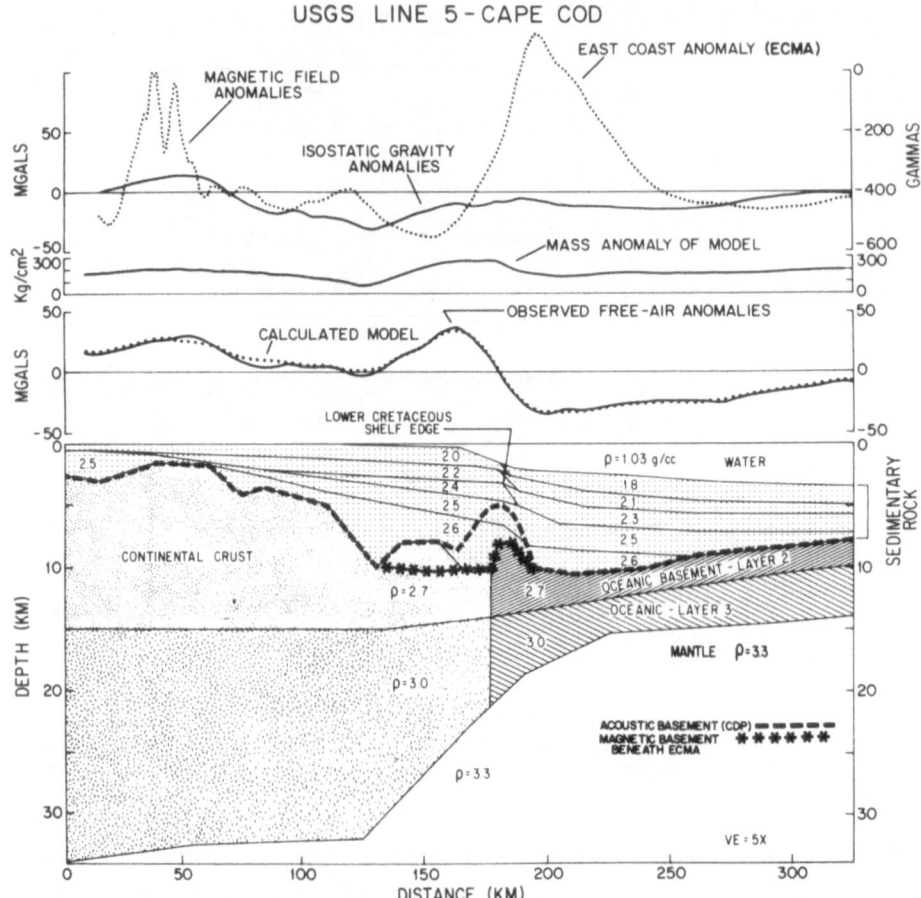

Fig. 11.16. Cross-section of the U.S. Atlantic Margin showing the densities of crust and the sedimentary sequence. (Grow et al. 1979)

from the cemented part of the rift. During the first part of the spreading phase basins with limited circulation will be formed so that evaporites and carbonates are often deposited. Upper Jurassic and Lower Cretaceous deposits of this sort are found extensively along the margin of the Atlantic Ocean (Fig. 11.16,17).

Subsidence Along Passive Margins

Passive margins develop from a rift phase to a spreading phase. The transition between continental crust and oceanic crust therefore consists of a thinned-out continental crust with listric faults and horsts formed during the rifting phase. As the ocean-floor spreading progresses, the geothermal gradients in this part of the continental shelf will decline and we will find thermal subsidence as a function of the age of the sea floor. Subsidence flexure will develop, with the most rapid subsidence on the outer parts of the continental shelf and slope near the oceanic

Model of the evolution of stable margins created by the divergence of two plates

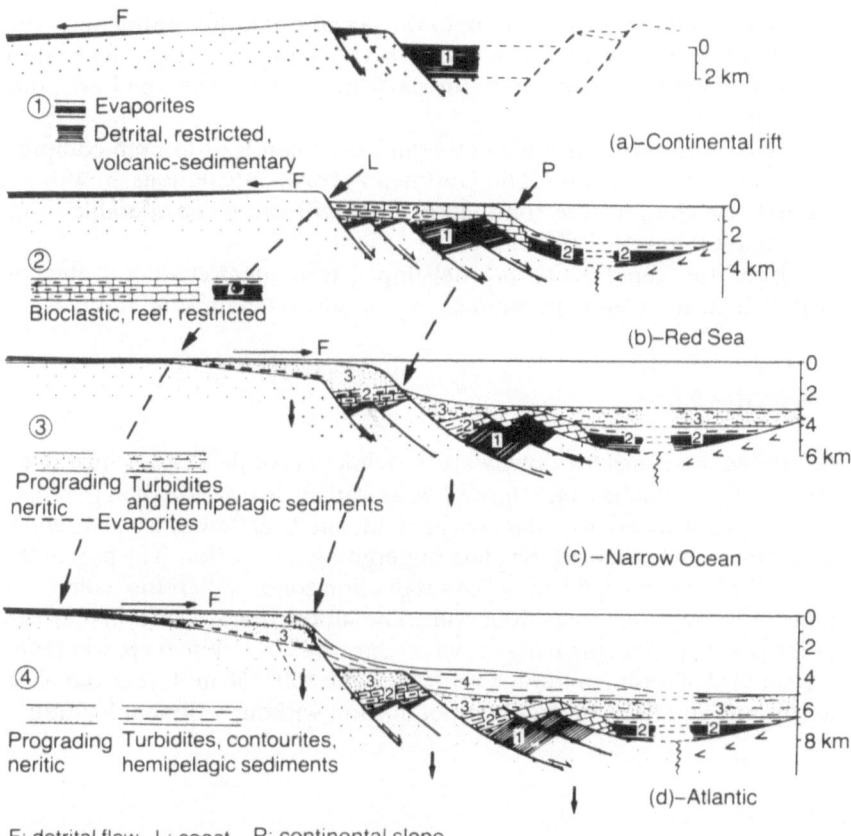

F: detrital flow   L: coast   P: continental slope

Fig. 11.17 Four stages in the development of an Atlantic margin. (Boillot 1981)

crust, and slowest in the interior of the continent. This will result in coastal onlap and transgression in areas where the rate of sedimentation is lower than the rate of subsidence. As sedimentation fills up the area between the oceanic crust and the continental crust, isostatic compensation due to the sediment load and the compaction will increase the rate of subsidence on the outer part of the continental shelf and slope in relation to the coastal plane. With cooling, the rigidity of the crust increases, so that the bending of the continental crust near the continental margin broadens and there is overall subsidence of the continental shelf, and in consequence transgression and onlap (Watts 1982).

Subsidence in Connection with "Strike Strip" Faults

We find "strike strip" faults in both the oceanic and the continental crust. Transform faults in the oceanic crust result in ridges with younger basaltic material which can help to limit the sediment basin, particularly in the early phases of the opening (see evaporites).

Strike strip faults in the continental crust can lead to both compression, i.e. thickening, and stretching of the continental crust. Calculations show that relatively modest stretching of the continental crust will cause considerable thinning and subsidence (Crowell 1974).

Here, too, relief which has developed through stretching of the continental crust will be made more pronounced by sediment loading and thermal subsidence.

Subduction Zones

In subduction zones the downward-moving part of the crust, which in most cases is ocean floor, will be characterised by extremely low geothermal gradients. This is because it is in most cases older ocean floor, and heat flow must move counter to the direction of movement of the plate undergoing subduction. The part of the oceanic crust which is carried down in the subduction zones is therefore colder and denser than the average old ocean floor crust. The subduction zones form the deepest parts of the ocean, producing troughs, which may be 10–11 km deep, where there is not a great deal of sedimentation. Ordinary, cold, 100–150 mill. year-old ocean floor is in isostatic equilibrium at a depth of 5–7 km without sediment loading.

12 Diagenesis in Clastic Sediments

What Happens to Sediments After Deposition?

The reactions which occur in sediments after deposition are called diagenetic processes. These processes convert loose sediments like sand and clay into solid rock-sandstone and shale, and involve both mechanical compaction and chemical solution-precipitation processes.

Diagenetic processes begin immediately after the sediments have been deposited, and continue as the thickness of overlying younger sediments increases, until the temperature reaches about 200–300°C. When sediments are subjected to higher temperatures, metamorphism commences, but there is no sharp boundary between diagenesis and metamorphism. We shall discuss here processes which occur at temperatures ranging from surface temperatures (0–25°C) up to 200°C, and at pressures of from 1 atmosphere to about 1500 bar (1.5×10^5 kPa), which corresponds to a depth of 6–8 km (Fig. 12.1).

A sediment consists of grains which may be mineral or rock fragments, amorphous material, e.g. volcanic glass, or biological material (fossils) which may be either crystalline or amorphous. Between the sediment grains there is porewater – or hydrocarbons – apart from in the vadose zone (above the groundwater table), where there is air between the grains of sediment.

Diagenetic reactions encompass reactions between sediment grains and the porewater between the grains. We can distinguish between mechanical compaction of the sediment, and chemical and mineralogical reactions. These vary very considerably according to the type of sediment, depending on the grain size and chemical and mineralogical composition, and we shall therefore deal separately with diagenesis in different types of sediment.

Circulation of Porewater in a Sedimentary Basin

The circulation of porewater in a sedimentary basin controls the transportation of ions dissolved in the porewater, and is of fundamental importance for the concentration of hydrocarbons, and also of a number of metals (uranium, lead, zinc etc.) in sedimentary rocks.

The flow of porewater is a function of potential fields which are defined by pressure gradients and the gravity potential of the porewater. Water flows from areas with a higher to areas with a lower hydrodynamic potential. Hydrodynamic potential can be most simply defined in terms of the *piezometric surface* (also known as the potentiometric surface), which is a function of porewater pressure in relation

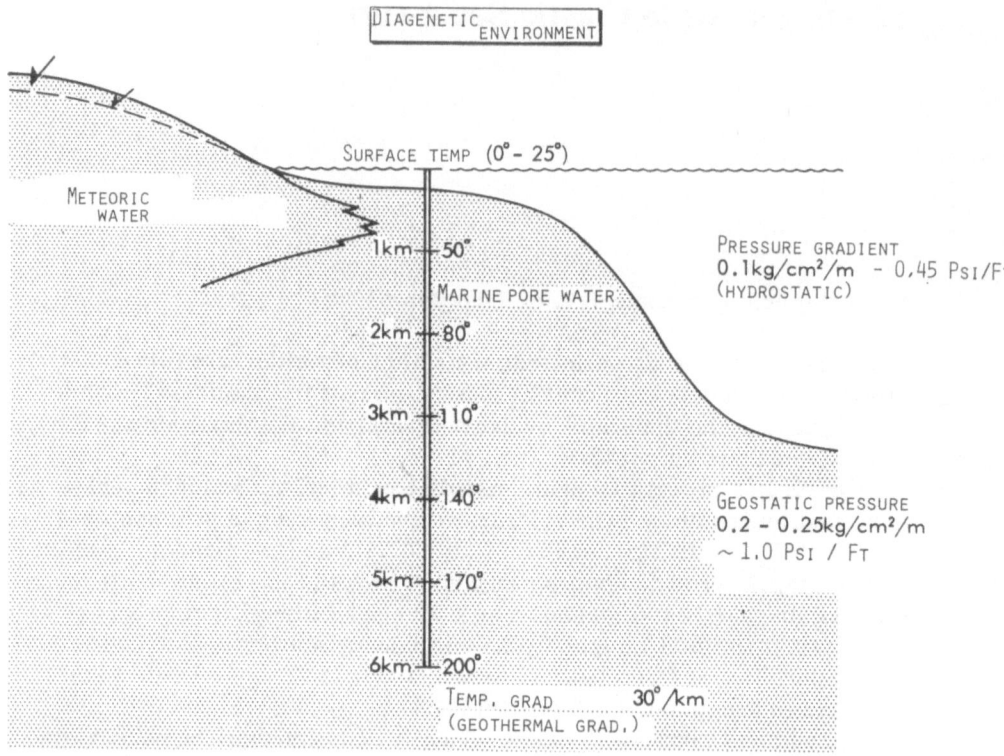

Fig. 12.1. Diagram showing different diagenetic environments and the temperature and pressure ranges for diagenetic reactions

to depth, and is defined as the height which the water would reach (in equilibrium with the atmospheric pressure) if a long pipe were stuck into the sediment or rock (Fig. 12.2). In sedimentary basins it will be useful to measure piezometric surface in relation to sea level, or the level of lakes or rivers. In marine basins porewater which is under normal hydrostatic pressure (not overpressure) will have a piezometric surface corresponding to sea level. This means that the pressure is equivalent to the weight of a column of water rising from the porewater at a particular place in the rock, up to sea level. If the porewater pressure is greater than the weight of the overlying water column, we have what we call *overpressure* or abnormal pressure. This corresponds to a piezometric surface above sea level in a marine sedimentary basin.

Variations in the density of the water due to dissolved salts and temperature must be taken into account. If the piezometric surface corresponds to a level above sea level, there is overpressure, i.e. the water has a higher hydrostatic potential, and is capable of performing work. Piezometric surfaces which are lower than sea level also occur, and the sediments then have a pressure lower than normal hydrostatic pressure, but this is rather rare. A piezometric surface can be contoured, i.e., we draw lines between points in a homogeneous matrix of similar hydraulic potential. Porewater will flow along lines with the maximum potential gradient. However,

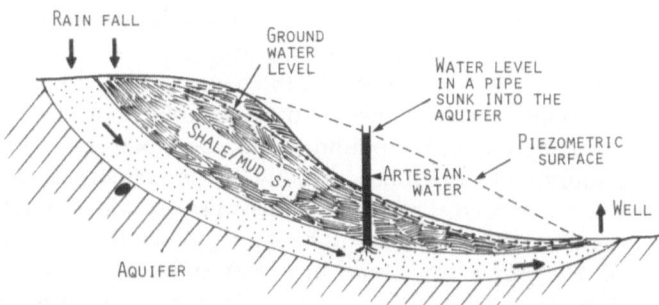

Fig. 12.2. Sketch showing how artesian water may develop in a confined permeable sand bed overlain by less permeable clay

sediments and rocks are not homogeneous (isotropic), and as a rule vary markedly in permeability from one bed to the next, and frequently also in different directions in the plane of the bed. This depends on sorting, grain packing, orientation of channels and cracks etc. In most cases, consequently, the porewater in sedimentary rocks does not simply flow vertically along equipotential lines, but is controlled to a large extent by beds with a high primary permeability or by permeable fractures and faults. Faults may often have a sealing effect, however, and a low permeability. When brecciation develops along the fault plane it will be highly permeable. Clay and softer sediments will tend to develop and seal the fault.

Porewater which is released through compaction will mainly flow with an upward component in relation to the subsiding bed. The average upward flow of porewater expelled by compaction is, however, of about the same order of magnitude as the subsidence rate, and if the porosity/depth curve is constant there is no net upward flow in relation to a fixed depth. We may say that the sediments are falling through a column of porewater. This is assuming an ideal homogeneous basin. In a real situation flow of porewater is to a large extent controlled by permeable beds and faults. It will only flow downwards in areas where the hydrostatic potential decreases downwards. When thicker beds of clay are compacted, porewater will flow down into an underlying sand bed if this has a lower hydraulic potential due to good communication with the surface.

The most important form of down-flowing porewater is fresh (meteoric) water, however. Freshwater has a hydrostatic potential which corresponds to the height of the water table above sea level. The excess pressure of a freshwater column above sea level is then $\rho \cdot g \cdot h$, where ρ is water density, g is gravity and h is the height above sea level, often referred to as the head. As long as there are variations in the hydrostatic potential in a basin, porewater will flow from areas with a high potential down to those with a lower potential. If these areas are separated by beds of rock with a very low permeability, however, this flow may take place very slowly, or virtually not at all. Permeable beds, e.g. sandstone, which end down-dip in pinch-outs, will not be very effective in channeling the flow of meteoric groundwater but significant amounts of porewater may flow through large cross-sections of poorly permeable mud or shale.

Meteoric water which flows downwards will initially have a high oxygen content, and will oxidise organic material and minerals. As a result, the oxygen content will be consumed and the oxygen will gradually decrease, until the water is reducing. Metals which are more soluble in an oxidised than in a reduced state will precipitate at the boundary between oxidising and reducing porewater (e.g. uranium and vanadium). As long as the meteoric water (groundwater) contains sufficient oxygen, it will be able to oxidise sulphides and sulphur in organic material to sulphate ions (SO_4^{2-}). Sulphate ions, which may also be derived from sea water, can exist for a long time in a metastable state in a reducing environment if sulphate-reducing bacteria are not present. The temperature will increase with depth and increase the solubility of most minerals, apart from carbonates. The solubility of carbonates tends to decrease with increasing temperature because of the increased solubility of CO_2. Porewater flowing out of overpressured structures may, however, precipitate carbonate if the reduced pressure causes dissolved CO_2 to form a separate phase (degassing).

Tectonic uplift of the land around the basin, and thereby the water table, will lead to an increase in the hydrostatic potential of the meteoric water. Lowering the sea level will have the same effect, and may act as a pumping mechanism for meteoric water in sedimentary basins along the continental shelf. There are examples of meteoric water having penetrated 2000–3000 m below sea level, for example in the Gulf of Mexico (Galloway and Hobday 1983), and on the continental shelf off the east coast of the USA there is freshwater just below the sea bed right out to the edge of the shelf, 100 km from the coast (Fig. 12.9). At greater depths (1–2 km) there is diffusion of saline porewater from underlying mesozoic evaporites.

Along the coast, and particularly in deltas, groundwater will flow through recently deposited sand-bodies into the ocean. Freshwater in the sediments will then flow in wedge into the underlying saline porewater, without much mixing. The meteoric water can be said to be floating on marine porewater like an iceberg in the sea.

If a porous sand bed which dips into the basin is exposed on land, the pressure on the freshwater column will correspond to the height above sea level of the water table where the sandstone is exposed. The freshwater column must therefore be in isostatic equilibrium with the salt water column.

If the height of the water table is H metres above sea level, the fresh water column may penetrate x metres below sea level down into the basin before isostatic equilibrium is achieved. At a depth of x metres the weight of the freshwater column should equal the weight of the marine porewater.

$$x \cdot \rho_{sw} = (x + H)\rho_{fw}.$$

Here ρ_{sw} is the density of saline water (1.025 g/cm³) and ρ_{fw} the density of freshwater (1.0 g/cm³).

$$x = \frac{H\rho_{fw}}{\rho_{sw} - \rho_{fw}},$$

if we use the above values for ρ_{sw} and ρ_{fw} we obtain $x = 40\ H$.

Theoretically, then, raising the ground water 100 m could result in freshwater right down to 4 km below sea level. The depth of freshwater penetration is usually less than what is calculated from the head, however, because the freshwater lens often meets with slight overpressures in the compacting sediments. In Pleistocene times the sea level was lowered about 100 m. This caused an increase in the pressure head of the freshwater lenses, so that the glaciations served as mechanisms for pumping freshwater into marine basins. If reservoir rocks are adjacent to freshwater lenses, hydrocarbons produced will be replaced by an equal volume of fresh water, and the pressure will remain constant during oil and gas production. This mechanism is called "water drive". Freshwater and salt water have very different chemical compositions, and this is of great significance for diagenetic processes. Clearly a powerful water drive requires a raised water table. This in turn depends on the topography around the basin, and whether the permeable beds are continuous from the land and out into the basin. Prograding sequences in a basin surrounded by areas undergoing tectonic uplift will provide the ideal situation for a water drive of this nature.

Conversely, in a flat-lying epicontinental environment, where there is low relief around the basin as well and almost horizontal bedding, there is not much potential for percolation of freshwater through marine sediments.

Diagenetic Processes in Sandstones

Immediately after deposition, sand has a porosity of about 35–50%. Poorly sorted sandstones with a high clay content may have a higher water content, and have many of the same properties as clays. If there are sand grains floating in a matrix of clay (matrix-home texture) they will respond to compaction in a manner similar to shales.

Diagenetic reactions will be controlled by the primary mineralogical composition, sorting and grain shape (texture) of the sand and by the external factors which affect the sandstone after deposition. We can distinguish between mechanical compaction and processes which include chemical and mineralogical processes.

Mechanical Deformation in Sandstone Diagenesis

Immediately following deposition, sand has a loose structure, because the sand grains are packed haphazardly and non-optimally. With increasing overburden, the sand grains will pack more closely and the porosity will be reduced. This may take place as a result of rotation of grains, and/or by grains breaking down mechanically. Porosity reductions require that the excess porewater between the grains be expelled. Grain packing may be gradual, or occur as a result of sudden collapse of the grain framework. Earthquakes may trigger such collapses. They cause the sand grain to be wholly or partially suspended in an upward flow of water for a short period, so that there is no friction between sand grains. The sand will thus behave like a liquid, and the process is called liquefaction. As the overburden increases, so will the stress at the contact points between sand grains. If we have well-sorted sand with round grains, the contact surface between the grains will be small, and the pressure (stress) per unit area (kg/cm^2) will be very great. This may cause deformation of the grains so that the stress due to the overburden is spread over a greater area through plastic deformation of minerals. By studying extinction under the microscope with crossed nicols, we can often see, for example, that the crystal lattice of quartz grains is deformed at contacts with other grains. Minerals may also suffer elastic deformation.

Minerals may also be crushed when they are subjected to stress. Mica flakes will often break between sand grains, and soft minerals, which are particularly abundant in volcanic rocks, e.g. chloritic basalt or gabbroic rocks, will be easily crushed or deformed in some other manner. We may also find that felspar grains split along the cleavage surface due to stress between sand grains.

As a general rule, we can say that relatively pure quartz sandstones have a high grain strength, and are relatively little subject to mechanical compaction, while sandstones which have been eroded from volcanic and basic source rocks have a low grain strength. This means that their porosity will be considerably reduced due to mechanical deformation, even with a relatively moderate overburden (1–2 km). This will reduce their potential as reservoir rocks.

What Are Effective Stresses?

The pressure in porewater is equally great in all directions, and we call this *pore pressure*.

In a porous sandstone we could consider porewater to be a water column in which the pressure increases as a function of the density of the water. A 1-m water column will weigh $100 \, g/cm^2$ and will produce a pressure of $0.1 \, kg/cm^2$, or $0.98 \, kPa$. A 10-m column of water corresponds to about 1 atmosphere. In the oil industry, pressure gradients are traditionally expressed as psi/ft (pounds per square inch/foot). Pure water with a density of $1.0 \, g/cm^3$ has a pressure gradient of $0.433 \, psi/ft$ or $0.1 \, kg/cm^2/m - 9-8 \, kgPa/m$. When the porewater at great depths and temperatures takes more salts into solution, the density will increase to about $1.1 \, g/cm^3$, in some cases more ($1.2 \, g/cm^3$). When the total dissolved solids amount to about 150,000 ppm, the density is about $1.1 \, g/cm^3$. The pressure gradient is then $0.476 \, psi/ft$ or $10.8 \, kgPa/m$. Immediately after deposition, sandstone will have a density of $1.9-2.0 \, g/cm^3$, and with increasing overburden and declining porosity the density will increase to about $2.5 \, g/cm^3$ (Fig. 12.3). The weight of the rocks, i.e. all sediment beds down to a certain depth, will exert a pressure which we call the geostatic pressure. At 2 km the geostatic pressure will be $440 \, kg/cm^2$ if we assume an average density of $2.2 \, g/cm^3$.

Geostatic pressure is a directional force or stress which is not uniform in all directions, and which will have a maximum vertical component in dense sediments which are not affected by tectonic stresses. This stress is transferred via the contact surfaces between sand grains, and we call the result the effective stresses (Fig. 12.3).

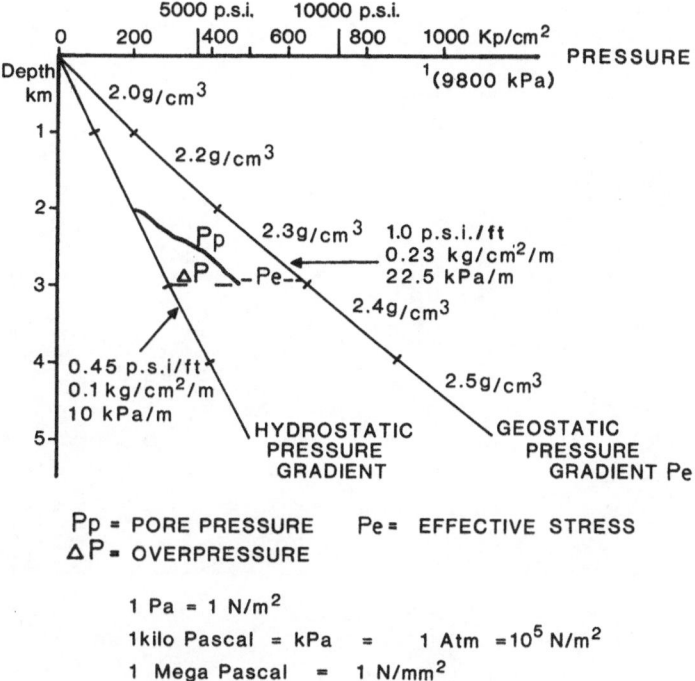

Fig. 12.3. Pressures and stresses in a sedimentary basin as a function of burial depth

The geostatic pressure (S_g) must be equal to the sum of the effective stresses (S_e) and the pore pressure (P_w).

$$S_g = S_e + P_w$$

or

$$S_e = S_g - P_w.$$

At a depth H the effective stress is: $S_e = H(\rho - \rho_w)$, where ρ is the average density of the rock column (overburden) and ρ_w the average weight of the porewater column. In the case of overpressure, the pore pressure equals the weight of the water column up to the piezometric surface.

We see here that the effective stresses at the grain contact are equal to the weight of the overlying rocks minus the pore pressure. When the pore pressure is greater than the normal hydrostatic pressure, the effective stresses are considerably reduced, and if the pore pressure equals the geostatic pressure, there are no effective stresses.

If the pore pressure exceeds the geostatic pressure, the rock will expand and fractures will form. We call this *hydrofracturing*. It may occur naturally, or be induced artificially by pumping water under high pressure into an oil reservoir. The exact pore pressure at which fracturing occurs depends on the maximum and minimum stresses in the rock and on the tensile strength of the rock.

Fig. 12.4. a Diagenetic quartz overgrowth on clastic quartz grains. To the left partly leached clastic felspar. From the Brent Group, North Sea. **b** Diagenetic quartz overgrowth in a pore partly filled with an aggregate of illite and smectite

What Determines the Mechanical Stability of Sand Grains?

When we want to assess the mechanical stresses to which sand grains are subjected, we must remember that it is the effective stresses which are of importance, and that these do not increase as a simple function of the depth of the overburden, but depend on the density of the rock and on the pore pressure. At a depth of 2 km the effective stresses with normal hydrostatic pressure and an average rock density ρ_r of 2.2 g/cm² will be:

$$S_e = H(\rho_r - \rho_w) \text{ where } \rho_r = 2.2 \text{ g/m}^3 \text{ and } \rho_w = 1.0 \text{ g/cm}^3$$
$$S_e = 240 \text{ kg/cm}^2.$$

If we have an overpressure corresponding to a piezometric surface at 1600 m (or a corresponding drilling fluid density of 1.8 g/cm³), the effective stresses will be:

$$S_e = H(\rho_r - \rho_w) - x\rho_w,$$

where x is the height at the piezometric surface. For H = 2000 m and x = 1600 m, $S_e = 80$ kg/cm².

We see that the effective stresses are reduced to a third.

The values express average effective stresses per surface unit of the rock, but the stress is in reality distributed over relatively small surfaces where the grains are in contact, particularly in the early phase of diagenesis, because cement which forms between the grains helps to distribute the stress.

Pressure solution means selective solution of minerals at grain contacts where the stress is great. Deformation of the mineral lattice under stress causes the solubility of minerals to increase, and we have solution of parts of the sand grains at contacts. The dissolved ions will cause a concentration of ions in the porewater next to the contact, and there will be transport of ions by diffusion away from this point and precipitation on other parts of the grain, where overgrowths will develop (Fig. 12.4). In this way the contact surface increases, and the pressure per unit of surface area decreases. Pressure solution causes reduced porosity due both to the precipitated overgrowth (mainly quartz) and to denser mechanical packing of the grains due to pressure solution. Pressure solution is very often localised along thin laminae of clay or clastic mica and may give rise to stylolites. Grains between stylolites will show little evidence of pressure solution and only traces of overgrowth. Mica seems to have some sort of catalytic effect on pressure solution, and we often see large sections of quartz grains dissolved against mica, even at shallow burial depths, 1–2 km (Fig. 12.5).

Conclusion: Mechanical compaction and deformation of sandstone is a function of mineral composition, grain strength, depth of overburden, sediment density and pore pressure. Pressure solution at the contact between grains is also a function of the effective stresses in the rock.

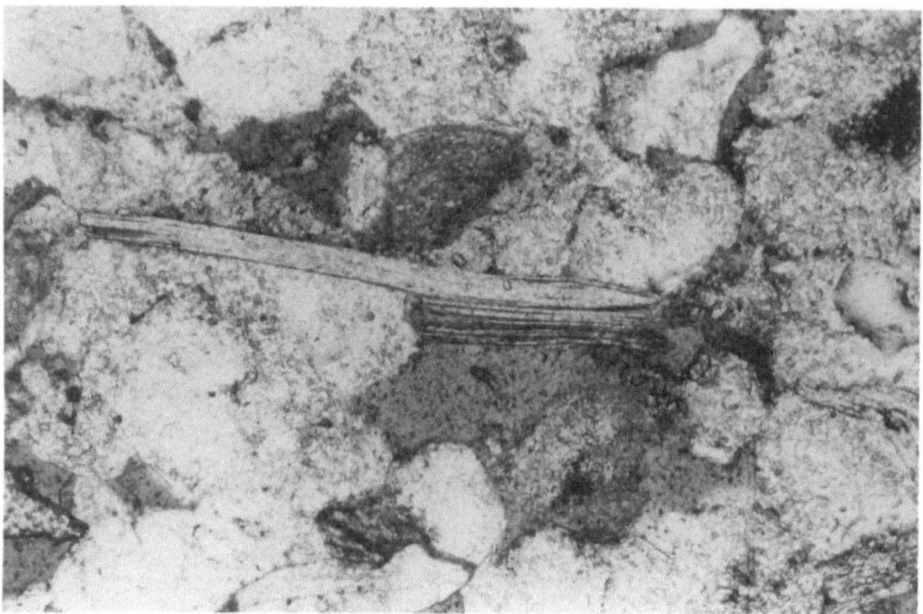

Fig. 12.5. Solution of quartz against mica. Notice that very large sections of quartz grains seem to have dissolved against the surface of the mica. The cement is poikilotopic calcite. From the Draugen field offshore Norway at 167° m burial

Chemical Precipitation and Solution of Minerals in Sandstone

What leads to the precipitation of cement, causing reduction of porosity and permeability? What leads to solution of a sand grain matrix or cement so that the porosity increases? Not only is the quantity of cement significant, but reservoir properties are also to a high degree a function of the various types of mineral growth (Fig. 12.6).

Diagenetic Reactions at Shallow Burial Depths (Early Diagenesis)

Sand is usually loose and uncemented if it has not been very deeply buried. However, we also find examples of thoroughly cemented sand which has not been buried at great depths. When relatively recently deposited sand undergoes rapid cementation, the cementing minerals are most frequently carbonates, iron oxides or amorphous phases, e.g. silica. Formation of cement in sandstone requires that the porewater is supersaturated with respect to a crystalline or amorphous mineral phase, which is then chemically precipitated. Sand which has been exposed for a long time to warm saline water which is supersaturated with carbonates may acquire a carbonate (aragonite) cement. Carbonate cemented sandstone (beach rock) is formed today on tropical shores where the waves wash over the sand. Finds

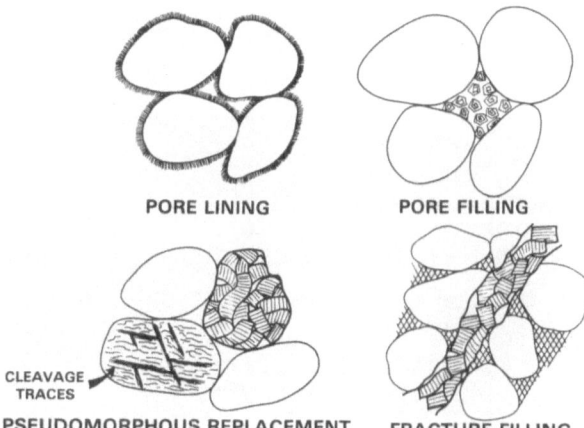

PORE LINING   PORE FILLING

Fig. 12.6. Types of cement formed by authigenic clay minerals in sandstones. Clay mineral grains are normally only 1–10 μ and their size is magnified in this drawing. (Wilson and Pittman 1977)

CLEAVAGE TRACES

PSEUDOMORPHOUS REPLACEMENT   FRACTURE FILLING

of beer cans etc. in these sandstones show that they may be formed over a very short period, geologically speaking, sometimes only 10–20 years.

Iron oxide cement is also common where rainwater in podzol profiles dissolves minerals containing iron, and where iron oxide is precipitated further down in the section where the water has acquired a higher pH after reaction with minerals. This sort of iron precipitation can form a permanent red layer (hardpan). Where water flows from reducing to oxidising environments near the sedimentary surface, ferrous iron will be oxidised to ferric iron, which will be precipitated because it is nearly insoluble.

Rapid formation of cement in the pores of sandstone requires that there be a local source of cement-like minerals undergoing solution or an amorphous phase resulting in almost constant percolation of supersaturated porewater through it. Each unit volume of water can carry only a very small quantity of iron or carbonate in solution, and 10^4–10^5 unit volumes of water are therefore normally required to precipitate one unit volume of cement. Circulation of porewater in the sediment basin therefore plays an important role in early reactions (Fig. 12.7).

Early Diagenetic Environment (Fig.12.8)

In the vadose zone above the water table the pores in coarse-grained sediments are filled with air. In areas with higher precipitation than evaporation there will be net downward transport of rainwater (meteoric water). This will dissolve the minerals nearest the surface, where the pH is lowest, and precipitate minerals or oxides, e.g. iron or quartz, as the pH rises further down the section. In drier areas where there is recharging of groundwater, e.g. from mountains, water will move up from the groundwater and evaporate. Ions in solution in the groundwater will then precipitate, particularly as carbonates (calcite) and silicates. In the vadose zone cement will only grow in water. The water accumulates in drops on the underside of sand grains, and *pendant* cement forms there. Below the water table there will be

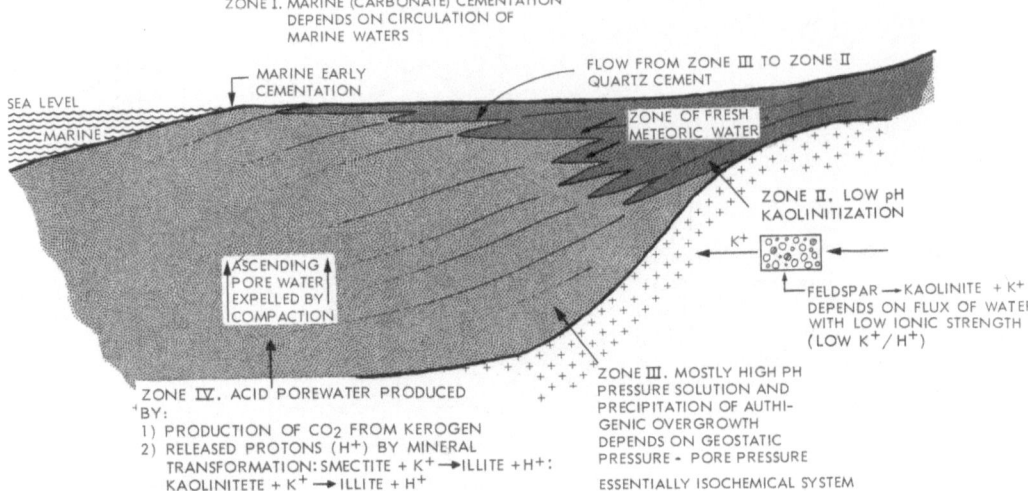

Fig. 12.7. Simplified model of porewater circulation in a sedimentary basin, and the diagenetic reactions which are typical of the various parts of the basin. (Bjørlykke 1983)

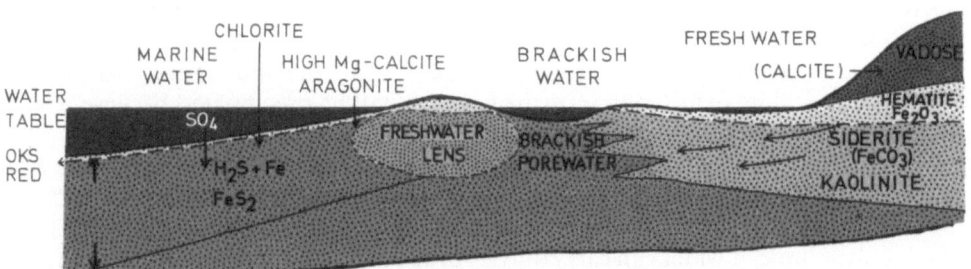

Fig. 12.8. Schematic overview of early diagenetic environment. In the sulphate-reducing zone all iron will precipitate as sulphides. Siderite will only form in the fresh or brackish water environment or below the sulphate-reducing zone

freshwater flowing through the sediments (Fig. 12.9). Fresh groundwater near the water table is normally characterised by a low content of dissolved salts and a high Eh. However, oxygen is gradually consumed by reactions with minerals and organic matter and the groundwater will become more reducing with time. The Eh may drop rapidly over short transport distances in some aquifers. In drier areas, and where the groundwater has dissolved evaporite minerals, meteoric groundwater may have a high salt content. In desert deposits with little organic matter, groundwater may remain oxidising for longer periods.

Diagenesis in ordinary, low-salt groundwater is in reality a continuation of weathering processes. Silicate minerals such as felspar, mica, pyroxene, amphibole etc. will be unstable and break down, and new minerals will be formed. Of particular importance is the formation of kaolinite at the expense of felspar and mica (see Figs. 12.10 and 12.11).

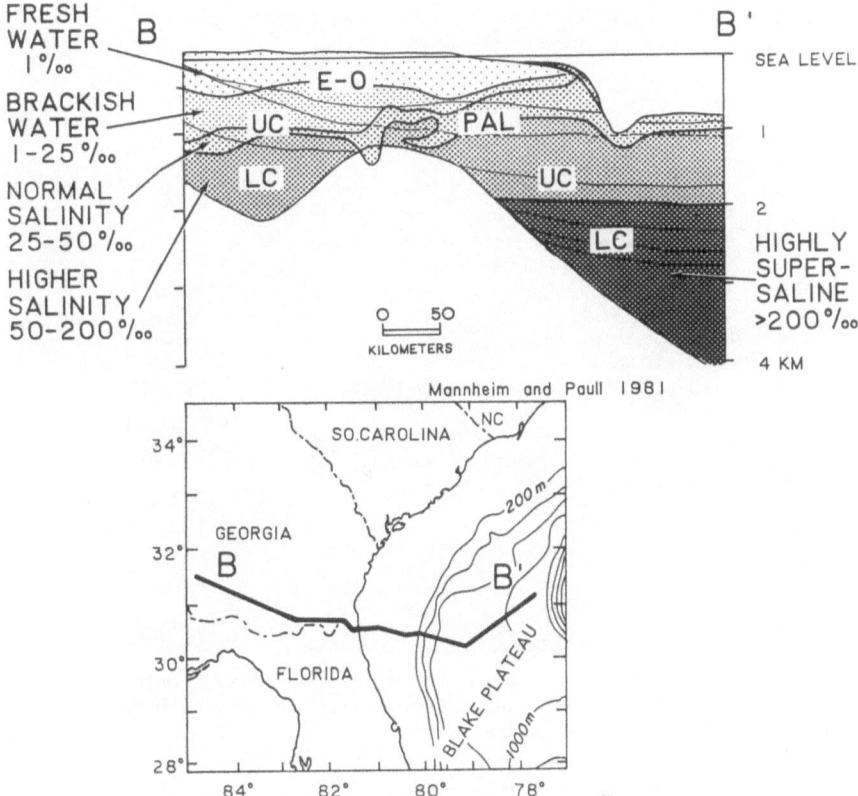

Fig. 12.9. Porewater composition in a section from Georgia to Blake Plateau off the east coast of the USA. Below the continental shelf there is fresh porewater right out to almost 100 km from the coast. At great depths the porewater is characterised by underlying Mesozoic evaporites (After Mannheim and Paull 1981). This section shows that porewater is stratified and that there is little vertical flow of porewater. A salinity gradient is probably established by diffusion from the supersaline porewater derived from Mesozoic evaporites

$$2KAlSi_3O_8 + 2H^+ + 9H_2O \rightarrow Al_2Si_2O_5(OH)_4 + 4H_4SiO_4 + 2K^+$$
$$\text{felspar} \qquad\qquad\qquad\qquad \text{kaolinite}$$
$$2KAl_3Si_3O_{10}(OH)_2 + 2H^+ + 3H_2O \rightarrow 3Al_2Si_2O_5(OH)_4 + 2K^+.$$
$$\text{mica} \qquad\qquad\qquad\qquad \text{kaolinite}$$

We see that protons (H^+) are consumed and alkali ions $(K^+, Na^+, Ca^{2+}, Mg^{2+})$ released, so that the pH increases.

Early diagenesis in the meteoric water zone is a kind of subsurface weathering, and the basic metal ions released must be constantly removed for the process to proceed. The reactions take place at low temperatures and consequently slowly, and the porewater is therefore not necessarily in equilibrium with the minerals. Groundwater is usually already supersaturated with respect to quartz (10–20 ppm SiO_2) and the silica released from felspar leaching may not precipitate directly as

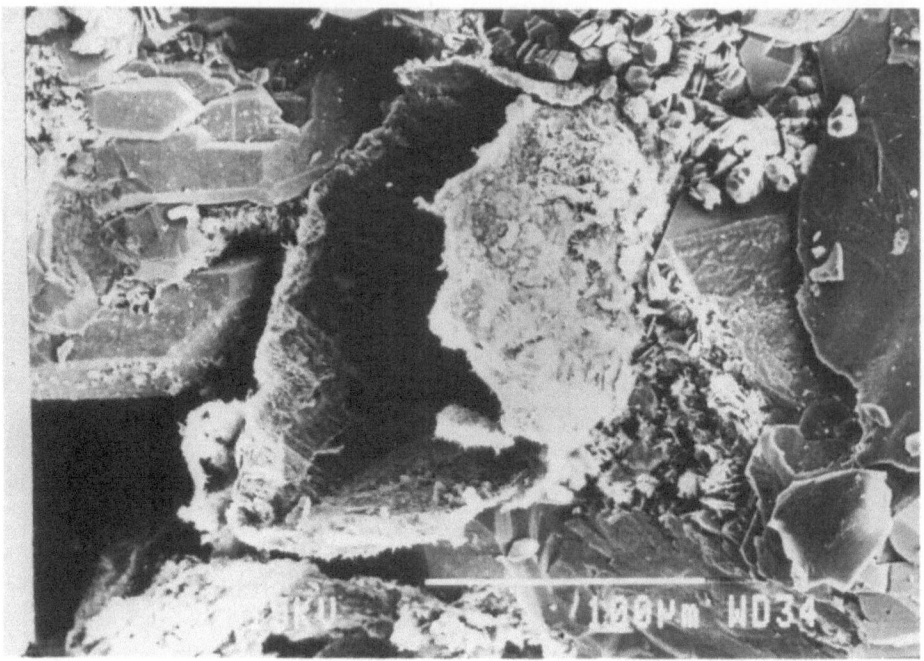

Fig. 12.10. Hole after dissolved feldspar. A clay coating on the feldspar is preserved and the surrounding porespace is partly filled with authigenic kaolinite. Brent Group (Jurassic), Huldra field, North Sea, 3750 m depth. By courtesy of Tor Nedkvitne

quartz for kinetic reasons. However, at higher temperatures ($> 80°C$) quartz precipitates more rapidly and the porewater tends to be in equilibrium with quartz. If the sediment contains amorphous silica (biogenic or volcanic) the porewater will be in equilibrium with this type of silica and have a higher silica content until the amorphous silica has dissolved.

Solution of felspars and other silicate minerals produces more pore space, so-called secondary porosity. However, the dissolved aluminium and silica will usually precipitate as kaolinite and quartz and occupy a similar volume so that there is little or no net gain in porosity (Fig. 12.10). In some cases kaolinite precipitates in the pore spaces of the dissolved felspar, but we also find examples where we see felspar solution but little kaolinite. We must then assume that silica and aluminium have been transported for some distance before precipitating. At low temperatures porewater may be oversaturated or undersaturated with respect to several minerals because the reaction rate may be very slow for kinetic reasons. It is very important to establish the degree of mobility of ions like Si^{4+} and Al^{3+} in porewater. It has been claimed that Al^{3+} and Si^{4+} can be complexed by organic acids in porewater and thus become more soluble. The fact that porewater analyses tend to show very low concentrations of Al suggests that it is not very soluble in natural formation water and therefore not very mobile.

Just as minerals will gradually neutralize meteoric water so that it approaches equilibrium with the minerals in sandstones, porewater will oxidise organic materials and a number of minerals. In this way its oxygen content decreases and it will

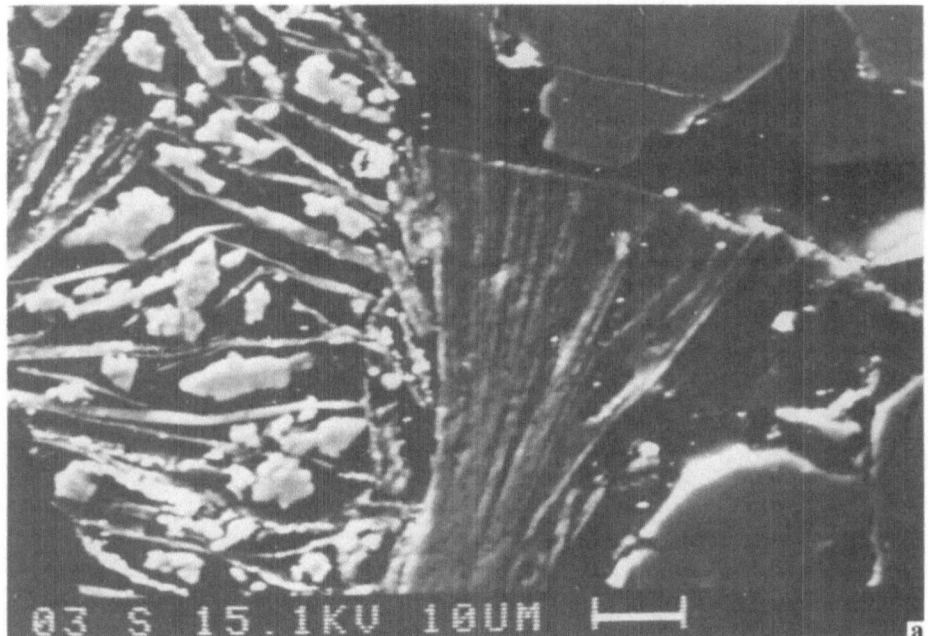

Fig. 12.11. a Electron backscatter picture of the Brent Sandstone in the Statfjord Field. In the centre of the picture is mica (muscovite) which is partially altered into kaolinite and is expanding due to increased water content. To the *left* altered biotite with authigenic siderite formed from the iron in the biotite. Iron-bearing minerals give intense backscatter because of the high atomic number. **b** SEM picture showing how authigenic kaolinite has grown between the sheets of dissolving mica. This results in an expansion of the clastic mica. From the Brent Group (Jurassic), Huldra Field, North Sea. 3700 m depth. (Courtesy of Tor Nedkvitne)

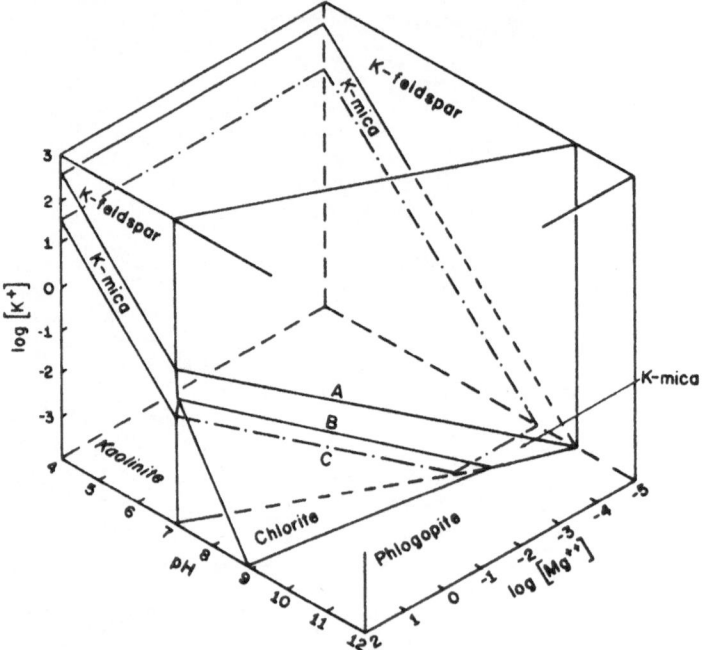

Fig. 12.12. Stability diagram for a number of layer silicates as a function of the concentration of magnesium, potassium and the pH. (Garrels and Christ 1965)

finally become reducing. At this point elements which are less soluble in the reduced state will be deposited. This is particularly true of uranium, which is released through the weathering of uranium-bearing rocks:

$$4e + 2UO_2^+ + 2H^+ \rightarrow U_2O_3 + H_2O$$
in solution   precipitated as uranium oxide.

Precipitation tends to take place at contacts between porous (channel) sand and clay or silt with a high organic content (roll-over fronts). Major deposits of sedimentary uranium ores are formed in this manner (e.g. in Texas and Colorado, USA) (Fig. 12.13).

Vanadium may be enriched in a similar manner when water flows from an oxidising to a reducing environment, since the vanadyl ion is soluble, while reduced vanadium oxyde, V_2O_3, is not very soluble.

$$2e + 2VO^{2+} + H_2O \rightarrow V_2O_3 + 2H^+ \text{ or}$$
$$2e + 2VO_2 + 2H^+ \rightarrow V_2O_3 + H_2O.$$

In shallow marine environments we may find early formation of carbonate cements, aragonite, high-Mg-calcite (beach rock) as mentioned above. Chlorite and chamosite are relatively stable in marine environments and may be formed early through diagenesis, often as a layer around sand grains (Fig. 12.14).

Marine sediments have reducing porewater only a few centimetres below the sea bed. This is due to the organic matter in the sediments which uses oxygen as it

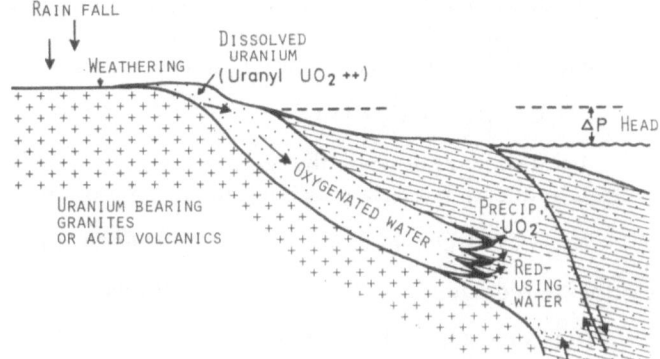

Fig. 12.13. Formation of sedimentary uranium ores (after Galloway et al. 1979). Uranium is most soluble in the oxidised state, and is precipitated as uranium oxide or hydroxide when the groundwater becomes reducing

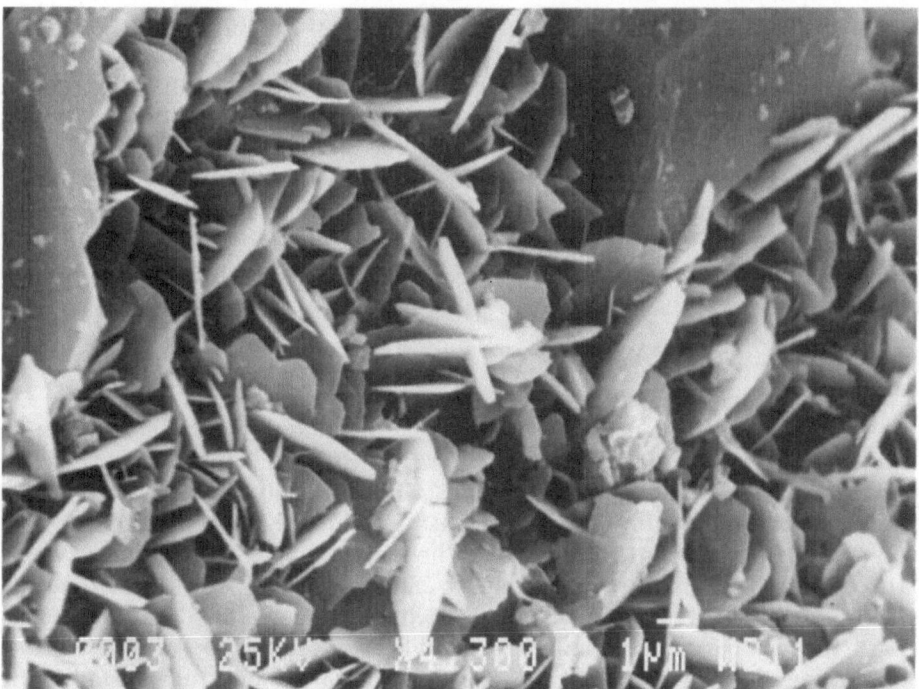

Fig. 12.14. SEM picture of grain-coating authigenic chlorite. In the upper parts of the picture the layer of authigenic chlorite is penetrated by authigenic quartz. (From The Brent Group, Haltenbanken, offshore Norway)

breaks down. Only in the uppermost layer will there be sufficient circulation of water from the overlying water to bring in new oxygen-rich water as the oxygen becomes used up through bacterial (aerobic) breakdown of organic material:

$$CH_2O + O_2 \rightarrow CO_2 + H_2O.$$

A little deeper down in the sediment, where there is little free oxygen, we find nitrate-reducing bacteria which use the oxygen bound up in nitrate ions:

$$2H^+ + 5CH_2O + 4NO_3^- \rightarrow 2N_2 + 2HCO_3^- + 3CO_2 + 5H_2O.$$

Sulphate-reducing bacteria function in a similar way:

$$2CH_2O + SO_4^{2-} \rightarrow H_2S + 2HCO_3^-.$$

The bicarbonate ions produced may cause precipitation of carbonates if suitable cations (Ca^{2+}, Mg^{2+}, Fe^{2+}) are available ($Ca^{2+} + HCO_3^- \rightarrow CaCO_3 + H^+$).

H_2S is a weak acid which may dissolve iron-bearing minerals, e.g. biotite, chlorite, hornblende, pyroxene and iron oxide, and forms iron sulphide. First a black monosulphide is formed (FeS — mackinawite) and then the disulphide (e.g. pyrite, FeS_2). Monosulphides are readily oxidised to sulphates again, while pyrite is more stable, but will also be oxidised during weathering. Iron-sulphide bonding is so strong that no other iron-bearing minerals form, e.g. iron carbonate, when there are sulphide ions present in porewater. Iron carbonate (siderite) therefore tends to form in freshwater where there is little sulphate to be reduced to sulphide, or in marine sediments outside the sulphate-reducing zone (more than about 10 m below the sea floor). Sulphide cement is common in marine sediments, particularly in organically rich sediments which provide nutrients for the sulphate-reducing bacteria. If we have a supply of other metals, e.g. Pb^{2+}, Zn^{2+}, Cu^{2+}, these will also form sulphides.

Burial Diagenesis

When sediments are buried by continued sedimentation in a sedimentary basin, increasing compaction and reduced porosity cause the water to be expelled upwards through the subsiding sediments. The average upward flow of porewater is of the same magnitude as the rate of basin subsidence, however, and the overall result is that there is little or no upward flow in relation to a fixed depth. We can regard the sediments as sinking through a column of water. Nevertheless, we often find more concentrated flow of water expelled by compaction out of a sediment basin through particularly permeable sand beds or faults. Flow of porewater in a sedimentary basin is very slow below the uppermost few metres and outside the area of meteoric (fresh groundwater) flow. The average flow of porewater upwards will be only a little greater than the average rate of sedimentation in the basin, i.e. from 1–0.001 mm/year. The porewater will then have ample time to reach equilibrium with the minerals in the sediments, particularly at high temperatures.

At grain contacts, however, the solubility of most minerals will be greater than it is on those parts of the surface which are adjacent to pores.

Pressure solution is common with quartz and felspar grains, and with rock fragments like carbonate grains. Cement forms a protective film on the grains, and the cement in turn may undergo pressure solution. When we observe grain contacts in thin sections under a microscope, we must remember that pressure solution may occur, even if the grains do not touch each other in the plane of the thin section. Pressure solution and precipitation of cement cause the geostatic pressure to be distributed over a larger surface, so that the pressure per unit area of surface decreases. Stylolites are continuous solution surfaces with an irregular surface, and are found not only in carbonate rocks, but also in sandstone. While stylolites in carbonate rocks start forming at depths of 1–2 km, sandstones normally require an overburden of about 4 km or more.

In felspathic sandstone we often find that clastic felspar grains have a thin, jagged overgrowth of authigenic felspar which has a slightly different mineralogical composition from that of the grain. Clastic potassium felspars from metamorphic or eruptive rocks always contain some Na^+, while the newly formed potassium felspar is a pure, low-temperature variety with a very low sodium content. In the same way, plagioclase will have an overgrowth of pure albite. The source of authigenic felspar may be pressure solution of felspar or precipitation from supersaturated porewater from other sources. There will always be a tendency for clastic minerals which are unstable at low pressures and temperatures to go into solution and precipitate pure, low-temperature albite or potassium felspar (adularia).

We often see evidence of plagioclase dissolving and albite precipitating instead, releasing Ca^{2+} (albitisation). Potassium felspar is often dissolved and replaced by albite.

The minerals may form finely divided aggregates which reduce the permeability considerably. In Triassic and Permmian sandstones, in the North Sea in particular, we often find development of threadlike aggregates of illite or montmorillonite which may stretch across pores, considerably reducing permeability, but without very greatly altering the porosity (Fig. 12.4). Aggregates of kaolinite, which are often formed in the middle of pores, (pore filler, Fig. 12.15), have the same effect. Kaolinite is the most common authigenic clay mineral in Jurassic reservoir rocks from the North Sea.

The solubility of carbonates decreases with increasing depth, as long as there are no chemical reactions in the sediments which produce acids. Increasing temperature reduces the hydration of the ions, and at depths of 2–3 km (60–90°) iron and magnesium carbonates form far more readily, because Fe^{2+} and Mg^{2+} are less surrounded by water molecules with their negative dipole towards the positive ions. Thus they can enter more easily into the carbonate structure. Siderite may form relatively early, but dolomite and ankerite are often late-diagenetic minerals in sandstones, e.g. the Brent Sandstone from the Statfjord Field (see Fig. 12.11).

Sandstone Which Contains Volcanic Material

The content of volcanic material is largely a function of the plate tectonic location of the basin. Sandstone deposited in the neighbourhood of island arcs will contain volcanic and basic eruptive material within the island arc (inter-arc basins), and in front of the island arc (fore-arc basins). Back-arc basins will be supplied largely with sediment fron the continent. Continental rift basins will also contain volcanic

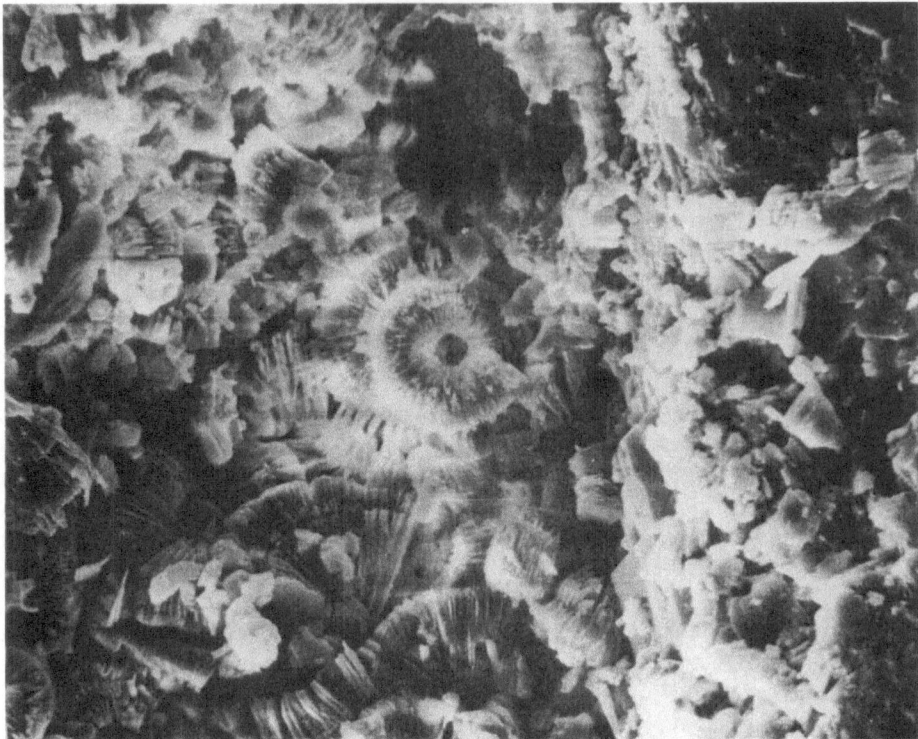

Fig. 12.15. Diagenetic pore-filling kaolinite in the Brent Group, Statfjord field, North Sea. The small pores between the kaolinite crystals may remain water-saturated also after oil emplacement due to capillary forces. Fig. = 0.1mm across

material, but these will normally make up a much smaller percent of the overall sediment mass.

A high content of volcanic fragments will reduce the mechanical strength of sandstones. A sandstone which is initially relatively well sorted may develop a high degree of "pseudo-matrix", through the grains being compressed and transformed into a matrix. Grains of basalt or other basic eruptive rocks will suffer mechanical deformation, and pyroxene and amphibole may alter into chlorite. It may be difficult to distinguish between chlorite which has formed in this manner and a primary clay matrix. Grains from basic and volcanic rocks are also less stable chemically, and will go into solution more easily. This means that porewater will have a greater content of silica and of cations which are released by these unstable silicate minerals (Na^+ K^+ Ca^{2+} Mg^{2+} Fe^{2+}). If amorphous silica is present in the form of volcanic glass or biogenic silica (diatoms, radiolaria), the porewater will acquire an even greater silica content (about 150 ppm at 20°C). This presupposes that the porewater circulates slowly enough to be in equilibrium with the amorphous phases. Then the composition of porewater will fall into the stability area for zeolites and smectite, which are authigenic minerals typical of volcano-clastic sandstones.

Heulandite is a zeolite typical of volcanic sediments, particularly if they also contain some carbonate so that the Ca^{2+} concentration is high. Montmorillonite (smectite) is stable in the stability fields between zeolites and kaolinite, and between felspar and kaolinite (Fig. 7.5).

Zeolites will dehydrate and then be altered to other minerals at higher temperatures or if the composition of the porewater changes. With lower concentrations of silica in solution, the following reaction may occur:

$$Ca_3K_2Al_8Si_{28}O_{72} \cdot 23H_2O \rightarrow 3CaAl_2Si_4O_{12} \cdot 4H_2O + 2KAlSi_3O_8 + 10SiO_2 + 11H_2O.$$
Heulandite   Laumontite   Felspar

At higher Na^{2+} concentrations:

$$8Na^+ + Ca_3K_2Al_8SiO_{28}O_{72} \cdot 23H_2O \rightarrow 8NaAlSi_3O_8 + 4SiO_2 + 23H_2O2K^+ + 3Ca^{2+}.$$
Heulandite   Albite

Volcanogenic sandstones are often characterised by rapid burial and the mechanically and chemically unstable grains will cause the porosity to decrease rapidly with increasing overburden. They are therefore not good reservoir rocks, but are often associated with high-temperature gradients and early migration of hydrocarbons before the reservoir rock has lost all its porosity.

Secondary Porosity in Sandstones

Secondary porosity is porosity which did not exist when the sediment was deposited. It is due to chemical solution of sediment grains or cement. Other forms of secondary porosity are biological solution (boring) and tectonic fractures.

Chemical solution of minerals may in many cases result in a considerable increase in the porosity of sandstones. In thin sections we often see partially dissolved felspar grains or large holes which cannot be primary, but are due to minerals which have dissolved. This is also often true of carbonate grains or cement. However, it is important to remember that what we are observing as solution porosity does not necessarily mean a net increase in porosity. When we see that felspar has dissolved out, some of the authigenic kaolinite and quartz present may be a direct result of its breakdown.

In the case of carbonate, too, dissolved carbonate cement or grains may be partially or entirely precipitated again somewhere else in the sandstone as pure carbonate cement. Nevertheless, we often have a secondary porosity which results in a net increase in porosity. But this requires percolation of great quantities of water which is undersaturated in the dissolving mineral. Meteoric (fresh) water will in most cases be undersaturated in carbonate or felspar. Solution of these minerals may often be due to groundwater percolating through sediments deposited in fluvial deltas, and in some cases also in a shallow marine environment.

At greater depths in sediments, the porewater will soon reach equilibrium with the most important mineral phases. In order for dissolution to occur here, apart from solution due to pressure, the porewater must become undersaturated through chemical reactions in the sediments. Organic material (kerogen) produces CO_2 gas when hydrocarbons are formed, and the kerogen becomes more and more rich in carbon. The quantity of CO_2 produced is determined by the oxygen in the organic compounds which kerogen consists of. The oxygen content is highest in humic kerogen, which produces about 10% CO_2, while sapropelic kerogen only produces 1% CO_2. Every gram of kerogen can therefore produce from 10–100 mg CO_2 which in turn can dissolve at the most 20–200 mg of carbonate (one mole CO_2 dissolves 1 mole $CaCO_3$). Chemical modelling suggests that an equilibrium will be established with dissolution of a considerable amount of carbonate. It is clear that even if we have a thick shale with about 1% kerogen, there are limits to how much carbonate or felspar can be dissolved by the CO_2 formed. In addition, the CO_2 formed, finely distributed in shales, will first dissolve any carbonate or felspar present in the shale before it dissolves these same minerals in the overlying sandstones. Organic acids released from source rocks will probably also be neutralised before they reach the reservoir rock.

Another source of acid is clay minerals. Through weathering and early diagenesis clay minerals become "charged" with protons (H^+), while the minerals which are weathered produce basic ions (K^+, Na^+, Mg^{2+}, Fe^{2+}). When the temperature increases, minerals which are rich in water (as H_2O or OH) will become unstable and the reactions will be reversed, so that cations are absorbed and protons produced.

$$3Al_2Si_2O_5(OH)_4 + 2K^+ \xrightarrow{120-150^\circ} 2KAl_3Si_3O_{10}(OH)_2 + 2H^+ + 3H_2O.$$
$$\text{Kaolinite} \qquad\qquad\qquad\qquad\qquad \text{Illite}$$

To neutralise the H^+ ions produced and add alkali ions (K^+), potassium felspar, for example, will go into solution

$$3KAlSi_3O_8 + 2H^+ + 12H_2O \rightarrow KAl_3Si_3O_{10}(OH)_2 + 6H_4SiO_4 + 2K^+,$$
$$\text{Felspar} \qquad\qquad\qquad\qquad \text{Illite}$$

so that the net reaction on adding the two equations becomes:

$$Al_2Si_2O_5(OH)_5 + KAlSi_3O_8 \rightarrow KAl_3Si_3O_{10}(OH)_2 + 2SiO_2 + H_2O.$$
$$\text{Kaolinite} \qquad \text{Felspar} \qquad \text{Illite} \qquad\quad \text{Quartz}$$

A survey of the mineralogy of borehole cores from the Gulf Coast (Hower et al. 1976) shows that the felspar content decreases markedly at approximately the same depth as the kaolinite content. Similarly, montmorillonite and mixed-layer minerals will take up K^+ and produce silica with increasing temperature.

Hower's investigations also show that carbonate disappears at the same depth, and this may be due to the formation of CO_2 from kerogen so that carbonates are dissolved, or to kaolinite reacting with calcium carbonate to form calcium illite.

Conclusion: Analyses of porewater and oilfield water in sedimentary basins have failed to reveal high concentrations of Al^{3+} and Si^{4+} in solution. This also suggests that it is difficult to transport silicates in solution over long distances.

At high temperatures (> 80°C) porewater will tend to be approaching equilibrium with the minerals present. The composition of porewater is therefore a function of the minerals in the sandstones, and porewater can only precipitate minerals without dissolving other minerals to a very limited extent.

Convection Currents

Several workers, including Wood and Hewett (1984), have suggested that convection currents play an important role in transporting solids in solution in sedimentation basins.

In homogeneous porous sediments or greatly jointed sedimentary rocks, one can imagine that convection currents might develop as a heat-transport mechanism. Convection currents are set in motion by very small differences in porewater densities resulting from the expansion of water as a function of temperature, but increased salinity as a function of deeper burial will increase the density and often compensate for the temperature effect. The quantity of soluble salts also plays a part. The solubility of most crystalline minerals and amorphous compounds (e.g. silicates) increases with temperature, whilst carbonates are less soluble with increased temperature. The small differences in density which we find in porewater in a sediment basin are probably not capable of driving convection currents, at least not in fine-grained sediments, or in sedimentary rocks with low porosity. It is possible that this process may have been of significance in very permeable rocks, and perhaps particularly in rocks with well developed joint porosity and in areas with high heat flows (high geothermal gradients).

The significance of convection currents in a sedimentary basin is still uncertain. Large-scale convection currents would make it possible to transport ions in solution so that we would obtain solution and precipitation with the same porewater. The quantity of porewater in a sedimentary basin would then be no limitation on the supply of cement or on leaching and development of secondary porosity (Fig. 12.14). Shale beds and cemented intervals in sandstones form low permeability barriers to convection flow and may reduce the height of the convection cell, and thereby also the driving force. Convection is probably only important around volcanic intrusions, where we have relatively steeply sloping isotherms, or where we have very thick sandstones without inter-bedded shales (Bjørlykke et al. 1988).

Relationship Between Diagenesis, Depositional Environment, Stratigraphy and Tectonics

We have seen that the nature of diagenetic transformation of sandstones depends on the original composition of the sediments, the pressure factor due to overburden (effective stresses) and percolation of porewater. All these factors can be related to depositional environment, the stratigraphy and tectonic development of the basin, and later deformation.

The original composition of the sandstones is a function of the source rocks, climate (weathering) and transportation (see p. 136). Arkoses typically develop in basins formed through faulting of the continental (granitic) crust (particularly rift basins), while orthoquartzites are formed in tectonically stable continental crust where chemically unstable sand grains break down slowly over a long period through weathering and transport. Unstable sand grains which form greywackes are found particularly where sediments have eroded, in part at least, from the ocean crust or basic eruptive rocks, i.e. in basins near subduction zones with island arcs.

Fluvial sediments, and also deltaic sandstones, will generally have large quantities of freshwater (groundwater) percolating through them after deposition, depending on precipitation, the slope of the water table and the permeability of the sand.

In this way we may obtain solution of carbonates and felspars and formation of kaolinite. Fluvial channels may provide good transport channels for ground-water after deposition. Delta-front sediments, which lie farthest out on the delta, have poorer contact with percolating groundwater, and any growth faults which form here may offset the permeable sand bodies and prevent percolation of fresh groundwater.

Shallow marine sediments will often have freshwater percolating through them after burial, depending on the slope of the water table or communication in the sandstone, while turbidites deposited in submarine fans seldom have meteoric water percolating through them. This meteoric flux depends greatly on landward continuity to recharge zones. Isotopic analyses for $\delta^{18}O$ in the porewater in sedimentary basins often show a composition in the shallow part which lies between meteoric and marine values.

Prograding offlap sequences will often have continuous sand beds from deeper parts of the basin to the surface. If there is no major faulting, porewater can flow upwards during compaction, and overpressure will not develop. This means that the effective stresses and consequently compaction and pressure solution, will be great in relation to depth.

Onlap sequences more easily form sand beds which are sealed by overlying shales. Overpressure may then develop, as is the case around the Viking Graben in the North Sea.

Tectonic uplift and exposure above sea level will result in renewed infiltration of meteoric water. We therefore often find increased leaching immediately below unconformities.

It is of great practical importance to be able to predict what kinds of diagenetic processes will be most important in different parts of sedimentary basins. These processes determine the final properties of reservoir rocks for petroleum — not only porosity and permeability but also pore-size distribution and pore geometry. These parameters are very important for planning efficient oil recovery.

Further reading: Parker and Sellwood 1983; McDonald and Surdam 1984; Gautier 1986.

13 Petroleum Geology

How Petroleum is Formed in Source Rocks, Migrates, and Accumulates in Reservoir Rocks

Introduction

Petroleum geology contains those disciplines which are of greatest significance for the finding and recovery of oil and gas. Since practically all petroleum occurs in sedimentary rocks, *sedimentology* forms an important part of the basis for petroleum geology. *Structural geology* is also important for understanding the mode and timing of oil traps and the general development of sedimentary basins. Oil and gas are derived from organic matter buried in a sedimentary basin. *Organic geochemistry*, which includes the study of organic matter in sediments and its transformation into hydrocarbons, has become an important part of petroleum geology. The possibilities of source rocks with a high organic matter content and reservoir rocks with high porosities being formed can to a large extent be predicted by means of sedimentological and palaeooceanographic models. For biostratigraphic correlation of strata encountered in exploration wells we use *micropaleontology* and including, *palynology*, fields which have been developed very largely by the oil industry. Due to the small size of the samples obtained during drilling operations, one cannot rely on macrofossils. A rock sample weighing a few grams, on the other hand, may contain several hundred microfossils or palynomorphs which also usually give better stratigraphic resolution than macrofossils.

In the infancy of the oil industry, more than 100 years ago, oil exploration consisted largely of looking for oil seepage at the surface and drilling in the vicinity.

Today we explore for structures which are seldom visible on the surface. To define subsurface structures, we use (1) geophysical measurements at the surface, and (2) correlation by means of logs and fossils from wells. Geophysical measurements may include gravimetry and magnetometry, but seismic measurements are definitely the most important.

Because of the rapid improvement in the quality of seismic data processing techniques, geological interpretation of seismic data has become an entirely new and expanding field. Seismic sections and other geophysical data are often the only information we have, particularly for exploration below the ocean floor, where drilling is very costly.

Geophysical well-logging methods have developed equally rapidly. Logs provide a continuity of information through long series of beds which one can seldom obtain from exposures or core samples. This information makes it possible to interpret not only the lithological composition of the rocks and the variation of porosity and permeability, but also the depositional environment.

It would be true to say that indirect methods of mapping rock types employing geophysical aids are becoming increasingly important in petroleum geology, but it is still necessary to take samples and examine the rocks themselves. A petroleum geologist should also have experience of mapping in the field and of sedimentology and petrography in order to be able to interpret indirect data like seismic records or well logs.

History

Bitumen produced from naturally occurring crude oil which seeped out of the earth has been used since ancient times, in medicine, for lighting, and by the Greeks even for warfare. Before 1859 oil was also recovered from coal for use in kerosene lamps. The first large scale production of oil took place in 1859 at Oil Creek near Titusville in Pennsylvania. Here Edwin Drake showed for the first time that it was possible to find oil in large quantities by drilling holes in the ground, although his wells weren't especially deep — only 20-25 m.

Drake's wells only produced 8-10 barrels per day. But by the end of 1860 there were already 74 production wells around Oil Creek, and the USA's oil production rose to 509,000 barrels per year. Production outside the USA only totalled 5000 barrels per year. In 1870 production was 10 times higher, with 5,251,000 barrels from the USA, and 538,000 barrels from other countries.

Oil production developed rapidly up to the end of the 19th century, and more systematic geological principles for prospecting were gradually developed. The principle that petroleum is less dense than water, and must therefore occur in sealed domes or anticlines, led to extensive geological mapping of exposed structures, particularly in the USA. It was also found that oil fields had a tendency to lie along structural trends, and this "rule" was used in prospecting. This is particularly often the case with salt domes, which became important prospecting targets. The geological information which one obtains at the surface, however, is often not representative of the structures deeper down. Methods had to be developed for recognising structures in about the upper 5 km of the earth's crust. One method was to measure the depth of particularly characteristic strata in various wells through analysis of cuttings. But the methods involving electrical measurements (logs) from wells, developed during the 1920s, made the whole effort much simpler because they provided continuous sections through the rocks, and the development of micropalaeontology made it much easier to determine the age of individual strata.

Geophysical methods made it possible to map structures without drilling. This could be done using gravity measurements, particularly when exploring structures like salt domes. However seismic methods, which can be used both for mapping structures and giving information about the sound velocity distribution of sub-surface strata, became the main prospecting aid.

In the 1960s and 1970s an increasing amount of prospecting took place offshore, and the high costs associated with drilling in such areas led to rapid development of better exploration methods to reduce the number of exploration wells. Improved reflection seismic methods and advanced well log technology in particular made it possible to glean maximum and optimal information from each well.

Rising oil prices made exploration feasible in areas which had previously not been very attractive economically, and there was also extensive research on methods for recovering a higher percentage of oil from reservoirs. However, the lower oil prices experienced recently have drastically reduced exploration activities worldwide. This has reduced the reserves which are commercial to produce, and this in turn will contribute to oil prices rising again.

Formation of Petroleum — Production of Organic Matter

It has long been known that hydrocarbons are formed from organic matter. In the past it was also assumed that hydrocarbons were formed in the crust of the earth through inorganic reactions. Today it is well documented that the bulk of the oil in all major accumulations is of organic origin and related to organic matter in sediments. Methane, however, can be formed inorganically, and is found in the atmosphere of several other planets. Inorganic methane from the interior of the earth is likely to be well dispersed, however, and will not form major gas accumulations.

Since petroleum is derived from organic matter, it is important to understand how and where sediments with a high organic matter content are deposited.

The total production of organic material in the world's oceans is now 5.10^{10} tonnes per year. Nutrients from erosion of rocks on land are carried out into the ocean. The supply of nutrients is therefore greatest in lakes and coastal areas, particularly near the deltas of rivers discharging sediment into the ocean. A good deal of plant matter is also added to the ocean from the land.

Biological production is greatest in the uppermost 30-50 m of the ocean, and all phytoplankton growth takes place in the zone above the depth where the sunlight becomes too weak for photosynthesis (about 100-150 m). Phytoplankton provides nutrition for all other marine life in the oceans. Zooplankton feed on phytoplankton and therefore proliferate only where there is vigorous phytoplankton production. Organisms which have died sink, and may decay so that nutrients are released and recycled at greater depths. Diatoms which have a silica skeleton are important producers of organic matter in the ocean.

In polar regions in particular, cold, dense water sinks to great depths and flows towards lower latitudes. In areas with prevailing land winds, e.g. on the west side of the continents, there is a strong upwelling of nutrient-rich water from the bottom of the sea which provides a basis for especially high primary organic production. The best example of this is the west coast of South America.

The energy in petroleum which we can later put to use is stored solar energy. Through photosynthesis low energy carbon dioxide and water are transformed into high energy carbohydrates (e.g. glucose, $CO_2 + H_2O \rightarrow CH_2O + O_2$).

This energy can be used directly by organisms for respiration, which is the opposite process, breaking carbohydrates down to carbon dioxide and water again. Oxidation of 100 g glucose releases 375 kcal. of energy. Some of the energy which plants accumulate through photosynthesis is lost through respiration. Any carbohydrates produced which are not used in combustion can be stored as glucose or cellulose in the cell walls. Photosynthesis is also the source of biochemicals for the synthesis of lipids and proteins. Nitrogen and phosphorus and many trace metals are essential for the formation of organic matter (protoplasm) in living organisms.

Sunlight →	Carbohydrates →	Organic → components	Living material
CO_2 ↗	Cellulose		
H_2O	Glucose	Proteins	Protoplasm
Nutrients	Starch	Lipids	
P, N and trace metals from water		*Zooplankton*	

Proteins are large, complex molecules built up of condensed amino acids (e.g. glycine, $(H_2NH_2 \cdot COOH)$).

Planktonic algae are the most important contributors to the organic matter which gives rise to petroleum. Among the most important are diatoms, which have silica skeletons. The organic part of diatoms consists of about 0–31% carbohydrates, 24–48% protein and 2–10% lipids (Tissot and Welte 1984). Dinoflagellates have similar compositions.

Zooplankton

The most important of the zooplankton which provide lipid-rich organic matter from which petroleum is derived are:

1. Radiolaria, silica shell, wide distribution, particularly in tropical waters.
2. Foraminifera (globigerina) which have shells of calcium carbonate.
3. Pteropods, pelagic gastropods (snails) which have a foot which has been converted into wing-shaped lobes. Carbonate shell.

In the marine food chain these zooplankton are eaten by crustaceans, which in turn are eaten by fish. We call each link in the nutritional chain a trophic level. Each step along the chain reduces the biomass by a factor of 10.

1. Trophic zone 1000 kg Phytoplankton	2. Trophic zone 100 kg Zooplankton	3. Trophic zone 10 kg Crustaceans	4. Trophic zone 1 kg Fish

In coastal swamps, and particularly on deltas, we have extensive production of organic matter in the form of plants and trees. The residues of these higher land plants may form peat, which with deeper burial may be converted into lignite and bituminous coal. But such deposits are also a potential source of gas and oil. Plant matter, including wood, also floats in rivers and is deposited when it sinks to the bottom, usually in a near-shore deltaic environment.

River water supplies not only inorganic nutrients, but also contains considerable amounts of organic matter. This is particularly true of humic acids

$(C_{20}H_{10}O_6)$, and similar substances which are formed by the breakdown of plant material and subsequent recombination of the degradation products. Humic acids are weakly soluble in water (about 0.01% in cold water at 5°C) and are transported to the ocean by rivers.

Other plant materials, like waxes and resins, are more chemically resistant to breakdown and are insoluble in water. Such organic particles tend to attach themselves to mineral grains and accompany sediments out into the ocean.

Recycling of Petroleum

Most oil reservoirs which have been formed since the Palaeozoic have been uplifted and eroded, and great quantities of oil have flowed out onto the land or into the sea. One might expect this to provide a source of recycled petroleum in younger sediments, but petroleum breaks down extremely rapidly when subjected to weathering. Clay minerals have a high adsorptive capacity, and on land bacteria will degrade the remainder, and evaporation will remove the lighter components. Fossil asphalt lakes consist of heavy substances which neither evaporate nor can be easily broken down by bacteria. In the ocean the lighter components will dissolve quite rapidly, while the heavier asphalt fraction will sink to the bottom and be degraded and recycled. Recent research has shown that naturally occurring petroleum seepage on the ocean floor is not as toxic to marine organisms as was previously believed.

Breakdown of Organic Matter

By far the greater part of all organic matter which is produced on land and in the oceans is broken down through direct oxidation, or by means of microbiological processes. On a global scale, this amounts to more than 99% of the organic matter produced by living organisms.

If oxygen is present, carbohydrates, for example, will be broken down in the following manner:

$$CH_2O + O_2 \rightarrow CO_2 + H_2O$$

and ammonia:

$$NH_3 + 2O_2 \rightarrow H^+ + NO^-_3 + H_2O.$$

These reactions are much more rapid when they are microbiologically mediated. Breakdown of this nature takes place on the surface of organic material suspended in sea water, and on the sea bed where there is a supply of oxygen. As organic matter sinks through the water column it consumes oxygen and if water circulation is restricted due to density stratification of the water column, the oxygen will be exhausted. We will then get stagnant bottom water conditions where sulphate-reducing and denitrifying bacteria decompose organic material without a supply of free oxygen. Instead, the bound oxygen in sulphates or nitrates is used.

Normally, however, the water in marine basins will be oxygenated, and the porewater of the uppermost 2-20 cm of the sea bed sediments will also be oxidising, while there will be reducing conditions below.

Sulphate reducing bacteria induce reactions whereby sulphides are produced as indicated below:

$$2CH_2O + 2H^+ + SO_4^{2-} \rightarrow H_2S + 2CO_2 + 2H_2O$$
$$NH_3 + 2H^+ + SO_4^{2-} \rightarrow H^+ + NO_3^- + H_2S + H_2O.$$

H_2S is liberated, giving stagnant water and mud a strong smell. Through denitrification we get

$$5CH_2O_4^{2-} + 4H^+ + 4NO_3^- \rightarrow 2N_2 + 5CO_2 + 7H_2O$$
$$5NH_3 + 3H^+ + 3NO_3^- \rightarrow 4N_2 + 9H_2O.$$

In the presence of organic matter, N_2 may be reduced to NH_3 or NH_4^+.

In summary, the carbon cycle starts with formation of organic matter from inorganic constituents, and this is followed by its breakdown into inorganic components after organisms die.

Petroleum is derived from organic matter which has ultimately been formed through photosynthesis, i.e. storage of solar energy (Fig. 13.1). Such energy is constantly accumulating in the Earth's crust, but the percentage of solar energy which is preserved as petroleum is very small. The annual growth of potential petroleum reserves is therefore miniscule in relation to the total organic production, so in practice petroleum must be regarded as a non-renewable resource.

Accumulation of Organic Matter

All marine organic material is formed near the surface of the ocean, in the photic zone, through photosynthesis. Some phytoplankton are broken down chemically and oxidised and some are eaten by zooplankton. Both types of plankton are eaten by higher organisms which concentrate the indigestible part of the organic matter into faecal pellets which may be incorporated into sediments. Plankton, which are very small organisms, sink so slowly that they are almost entirely degraded in the water column. Pellets, on the other hand, are the size of sand grains, and sink more rapidly to the bottom.

On the bottom organic matter will be subjected to breakdown by microorganisms (bacteria). It will also be eaten by burrowing organisms which live in the top portion of the sediments. The activity of these organisms contributes to reducing the organic content of the sediments because most of the organic matter is digested when the sediment is eaten. Bioturbation results in stirring up and exposure of the sediments to the oxygen-bearing bottom water. If the bottom water is stagnant (contains little oxygen) there will be no bioturbation, due to the toxicity of H_2S, and more organic matter will be preserved in the sediments.

Unless the water is entirely stagnant, a slow rate of sedimentation will lead to each sediment layer spending a longer time in the bioturbation and microbiological

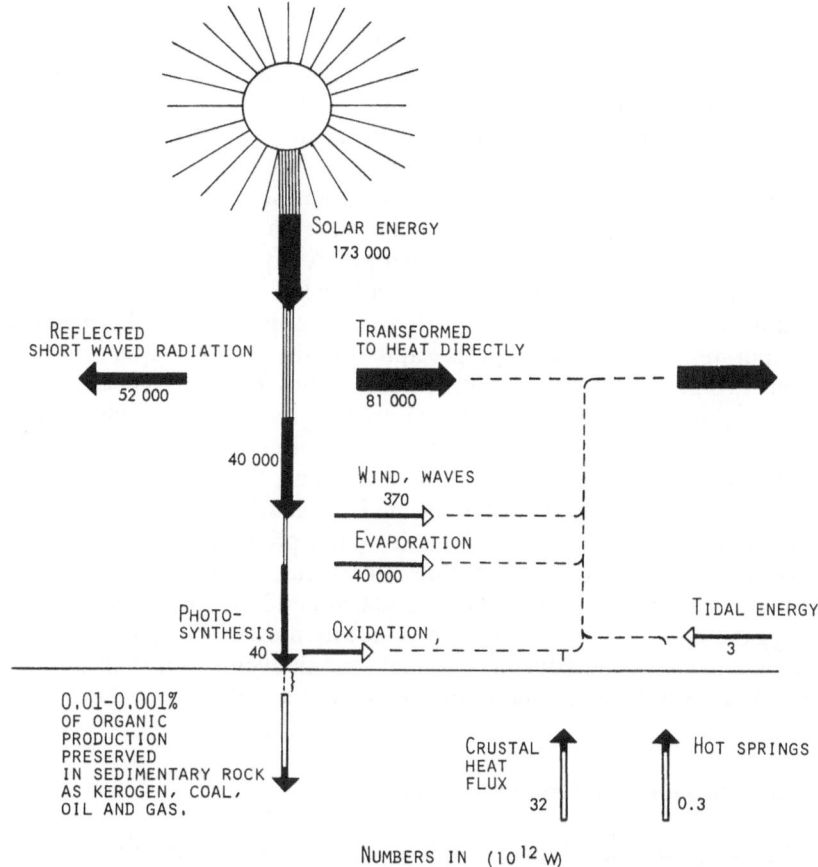

Fig. 13.1. Diagrammatic representation of the energy budget on the earth. We see that only a very small percentage of light energy is used for photosynthesis, and most of this (more than 99%) is broken down, with only a few ppm being converted into fossil energy

breakdown zones, and consequently less organic matter will be preserved in the sediment. Rapid sedimentation leads to more of the organic matter deposited being preserved, and from the start it will be very diluted with mineral grains. Consequently a sedimentation rate which is intermediate in relation to organic production (10-100 mm/1000 years) results in the formation of the best source rocks.

Since bioturbating organisms cannot live in reducing conditions, we get sediments with perfect, undisturbed lamination preserved. Sulphate-reducing bacteria, however, will use a good deal of organic matter, and precipitate sulphides (e.g. FeS_2). If the sediments contain insufficient free iron or other metals which could precipitate sulphides, more sulphur will be incorporated in the organic matter and will eventually be enriched in the oil derived from such source beds.

As we have seen, the net accumulation of organic matter in sediments is to only a limited degree a function of the total productivity. Rather it is a function of the relationship between productivity and biogenic breakdown and oxidation. In areas with powerful traction currents, most organic matter will be oxidised. An important source of oxygen-rich water is the cold surface water which sinks down in polar regions and flows towards equatorial regions - over the bottom of the ocean basins. Limited water circulation in semi-closed marine basins due to restricted outflow over a shallow threshold is one of the most important causes of stagnant water bodies. The Black Sea is a good example. A supply of fresh water which is large compared to the rate of evaporation leads to low salinity in the surface layer and to reduced circulation due to the density stratification in the water column. In evaporation basins with little precipitation the surface water will have a greater density than the water below it, and sink down so that we get circulation (see Fig. 7-10). Areas with coastal upwelling have the greatest production of organic matter.

Lakes or semi-enclosed marine basins often have a temperature or salinity-induced density stratification so that oxygenated surface water does not mix with water in the deeper part of the basin. This leads to anoxic conditions and a high degree of preservation of the organic matter produced in the surface waters. This aspect is therefore of considerable interest in exploration for petroleum in fresh water basins, particularly in Africa and China. The open oceans have normally had oxygenated water, but during the Cretaceous most of the Atlantic Ocean is believed to have been stagnant during so-called "anoxic events", and substantial amounts of black shale were deposited in the deeper parts of the ocean during these periods.

Early Diagenesis of Organic Matter

Microbiological breakdown of organic matter is due to the activity of bacteria, fungi and protozoa, etc. Under oxidising conditions these are extremely effective in breaking down organic matter in sediments. Oxygen may penetrate to depths of 5-20 cm in relatively coarse-grained sediments (sand), while in clay and fine-grained carbonate muds the pores will be so small that water circulation and diffusion is insufficient to introduce new oxygen as the original oxygen in the pore water gets used up. Each pore in such sediments is soon a relatively closed oxygen-poor system. Aerobic breakdown is therefore much more effective in coarse-grained sediments than in fine-grained. In anaerobic transformation bacteria use organic matter, e.g. short carbohydrate chains. Cellulose is broken down by fungi, and finally by bacteria. The final products are methane (CH_4) and carbon dioxide (CO_2). Methane, however, is the only hydrocarbon produced in any quantities by bacteria at low temperatures right near the surface of the sediment. Gas occurring at shallow depths (shallow gas) therefore consists largely of methane (dry gas) unless there has been addition from much deeper strata. Biogenic gas may form commercial accumulations, as in Western Siberia. The presence of abundant shallow gas, as on the Norwegian continental shelf, may also represent a hazard in the form of blowouts and fire during drilling.

Kerogen

Kerogen is a collective name for the fraction of sedimentary organic matter which is not soluble in organic solvents and has a polymer-like structure. The organic fraction which is soluble in organic solvents is called *bitumen*. Kerogen is formed gradually within the upper few hundred metres of the sediment column after deposition from precursor products like humus, humic and fulvic acids. Kerogen may also include organic particles of morphologically recognisable biological origin such as vitrinite (derived from woody tissues and liptinite materials (e.g. algae pores, cuticles etc.) Because of its resistance to strong oxidising acids, kerogen can be recovered from sedimentary rocks by dissolving most of the rock away with HCl or HF.

It is also possible to separate kerogen by a density method, using heavy liquids, because kerogen is lighter than minerals. The resulting kerogen concentrate can be studied microscopically using transmitted and reflected normal light, to identify the biological or diagenetic origin of the various precursor products. It is also useful to use ultraviolet light microscopy, since certain components, i.e. liptinites, display characteristic fluorescence colours. Infra-red spectroscopy and ^{13}C-NMR spectroscopy can be used to investigate the chemical composition and structure of kerogen.

Being a complex of very large molecules, kerogen is difficult to analyse, but upon heating in an inert atmosphere (pyrolysis) it will break down into smaller units which can then be analysed by means of gas chromatography and mass spectrometry.

Sapropelic kerogen is formed from lipid-rich organic matter which contains abundant aliphatic structural units. It consists of the products of microbial breakdown of spores, planktonic algae, and also animal organic matter. Sapropelic matter, which is enriched in lipids like fatty acids, oils, alcohols, wax etc., has a high H/C atomic ratio — usually between 1.3 and 1.7 (Fig. 13.2). This type of kerogen, often called Type I kerogen (Tissot and Welte 1984), contains little oxygen (O/C less than 0.1). Upon burial it will produce mainly oil during thermal maturation. Type I kerogen is typical of some oil shales, particularly in freshwater basins such as the Green River basin in Colorado, Wyoming and Utah, but is also found in marine basins.

Type II kerogen is a mechanically and chemically complex mixture, the composition of which varies considerably, depending on the initial organic precursor materials.

Humic kerogen is derived from organic matter from land plants such as lignin, tannins and cellulose. This type has a low initial H/C ratio, and a high initial O/C ratio reflecting the composition of the precursor plant matter. In maturing (through the effect of temperature) this kerogen, which is often called Type III (Tissot and Welte 1978) generates abundant CO_2 and methane (CH_4). Most coals have a composition and structure similar to Type III kerogens.

Type II kerogen has a composition intermediate between Type I and Type III, but it does not represent a mixture of these end members. It has a relatively high initial H/C and low initial O/C ratio, but contains more oxygen-containing compounds (ketone and carboxyl acid groups) than type I. Alipathic and naphthe-

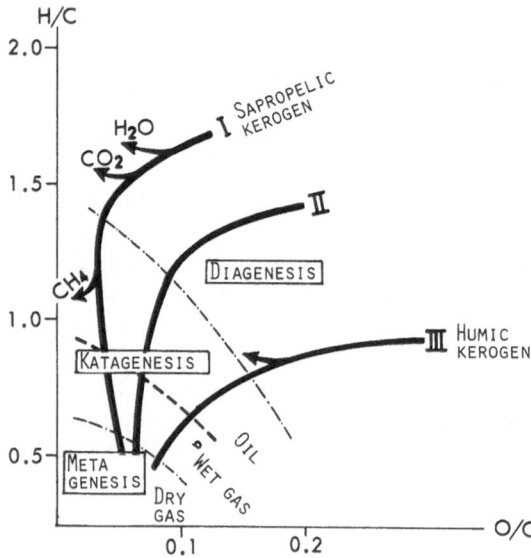

Fig. 13.2. The Van Krevelen diagram. Chemical classification of kerogen types and their change in elemental composition with increasing temperature

nic structures which have hydrocarbon potential are also common. This type of kerogen is common in marine basins where a mixture of phytoplankton, zooplankton and other marine organisms has been deposited under reducing conditions, sometimes along with land-derived plant material. The most common and richest source rocks for oil are sediments of this type (Fig. 13.2).

Transformation of Kerogen with Burial and Temperature Increase

Conversion of kerogen to oil and gas is a process which requires higher temperatures than one finds at the surface of the earth and a long period of geological time. Only at temperatures of about 50°–70°C, i.e. 1–2 km of sediment overburden does the conversion of organic plant and animal matter to hydrocarbons very slowly begin to take place. 80°–130°C is the ideal temperature range for this conversion to oil, which is called maturation. This corresponds to a depth of 3–4 km with a normal geothermal gradient (about 30°/km). In areas with higher geothermal gradients (e.g. high heat-flow areas) organic matter may mature at much lesser depths. In large intracratonic sedimentary basins or along passive margins, however, the geothermal gradient may be only 20–25°/km and the minimum overburden required to initiate petroleum generation will be correspondingly greater.

In general one can say that petroleum cannot be generated near the surface except locally through the influence of hydrothermal and igneous activity. We find shallow deposits of oil and gas today because they were actually formed at great depths, and the overburden was later removed by erosion or because hydrocarbons migrated considerable vertical distances. As we have already seen, however, natural gas, chiefly methane (CH_4), may be formed by biochemical processes near the surface.

Temperature increases with increasing overburden, and causes the carbon-carbon bonds in the organic molecules of which kerogen consists to rupture. Gas is also formed in this phase. As the temperature rises, more and more carbon-carbon bonds are broken, both in the kerogen and in the hydrocarbon molecules which have already been formed. This "cracking" leads to the formation of lighter hydrocarbons from the long hydrocarbon chains and from the kerogen. The removal of CH_4 and other hydrocarbons leaves the residual kerogen relatively enriched in carbon, because at the outset kerogen (Type I and II) has a H/C ratio of 1.3–1.7. Humic kerogen (Type III initially) which has a high initial oxygen content, produces mainly CO_2 gas, and so its oxygen/carbon ratio gradually diminishes. This diagenesis begins at 60–70°C and continues until the H/C ratio is about 0.6 and the O/C ratio is less than 0.1 at about 150°C. At higher temperatures all longer hydrocarbon chains will already have cracked, leaving us only with gas – mainly methane (dry gas). The kerogen composition will gradually move towards pure carbon (H/C = 0) (Figs. 13.2, 13.3).

When kerogen maturation reactions are completed, the kerogen's aliphatic structural components, which may be derived from lipids, fatty acids and proteins etc. have been removed and converted into hydrocarbons.

During this transformation of organic matter, water and oxygen-rich compounds are liberated first, and then compounds which are rich in hydrogen, e.g. hydrocarbons. This conversion results in enrichment of carbon and the colour of the residual kerogen changing from light yellow to orange, brown and finally black. For application in exploration a rapid semi-quantitative method has been developed whereby these colour changes are estimated from smooth spores examined under transmitted light and compared with a standard colour scale. This parameter is called the "Thermal alteration index" (TAI) and will give a rough idea of the thermal maturity of the sediments and their temperature history.

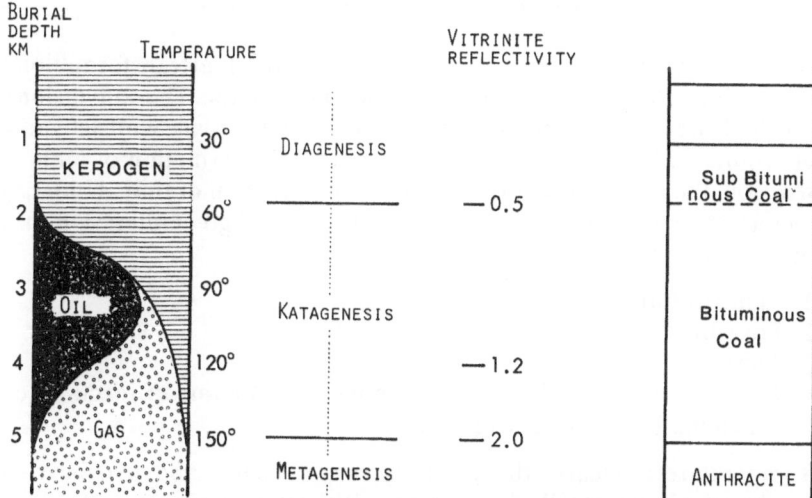

Fig. 13.3. Conversion of kerogen to oil, gas and coal as a function of temperature. The reflectivity of vitrinite (a type of humic kerogen) is a measure of maturation. The temperature indicated assumes burial and heating for tens of millions of years. With shorter burial times, higher temperatures are required

Another more quantitative way of determining the temperature at which alteration of sedimentary rocks takes place is to record the degree of carbonisation of the plant remains which are present in most sediments. Vitrinite derived from woody tissues is a common component of coal. It becomes shinier and reflects light better as the degree of carbonisation increases. By measuring the reflectivity of vitrinite particles under a reflected light microscope an exact value can be obtained for this maturity parameter. For certain source rocks the ratio between extractable alkanes (paraffins) with an even number of carbon atoms per molecule and those with an odd number may also be an expression of maturity. Apart from in marine algae, one finds a higher abundance of alkanes with an odd number of carbon atoms in recent organic matter from plants (mainly land plants). The decrease in this predominance of odd over even in source rocks with increasing maturity is due to the dilution of the original biologically derived n-alkane mixture with a newly generated mixture which has a regular carbon number distribution. This odd/even ratio is normally expressed by means of an index called the Carbon Preference index (CPI).

$$CPI = \frac{\text{concentration of n-alkanes with an even number of C atoms}}{\text{concentration of alkanes with an odd numbers of C atoms}} .$$

In addition the maturation process will cause a shift in the carbon number distribution towards smaller molecules, particularly in the range C_{13}-C_{18}.

There is isotopic fractionation of carbon, and when kerogen is releasing petroleum, this phase is somewhat enriched in ^{12}C corresponding to the precursor kerogen. Gases, particularly methane, normally have lighter carbon isotopes than kerogen and oil. When methane is formed from larger hydrocarbon molecules by thermal cracking, the ^{12}C-^{12}C bond is less stable than the ^{13}C-^{12}C bond and the product becomes enriched in ^{12}C.

What Factors Influence the Maturation of Kerogen?

The term "maturity" refers here to the degree of thermal transformation of kerogen into hydrocarbons and ultimately into gas and graphite. The conversion of kerogen into hydrocarbons is a chemical process which takes place with activation energies of around 50 kcal/mol. It has been assumed that formation of oil is a first order reaction, the rate of which is an exponential function of time. Understanding the factors which influence the rate of this reaction is of great interest. Four factors are thought to contribute:

1. Temperature
2. Pressure
3. Time
4. Minerals or other substances which increase the rate of reaction (catalysts) or which inhibit reactions (inhibitors).

Temperature is clearly the most important factor, and hydrocarbons can be produced experimentally from kerogen by heating it. However the reaction is also time-dependent. In laboratory experiments, where time is more limited than it is in nature, fairly high temperatures (200–500°C) have to be used. Pressure appears to play a minor role, although one might have expected that increasing pressure would reduce the rate of the reaction because of the increase in volume involved in the formation of hydrocarbons (Le Chatelier's rule). For a long time it has been believed that minerals, particularly clay minerals, might have an effect on the rate of formation of hydrocarbons. A number of laboratory experiments have been carried out in which kerogen is mixed with various minerals. The results are still fairly ambiguous.

The conversion of organic matter begins at 50–60°C, given geological time periods. At lower temperatures we do not get hydrocarbons formed thermally, regardless of the duration of heat exposure. Between 50° and 70°C the transformation of kerogen proceeds very slowly, and most of the maturation occurs between 70° and 130°C. The degree of transformation of kerogen is also a function of time. This means that rocks which have been subjected to 90°C for 50 mill. years are more mature than rocks which have been exposed to this temperature for 10 mill. years. As the organic-rich sediment (source rock) is buried in a sedimentary basin, it will normally be subjected to increasing temperature as a function of increasing burial depth. If we know the geothermal gradient and the subsidence curve we can calculate its temperature history. If the geothermal gradient can be assumed to have been constant throughout the relevant period, this is relatively simple. However, if the geothermal gradient has varied considerably through time it is a lot more complicated. A theoretical maturity parameter (P) can be calculated by integrating temperature with respect to time.

$$P = Ln \int_0^t 2^{T/10} \cdot dT$$

t = geological time (mill. years) and T = temperature (°C).

We see that a doubling of the reaction rate for every 10°C is built into this expression (Geoff 1983). This is an expression which is very similar to Lopatin's Time-Temperature Index (TTI) (Waples 1980) which integrates the temperature the source rock is subjected to with respect to the burial time. When the temperature rises above about 130°–140°C, maturation proceeds very rapidly, and then time is not so crucial.

There are differing views as to how much emphasis should be placed on time in relation to temperature in the matter of maturation. Different formulae are used for calculating these temperature factors and the 10-degree rule is now found not always to be valid, particularly for very young sedimentary basins with high geothermal gradients.

The rate of sediment heating, i.e. temperature increase per unit time, can be expressed as dT/dt. The rate of subsidence is dz/dt, and the geometrical gradient dT/dz. We then get:

$$\frac{dT}{dt} = \frac{dz}{dt} \cdot \frac{dT}{dz}$$

Rate of heating = rate of subsidence × geothermal gradient

If we know the stratigraphy of the overlying sediments and the geothermal gradient, temperature can be calculated as a function of time.

Calculating the maturity of source rocks is important not only for predicting whether source rocks are sufficiently mature to produce oil, or perhaps overmature so that only gas remains. It is also possible to estimate the timing of the generation and migration of oil. Calculation of source rock maturity lends itself to computerised calculation and modelling and has become an important part of basin analysis for oil exploration. The most important input into the calculation is the subsidence history as derived from the stratigraphic record and the estimated geothermal gradient as a function of geological time. The success of the basin modelling depends as always very much on the quality of the input data.

Further reading: Hunt 1979; Bjorøy 1981; Geoff 1983; Tissot and Welte 1984; Leythaeuser and Rullkötter 1986.

Composition of Petroleum

Naturally occurring petroleum has a very complicated chemical composition. The most important hydrocarbons occurring in oil are alkanes (paraffins) (C_nH_{2n+2}). These are saturated (aliphatic) chains with from 1 to over 60 carbon atoms in every molecule. Important gases are methane (CH_4), ethane (C_2H_6), propane (C_3H_8) and butane (C_4H_{10}). Paraffins from carbon numbers 5, pentane (C_5H_{12}) to 15, pentadecane ($C_{15}H_{32}$) occur mainly as liquids at room temperature. Paraffins dominate the gasoline fractioe oil. Naphthenes (cycloparaffins) are a series of cyclic, saturated hydrocarbons with the general formula (CH_2)$_n$. Cyclopentane (C_5H_{10}) and cyclohexane (C_6H_{12}) are important members of the naphthene series which are found in all types of petroleum. Naphthenes with lower carbon numbers, C_3H_6 and C_4H_8, are gases. The percentage of naphthenes in petroleum varies from 7 to 31. Hydrocarbons of the type described above are also referred to as saturated hydrocarbons. The higher paraffins are wax-like in the solid phase.

Aromatic hydrocarbons are unsaturated, cyclic, and have the general formula C_nH_{2n-6}, for example benzene, C_6H_6. Aromatic hydrocarbons have a strong odour and a lower boiling point than aliphatic hydrocarbons with the same number of carbon atoms. Important compounds in petroleum are toluene (methylbenzene, $C_6H_5CH_3$) and xylene (dimethylbenzene, $C_6H_4CH_3CH_3$). Aromatic compounds make up 10–39% of crude oil. Refinement of crude oil with a high content of aromatic compounds results in a high octane product.

Asphalt is the high molecular weight, high boiling point fraction of crude oil, and consists of resins and asphaltenes, which are rich in sulphur and nitrogen.

API Gravity is an expression of the density of hydrocarbons. A high API gravity means a low density.

$$\text{API gravity} = \frac{141.5 - 131.5}{\rho},$$

where ρ is the density of the petroleum at 60°F. A density of 1 g/m^3 corresponds to an API gravity of 10. Light oil has an API gravity higher than 40, and heavy oil has an API gravity of about 15–20.

The sulphur content of crude oil varies from 1% to 5–6%, and sulphur occurs as elemental sulphur, hydrogen sulphide, H_2S, and organic sulphur compounds.

Elemental sulphur occurs in great quantities in a number of oil fields, and most of the world's sulphur production comes from sources of this type. In most oil reservoirs, however, sulphur compounds are undesirable components, since they cause corrosion problems, problems in the refining process, and burning oil of this type releases SO_2 into the atmosphere.

Table 13.1 shows the composition and density of a number of hydrocarbons. A number of definitions of crude oil are given in Table 13.2.

Table 13.1. Composition of oil and kerogen as % by weight of the constituent elements. (Hunt 1979)

	C	H	S	N	O
Oil	84.5	13	1.5	0.5	0.5
Asphalt	84	10	3	1	2
Kerogen	79	6	5	2	8

The N.S.O. compounds include resins and asphaltenes which are enriched in oxygen, nitrogen and sulphur. In addition hydrocarbons contain a number of trace metals. The most important are Ni and V.

Hydrocarbon	Molecule type	Formula	H/C ratio	API Gravity	Density g/cm3
Hexane	Paraffin	C_6H_{14}	2.3	82	0.6594
Cyclohexane	Naphthene	C_6H_{12}	2.0	50	0.7786
Benzene	Aromatic	C_6H_6	1.0	29	0.8790

Hydrocarbons have a density which is largely an inverse function of the H/C ratio.

Table 13.2. Classification of crude oil

1. *Low shrinkage oil*: Oil which after production and separation at ordinary pressure and temperature contains a high percentage of fluid (> 80%) and little gas.

2. *High shrinkage oil*: Oil which after production and separation at ordinary temperature and pressure contains a smaller percentage of liquid (less than 70%) and a large amount of gas.

3. *Condensate*: Gas which turns into a liquid with pressure reduction. This is gas which has a temperature higher than the critical point. Condensates have low densities (about 40–60° API).

4. *Dry gas* consists almost exclusively of methane.

5. *Wet gas* is gas which in addition to methane also contains significant amounts of alkanes with high carbon numbers. It may be in the gaseous phase in the reservoir, but after production and separation may form a good deal of liquid at normal pressure and temperature.

6. *Undersaturated crude* is oil which contains less gas than is potentially soluble at reservoir temperature and pressure.

7. *Saturated crude* is oil which contains as much gas as can be dissolved at reservoir pressure and temperature.

Gas

Methane is a thermodynamically stable compound, even at temperatures of 500–600°C. As a result we may find methane gas at very great depths. The limiting factors for economic exploitation of gases of this type are the greatly reduced porosity and permeability of reservoirs at great depths (6–8 km), and the expenses of drilling to such great depths. In rocks which contain sulphur, either as free sulphur or as sulphate, which breaks down, sulphur will react with methane and form mercaptanes.

The main component of natural gas is normally methane (CH_4), but ethane (C_2H_6), propane (C_3H_8) and butane (C_4H_{10}) may also be important. In addition we have varying amounts of CO_2, H_2O, nitrogen, hydrogen and inert gases such as argon and helium.

Methane produced by bacterial breakdown of organic matter at low temperatures ($< 60°C$) is characterised by a light carbon isotope composition. This kind of gas is often referred to as "shallow biogenic gas".

Migration of Petroleum

It has been evident for a long time that petroleum occurs in oil reservoirs which cannot have been the source rock. This means that petroleum cannot have been formed in the reservoir rock, but must have *migrated* from a *source rock* to a *reservoir rock* (Fig. 13.4).

In some special cases the source rock may fracture and become permeable enough to become a reservoir rock. This is the case with the Miocene Monterey Formation in California, which consists of an organic-rich, shaly, calcareous phosphatic chert which later formed a fractured reservoir.

Most of the typical petroleum reservoir rocks were deposited under extreme oxidising conditions, and must originally have contained very little organic material. This is particularly true of many sandstones, oolitic limestones, reefs etc.

Consequently there is no doubt at all that oil migrates to reservoirs from another source rock. We do not as yet know precisely how this occurs.

Kerogen occurs finely dispersed or concentrated in organic-rich laminae in the source rock. When oil is generated in the source rock there may be an increase in volume which may build up overpressure. This in turn helps to produce fractures along which oil can flow out of the source rock. Overpressure is also developed due to retardation of oil expulsion from tight rocks, particularly shales. In the case of gases this expansion is much greater. A build-up of pressure in the source rock may not be necessary to permit oil to escape from the source rock, but it will make the oil expulsion more efficient. Oil must, however, saturate the pores and fractures of the source rock before it will be expelled into the adjacent sediment. A minimum quantity of hydrocarbons must therefore be generated before the source rock will expel oil that can migrate further. Source rocks must therefore contain more than about 1.0% organic carbon, preferably more than 2-3%.

It was earlier suggested that oil was transported dissolved in porewater. However hydrocarbons have a low solubility in water, varying at room temperature

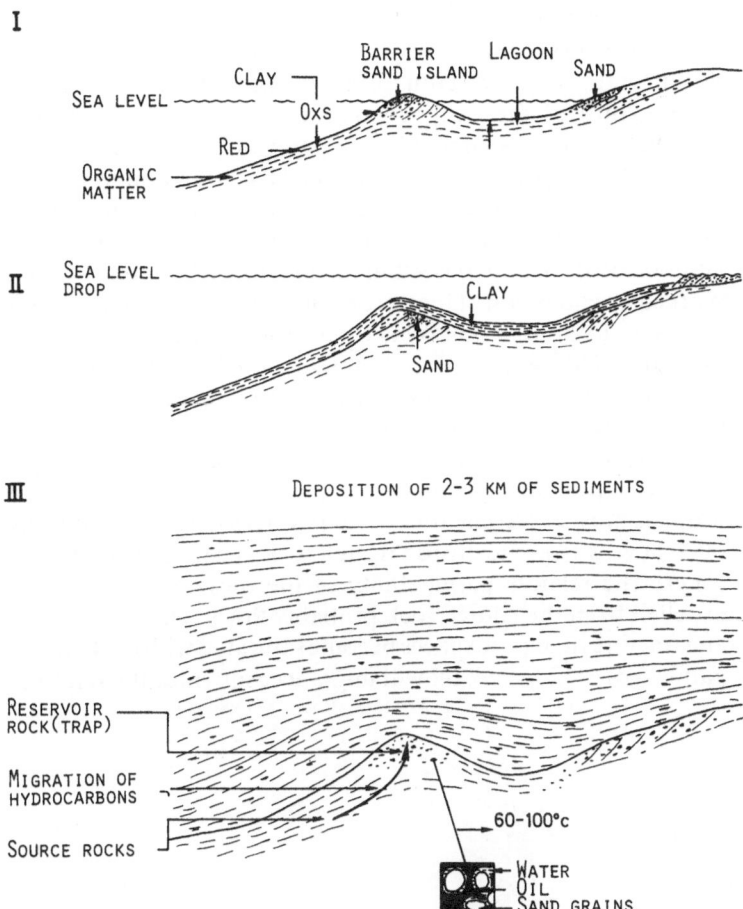

Fig. 13.4. Example of reservoir rocks and source rocks. When the source rock is "mature", oil can migrate into the reservoir rock

from only 24 ppm for methane to 1800 ppm for benzene. Other compounds, e.g. pentane, are even less soluble (2-3 ppm). Many hydrocarbons have solubilities of less than 1 ppm in water. The presence of colloidal electrolytes or surfactants, however, could cause formation of molecule aggregates (micelles) which can greatly increase the solubility of hydrocarbons. Solubility increases with temperature, but is little affected by variation in pressure.

The solubility of hydrocarbons will, however, increase greatly when we have large aggregates of molecules (micelles) present in colloidal electrolytes. However micelles are not present in high concentrations in source rocks, and are not considered a viable mechanism for transport of hydrocarbons. The principle reason for discounting aqueous transport of hydrocarbons is that the most soluble types are not most abundant in petroleum, and the distribution of hydrocarbons is not what would be expected from a possible precipitation mechanism.

The greatest problem with migration of hydrocarbons in aqueous solutions, however, is explaining where one gets large enough quantities of water to flow through the rock. Petroleum forms at depths of at least 2 km, and most of the water released through the compaction of sediments is expelled from the rock very early.

One potential source of extra porewater is that clay minerals, e.g. montmorillonite (smectite), release water which has been bound up in the lattice structure. But this will still not produce sufficient water to transport large quantities of soluble hydrocarbons and many sediments contain rather a small amount of smectite. In addition, with an aqueous solution of hydrocarbons one would expect to get a very marked fractionation of hydrocarbons according to their solubilities. It is therefore necessary to assume that oil is transported essentially as a separate phase during migration. Oil is lighter than water, and oil droplets will be able to move through pores in the rock. This is facilitated by the fact that in most cases water moistens the surface of sediment grains far more easily than oil. Oil droplets will therefore often occur in pores surrounded by a thin membrane of water, if the rock is water-wetting. In order to pass through the narrow passage between two pores (the pore throat) oil droplets must overcome the capillary forces (Fig. 13.5). It is therefore difficult for oil to flow as separate drops through rock in which most of the pores are filled with water. The relative permeability of the rock to oil is therefore rather low. Oil expelled from the source rock probably flows selectively along special pathways, and fractures and pore spaces become saturated with oil. Once such an oil-saturated pathway is established, oil will migrate easily since the relative permeability of oil

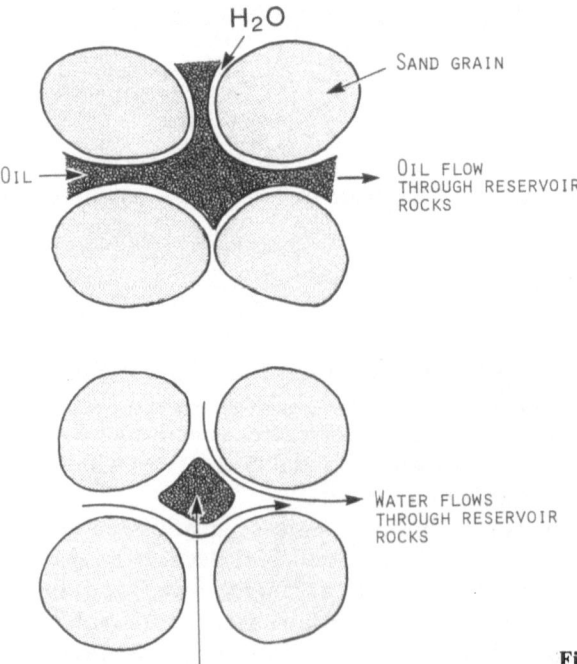

Fig. 13.5. High and low oil saturation in water wetting sandstone reservoir

is very much higher here than in the water-saturated parts of the rock. Oil migration is therefore not likely to be accompanied by movement of a great deal of pore water, since the oil is likely to follow special pathways.

The cap rock is a layer of rock with a low permeability which prevents hydrocarbons from migrating further upward. The cap rock may consist of evaporite (salt) layers but is most commonly a shale. Well compacted shale, perhaps partly carbonate-cemented, has very small pores, and the oil will not be able to overcome the capillary forces acting against oil infiltration. A cap rock may be permeable to water and not to oil, but if a thick column of oil is trapped beneath the cap rock the pressure in the oil phase may build up sufficiently for the oil to flow slowly through the cap rock as well. Gas may also leak through cap rock by diffusion, or more easily along special zones of weakness.

Further reading: Roberts and Cordell 1980.

Biodegradation

Hydrocarbons in the reservoir can be broken down by microorganisms (bacteria, yeast and fungi). This is a form of biological oxidation by which hydrocarbons are oxidised to alcohols, ketones and various acids. The biological breakdown of hydrocarbons proceeds far more rapidly with smaller molecules (low carbon number, less than 20). When it comes to molecules with the same carbon number, n-paraffins will break down first, and then isoparaffins, naphthenes and aromatics. Isoprenoids, steranes and triterpanes show the greatest resistance to biodegradation. Biodegradation of aliphatic hydrocarbons is dependent on a supply of oxygen, normally dissolved in water. If we have dissolved sulphate, hydrocarbons may also be broken down by sulphate-reducing bacteria, but this takes longer.

Biodegradation occurs as soon as the oil flows out at the surface as oil seepage. Bacteria can then use oxygen from the air. Asphalt forms from oil of this kind not only because the lighter components of the oil evaporate more readily, but also because the remaining petroleum components are rapidly broken down by bacterial processes, so that the concentration of the larger molecules increases, particularly the asphaltenes, which are the most important components of asphalt.

In order for the biodegradation process to continue, we must have a continuous supply of oxygen in the porewater. Only surface (meteoric) water contains fairly large quantities of oxygen, but free oxygen is rapidly lost when water flows through sediments through oxidation of organic and inorganic compounds. However the sulphate ion is more stable, and will not easily be reduced, except by sulphate-reducing bacteria. In most cases biodegradation occurs in relatively shallow reservoirs (< 1500 m, $< 60-70°C$), which are in contact with circulating meteoric water. In special cases meteoric water can penetrate more than 2 km down below sea-level. In reservoirs on land along mountain chains, the probability of fresh water percolating through is particularly great. Biodegraded reservoirs will generally have normal hydrostatic pressure. As long as there is overpressure in the reservoir, meteoric water will not penetrate down to it.

Sulphates, e.g. gypsum, are also a source of oxygen which can cause biodegradation through the action of sulphate-reducing bacteria. Biodegraded hydrocarbons are rich in asphalt, and have a high density (low API gravity and high viscosity). As a result the oil is less valuable and difficult to produce. Injection of steam to heat the oil and reduce its viscosity is the most effective method of increasing production from reservoirs containing heavy, biodegraded oil.

Properties of Reservoir Rocks

Packing of sediment grains — porosity and permeability.

When sediment is deposited, it will pack with a certain ratio between the total volume, the volume of open pores (possibly filled with fluid) and the volume of sediment grains. We normally express this as porosity.

$$\text{Porosity } (\varphi)\ \% = 100 \times \frac{\text{volume of pores in the sediment}}{\text{total volume of the sample}}.$$

We can also use the ratio between pore volume and the volume of the grains (void ratio).

Well sorted, rounded sand grains are almost spherical in shape. If we have grains of the same size, which are all quite well rounded, with a high sphericity, we will be able to pack the grains so as to get minimum porosity. Rhombic packing results in 26% porosity. Cubic packing, where the grains are packed directly one above another, results in about 48% porosity. Most well sorted sandstones have a porosity which lies between these two values.

Clay-rich sediments have a much greater porosity immediately after deposition: it may well be around 60–80%. This means that a sand bed is denser than a bed of clay or silt immediately following deposition. However, clay and silt lose their porosity more rapidly with burial.

If we assume that we know the density of the grains (minerals), the porosity can be measured by measuring the density of a certain volume of the sediment. Using a pyknometer is one of the most accurate ways of determining porosity. Porosity may also be determined on a dry sample through gas expansion, employing Boyle's Law.

The *absolute porosity* is the total volume of pore spaces in the rock. However some of the pore spaces may be isolated, so that they are not connected with other pore spaces.

The *effective porosity* is the designation for pore spaces which are connected so that fluids can flow through the rock. The effective porosity is measured by pumping fluid under pressure into the rock, to find the volume which the rock can absorb.

Absolute porosity can be determined by measuring the area of pores in thin section, and the relative density. Some rocks, e.g. pumice, have a very high absolute porosity, while the effective porosity is very small. Porosity is determined in the laboratory by forcing a liquid, e.g. mercury or a gas, into the sample. One can gain an idea of the porosity of a hand specimen by pouring a drop of water onto a dry sample of the rock, and seeing how rapidly the drop is sucked in. In wells porosity is most effectively measured by means of logs. In most cases porous strata give the highest self potential. Neutron logs register high contents of oil or water, and consequently express porosity. Sonic and density logs also provide an expression of porosity (see Chap. 14).

Permeability and Porosity

Permeability is an expression of the ease with which fluids flow through a rock.

It will depend on the size of the pore spaces in the rocks, and in particular the connections between the pore spaces. Even thin cracks will contribute greatly to increasing the permeability.

Permeability can be measured by letting a liquid or gas flow through a cylindrical rock sample under pressure. The difference between the pressures at the two ends of the cylinder is P, the length of the cylinder, L, and the volume of e.g. water which flows through the cylinder, V cm^3/s. A is the area of the cylinder, and μ the viscosity. According to Darcy's Law:

$$V = \frac{k \cdot A \cdot \Delta P}{L \cdot \mu}.$$

k is the permeability factor which is expressed in darcies, the permeability unit (Fig. 13.6). A rock which has water flowing through it with a velocity of 1 cm^3/s. per cm^2 (cross-section) with a fall in pressure of one atmosphere per centimetre when the viscosity is 10^{-3}N/s/m^2, has a permeability of 1 darcy. Permeability is normally measured in 1/1000 darcy, i.e. a millidarcy.

A well-sorted sandstone has a permeability of between 100–1000 md, which is very good. Permeabilities of 10–100 md are also considered to be good values for reservoir rocks. Permeabilities of 1–10 md are typical of relatively dense sandstones and limestones, so-called "dense" reservoirs. There are also examples of rocks with lower permeabilities being exploited commercially for oil production (see the

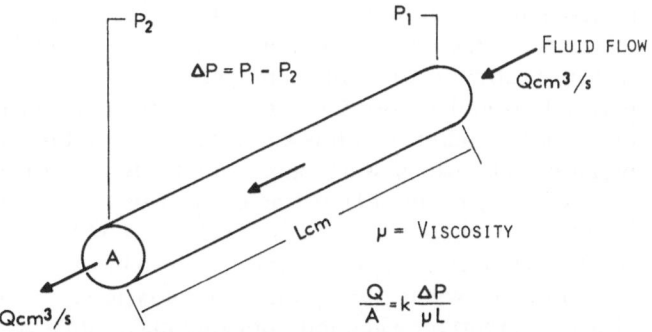

Fig. 13.6. Definition of permeability illustrated by means of flow through a cylindrical sample of rock. ΔP is the difference in pressure over a distance L. The flow velocity through cross section A is Q cm^3/sec. The coefficient of permeability is k and μ is the viscosity of the fluid

Ekofisk Field). In these the generally low permeability of a chalk matrix is enhanced by fractures which increase the overall permeability. The speed with which a liquid flows at a particular pressure gradient and permeability is called its *mobility*. Mobility is thus a function of viscosity.

Primary Porosity — Porosity Existing from the Time of Deposition

1. Intergranular porosity.
   Porosity between sediment grains, e.g. sand grains in sandstone or in a calcitic sandstone.
2. Intragranular porosity — porosity within the grains.
   Cavities in fossils, e.g. foraminifera, molluscs etc. Partly and entirely dissolved felspar and rock fragments.
3. Porosity in primary cavities in biogenic sediments, e.g. reefs (framework porosity).
4. Solution porosity or secondary porosity. Cavities will form through selective solution of sediment grains or fossils (mouldic porosity). This is particularly true of selective solution of carbonates, or of aragonite rather than calcite or dolomite. Chemical solution of a matrix (e.g. a carbonate matrix) may give secondary porosity. Sediment grains may also be subjected to chemical solution and corrosion if the porewater changes its composition so that they are no longer stable. Felspar grains and rock fragments may go partially or wholly into solution and cause secondary porosity.
5. Fracture porosity.
   This is porosity due to the fracturing of rocks. Fracture porosity may develop in addition to other types of porosity, and will then increase the permeability very considerably. Normally dense sedimentary rocks, and igneous and metamorphic rocks may develop considerable porosity (5–15%) after fracturing and tectonic brecciation.

Most rocks are far from homogeneous. We may measure the porosity and permeability of a hand specimen or core plug, but it is not certain that these are representative of the rock as a whole. Fractures occur at varying intervals, and vary in size from large, open joints down to microscopic cracks which can barely be seen with an electron microscope. In many cases joint systems represent a considerable part of the overall porosity, and there are a number of oil reservoirs which are based on rocks with porosity of this kind. It is important to be aware of this possibility, because rocks which were once completely impermeable may form large oil reservoirs as a result of later fracturing. This may be the case for well-cemented limestones or chert, which easily develop a good fracture system.

Existing tectonic joint systems may also be expanded through chemical solution caused by water passing through the system. Fracturing of reservoir rocks is also very important when joint porosity makes up a very small percentage of the overall porosity, because the permeability increases greatly and microjoints form connections between otherwise isolated, oil-filled pores in impermeable reservoir rocks. This is true, for example, of the Ekofisk Field in the North Sea, where the dense Cretaceous rock is only productive because of microjointing. The measured permeability in core samples is low ($<$ 1 md).

Jointed reservoirs may occur in rocks which we do not otherwise think of as reservoir rocks, e.g. shales, micritic limestones, chert and metamorphic and igneous rocks.

Capillary Pressure

Capillary pressure (P_c) is defined as the difference in pressure on opposite sides of the interface between two immiscible fluids, e.g. between gas and liquid, or between two immiscible liquids like oil and water. This pressure difference (P_c) is a function of the interfacial tension γ (dynes/cm) and the curvature of the interface, expressed as a radius (r). $P_c = 2\gamma/r$. The angle θ between a drop and the wall of a tube (Fig 13.7b) or a mineral surface (Fig. 13.7a) is called the contact angle. If the contact angle between water and oil is less than 90° the surface is said to be water-wet.

The capillary pressure can be expressed:

$$P_c = \frac{\cos \theta}{R}.$$

With a cylindrical capillary tube with a radius r, the height h that water rises in the tube due to capillary forces can be calculated. Assuming that the capillary forces equal the gravitational forces (Fig. 13.7b) we get:

$$h = \frac{2\gamma \cos \theta}{r \cdot g(p_w - p_o)}.$$

Capillary pressure plays a major role during the migration and accumulation of hydrocarbons. When the pore diameter and the pore throat decrease, permeability falls off rapidly, while capillary pressure increases. This capillary pressure is capable of making water rise up through the fine-grained sediments above the oil/water contact until equilibrium is reached.

When we have two different liquids which are immiscible (oil and water), it will not be possible for oil to flow through a fine-grained rock, like shale, for example, which is saturated with water (more than 70–80% water). Conversely, if the rock is saturated with oil, it will act as a barrier to the flow of water (Fig. 16.2).

Cap rocks (or roof rocks, or seals) always have very small pores and high capillary pressures, while reservoir rocks normally have low capillary pressures. It is the high capillary pressure in the cap rocks which prevents oil from flowing through from the reservoir rocks, because the capillary force is stronger than the buoyancy of oil in water. The fact that the cap rocks have a low permeability does not prevent flow, but reduces the velocity. In fine-grained rocks, high capillary pressure is more or less correlatable with low permeability. Shales or siltstones which occur as lenses in sandstone reservoirs, may contain water even if they are surrounded by oil because the oil has not overcome the capillary forces required to displace the water in the small pores. If we have both fine-grained and coarse-grained sandstones in

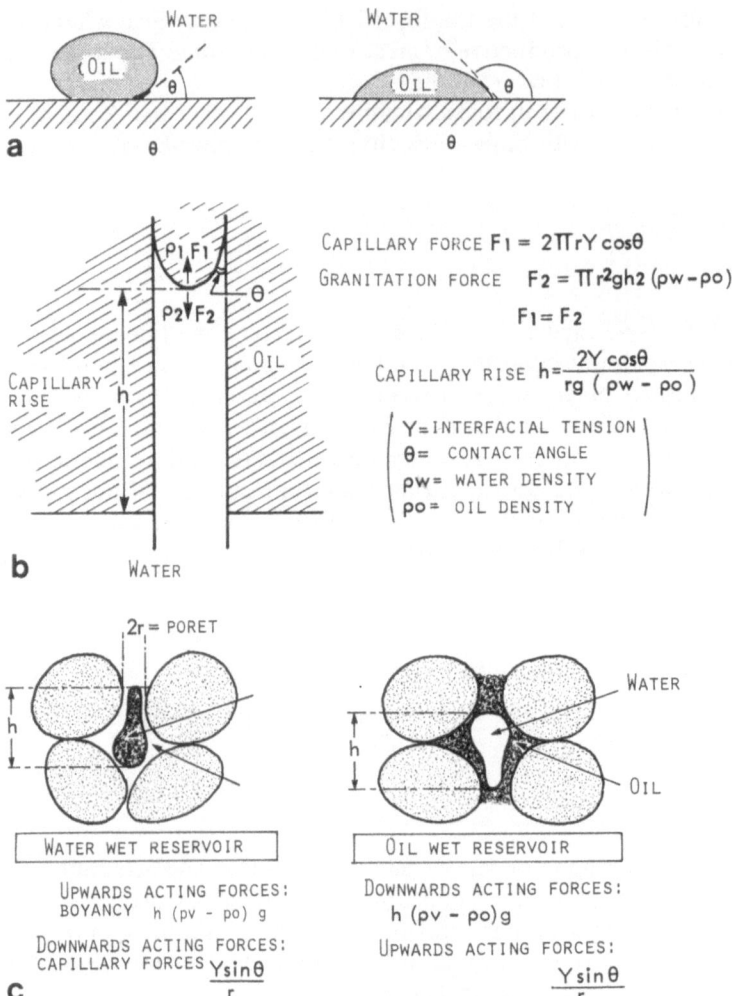

Fig. 13.7. a If the interfacial tension between oil and the mineral surface is greater than that between water and the mineral surface, the contact angle will be less than 90°, and we will have a water-wet system. If the contact angle is greater than 90, the system is oil-wet. The contact angle is therefore an expression of the degree of water-wetting or oil-wetting. **b** Diagram showing capillary forces between oil and water in a tube of radius r. The capillary head is inversely proportional to the pore radius. **c** Capillary forces counteract the buoyancy of oil droplets in water (water-wet reservoir) and the opposite is the case for droplets of water in oil (oil-wet reservoir). The force of gravity will cause the oil to move upwards because of its lower density, but in fine-grained sediments there are capillary forces which work against the movement far more strongly because the pore radius (r) is small. Consequently the oil/water contact will defined by a fairly broad mixed zone with different oil/water saturation in fine-grained sediments

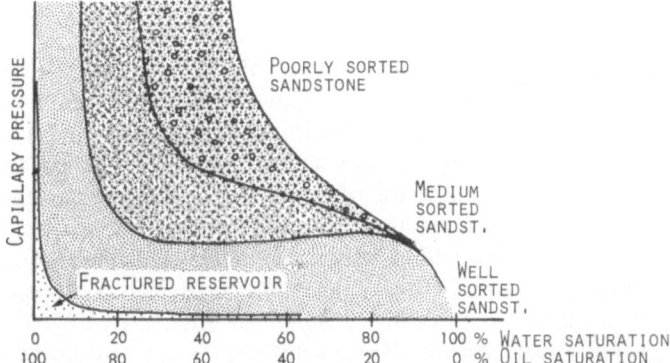

Fig. 13.8. Capillary pressure as a function of water/oil saturation in different types of clastic reservoirs. A moderate capillary pressure must be overcome in well sorted sandstones in order for oil saturation to be achieved. In finer-grained sediments the capillary pressure will be higher

the reservoir rock, the oil-water contact could slope. This could also happen if the water is moving horizontally in the reservoir, producing a fluid drag which tilts the oil/water contact.

In a reservoir we will have a transition zone between the largely oil-saturated part of the reservoir, and the water-saturated section. Because water normally "wets" mineral surfaces better than oil, we will more often have some water saturation in the sector of the reservoir above the oil/water contact than the reverse.

When we have a droplet of oil in water, there will be a difference in pressure between the droplet and the water which is a function of the surface tension (γ) and the radius of the droplet.

$$P = 2\gamma/R.$$

The difference in capillary pressure between the top and the bottom of the mixed zone is

$$2\gamma/R_1 - 2/R_2.$$

R_1 and R_2 are the radii of the meniscus in the pores in the rock at the top and bottom of the mixed zone. This difference must correspond to the upward force of oil above the height (h) in the mixed zone hg (ρ_w-ρ_o) where ρ_w is the density of water and ρ_o is the density of oil (Fig. 13.7).

If the grain size and the pore size decrease, the capillary pressure increases because R_1 gets smaller, and we obtain a broader zone, h, where oil is partly displaced by water due to capillary forces (Fig. 13.7). When calculating oil reserves it is important to take into account oil and water saturation and the thickness of the transition zone. In fine-grained reservoir rocks the transition zone could be greater than the height of the reservoir, in which case water will always be produced along with oil from the reservoir. In fine-grained sediments with small pores oil must overcome much higher capillary pressures than in coarser sediments and fractured sediments (Fig. 13.8).

Osmosis

In fine-grained sediments clay particles form a semi-permeable medium in which the pores are so small that they function as membranes. Osmotic forces may therefore be active in clay sediments. If the salt content of porewater is different in two sand beds separated by a clay bed, water will flow from the sand bed with the lowest salt content to the bed with the highest salt content, until the pressure in this bed has increased so much that an osmotic equilibrium is reached.

The osmosis effect is greatest where we have a great contrast in the salt content of the porewater in the sedimentary basin. This could apply where fresh (meteoric) water flows down in the sediment basin, driven by hydrostatic pressure from the groundwater (see p. 248). Osmotic pressure will build up between the meteoric groundwater with low salinity, and the marine porewater in the surrounding clay sediments. A thin clay bed between two sand beds with different salinity could function as a membrane.

Cap rocks over oil and gas reservoir rocks can also act as membranes and contribute to the pressure in the reservoir. Osmotic pressure will also build up in salt deposits and domes in relation to sediments with normal "marine"salinity. However, it is uncertain how important osmotic pressure build-up is in sedimentary basins. Few examples demonstrating the effect of osmosis have been demonstrated yet.

Pressures and Stresses in Sedimentary Basins

Sediments on the sea floor are already being subjected to a pressure which corresponds to atmospheric pressure plus the weight of the water column above. This is hydrostatic pressure, i.e. it is equal in all directions. Pressure is force per unit area, and is often expressed in Kp/cm^2 or pascal. In the oil industry, P.S.I. (pounds per square inch) is still used. The pressure in a column of water increases by 9.8 k Pa/m ($0.1 Kp/cm^2/m$ or 0.443 P.S.I./ft.) ($1 Pa = 1 N/m^2$, kg Pa $= 10^3 Pa$, MPa $= 10^6 Pa$) 1 atmosphere $= 1.033\ Kp/cm^2$, 14.7 P.S.I. or 98 KPa/m (Fig. 12.3).

As sedimentary beds subside and become overlain by younger sediments, they are subjected to increasing pressure from the overburden.

In plastic clayey sediments there will also be hydrostatic pressure, while sandstones and more indurated sediments will be sufficiently firm for the pressure or stress to be of different intensity in different directions. Stress is force per unit area. The distribution of stresses in solids can be represented by a triaxial ellipsoid, where S_1, S_2 and S_3 represent the greatest, medium and smallest compressive stress directions.

S_1-S_2, S_2-S_3, S_1-S_3 etc. are the differential stresses. Thus hydrostatic pressure corresponds to $S_1 = S_2 = S_3$, as in liquids or plastic material. The designation "hydrostatic pressure" is also used of pressure which corresponds to the weight of the water column up to sea level or the water table (see Chap. 12). Overpressure is thus a pore pressure which is greater than the hydrostatic pressure, i.e. the piezometric surface is above sea level or the water table.

Overpressure in shales, for example, leads to them having a higher water content and greater porosity than other rocks at the same depth. The resultant lower density may under certain conditions cause shales to move upwards in the same way as a salt dome (clay diapir).

Seismic waves will propagate themselves more slowly in rock with overpressure due to decreased density and rigidity when the effective grain-to-grain stress is reduced. High pressure zones can therefore be recognized on seismic profiles due to the decreased seismic velocities in the sediments. Clay diapirs may develop if large volumes of sediment are sufficiently undercompacted. Clay diapirs will generally show smaller contrast acoustic impedance and therefore weaker reflections. Flow of clay will also tend to produce a chaotic seismic signature.

Shortly after deposition, sediments are extremely porous and contain a lot of water. In the case of clay sediments this may be up 70%, and for sandstones about 40%. Sandstones are therefore denser than clay-rich sediments just after deposition. This may lead to the formation of deformation structures.

The sediments will undergo compaction due to the overburden, and the porosity of clay-rich sediments in particular will be rapidly reduced. Excess water can in reality only escape upwards, and this leads to the fact that one in principal has a stream of water moving upward through sediments in a basin. Water flows from areas with a higher to areas with a lower potential than the hydrostatic potential.

The pore pressure in sedimentary rocks can thus reveal a great deal about the development of the basin. Normal hydrostatic pressure indicates that there is free communication of water through relatively permeable rocks; overpressure shows that we have structures surrounded by strata with low permeability. By testing whether the pore pressure varies from one sandstone to another, one can determine whether there is continuity between sandstone strata. This is particularly important with reservoir rocks, and in connection with assessment of reservoirs and production planning.

Hydrocarbon Traps

Hydrocarbon traps are structures resulting from permeable reservoir rocks being covered by rocks with a low permeability (cap rocks) which are capable of preventing the hydrocarbons from migrating further upwards.

The cap rock may not be 100% effective in preventing upward flow of hydrocarbons. If loss of hydrocarbons upwards is less than the supply of hydrocarbons to the trap from the source rocks, hydrocarbons will still accumulate.

Traps can be classified according to the type of structure that produces them.

We distinguish between traps that are formed by structural deformation of rocks (structural traps) and traps related to primary features in the sedimentary sequence (stratigraphic traps).

1. *Anticlinal — dome traps*
Conditions: Impervious cap rock — porous reservoir rock — source rock. Closure in all directions (four-way closure).
a) Fold
   anticlinal with axis culmination (fold axis dipping in both directions). Fig 13.9.

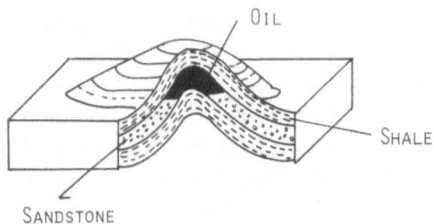

OIL

SHALE

SANDSTONE

Fig. 13.9. Simple anticlinal trap (dome) with oil in sandstone and a cap rock of shale

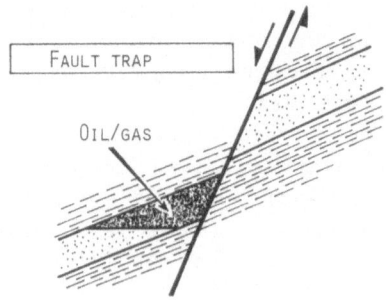

FAULT TRAP

OIL/GAS

Fig. 13.10. Simple fault trap. The condition for formation of such faults is that the fault plane is impervious to oil (gas). The fault plane may in some cases be sealed, in others open. A clay filling, which is impervious to oil because of capillary forces, often forms in the fault plane. When we have sandstones on both sides of the fault plane it is often permeable.

b) Salt domes.
   Strata around the dome curve up creating traps against the sealing salt layers (cap rock).
c) Growth dome.
   Domes which are formed simultaneously with sedimentation through the area subsiding more slowly than the surrounding areas.

2. *Fault traps*
The fault plane must have a sealing effect so that it functions as a barrier for reservoir rocks. (Fig. 13.10).

a) Normal fault — often in connection with graben (rift) structures.
b) Strike-slip faults — these will often not be sealed due to continued movements, but basement-controlled strike-slip faults often produce good anticlinal structures in overlying softer sediments.
c) Upthrust faults.
d) Growth faults.

3. *Stratigraphic traps*
Traps which are partially or wholly due to variation in facies or unconformities, and which are not primarily a result of tectonic folding or faulting of the strata (Fig. 13.11).

a) Primary pinch out of strata. Porous strata, for example sand strata which pinch out updip in less permeable rocks, e.g. shale, (Fig. 13.12).
b) Fluvial channels of sandstone. Fluvial channels may be isolated and surrounded by impermeable clay-rich sediments, or they may be folded so that we obtain a combination of stratigraphic and structural traps (Fig. 13.13).

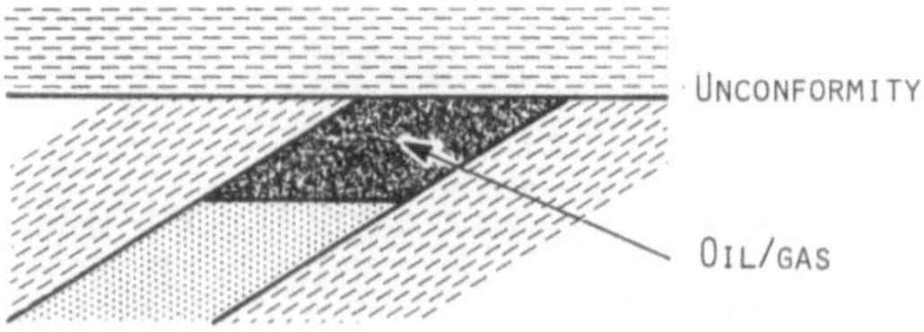

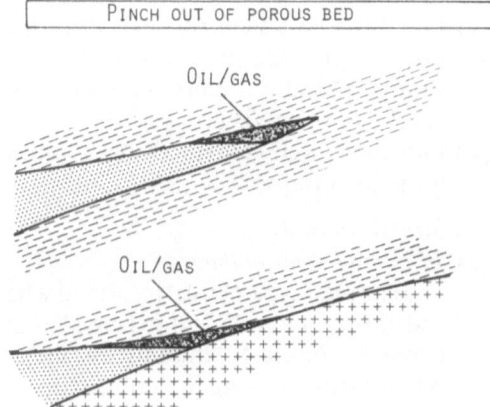

Fig. 13.11. Diagram of stratigraphic trap formed by unconformity due to tectonic tilting, erosion and renewed sedimentation

Fig. 13.12. Stratigraphic traps. Primary development of a porous bed

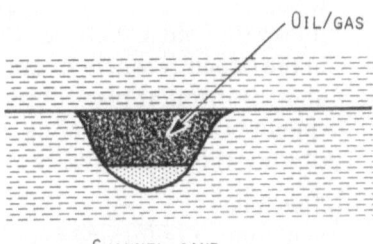

Fig. 13.13. Sandstone deposited in fluvial channels on a delta which forms a partial stratigraphic trap. In order to get sealing in all directions, the sandstone must also be folded or faulted

c) Submarine channels and sandstone turbidites in strata rich in shale. Here we will often find pinch-out of permeable layers updip from the foot of the continental slope. This will result in stratigraphic traps without any further folding being necessary.

d) Reefs often form stratigraphic traps. A reef structure projects up from the sea bed, and often has shale sediments surrounding it, so that oil could migrate from the shale into the reef structure (Fig. 13.14).

e) Traps related to unconformities. Sandstones or other porous rocks may be overlain with an angular unconformity by shales or other light sediments, forming a trap underneath the unconformity.

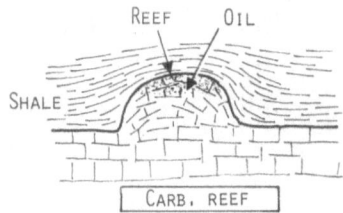

Fig. 13.14. Two types of primary stratigraphic trap, reef structure, and upward projecting basement. The oil migrates from the shale into the projecting structure

Fractured Reservoirs

a) Fractured basement rock which projects upward locally into overlying shales can also provide good traps. Remember that oil can migrate upwards into stratigraphically underlying rocks (Fig. 13.15). In China there are many examples of this type of trap.
b) Fractured, well-cemented rocks (limestones and cherts) along faults may also form good traps.

A. Structural traps
a) *The anticline or dome trap*
   The simplest type of trap is formed when a sandstone is overlain by a tight shale and this sequence is folded into an anticline. A regular anticline is no trap, however. The crest of the anticline must have a culmination so that hydrocarbons can be trapped along the fold axis.
   Dome-shaped structures can be formed by salt diapirs, horsts or by differential compaction. Anticlinal or dome-shaped structures can be mapped using contours displaying the relief of the structures, similar to contours on topographic maps. The traps appear as mountains or hills on the contours which follow the contact between the reservoir rock and the cap rock. The lowest point in the structure where hydrocarbons could be contained is called the *spill point* and a horizontal plane through that point is called the *spill plane*. This corresponds to the lowest sealing contour. The vertical distance from the highest point (the crest) of the structure to the spill plane is called the *closure* (Fig. 13.16).

b) *Salt domes*
   Salt domes are formed because salt (specific weight about 1.8–2.0) is less dense than the overlying rock, and the salt "flows" upwards due to buoyancy.

   In order for a salt deposit to move upwards and form a salt dome, we have to have a certain minimum overburden and the thickness of the salt deposits must be more than about 100–200 m. The movement of salt up through the strata, and the deformation of all the sedimentary strata this involves, is called halokinetics, or salt tectonics. The movements are very slow, and may continue for several hundred million years. They may lead to the salt breaking through the overlying rocks and flowing up to the surface, or forming intrusions into younger rocks. If the salt contains anhydrite, this mineral will expand and form gypsum when it comes into contact with groundwater as the salt approaches the surface. Traps may be formed (1) in the strata above the salt dome, (2) in the top of the salt domes (cap rock), (3)

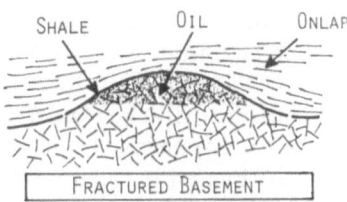

Fig. 13.15. Diagram showing a fractured basement rock acting as a trap oil migrates from the shale into the trap

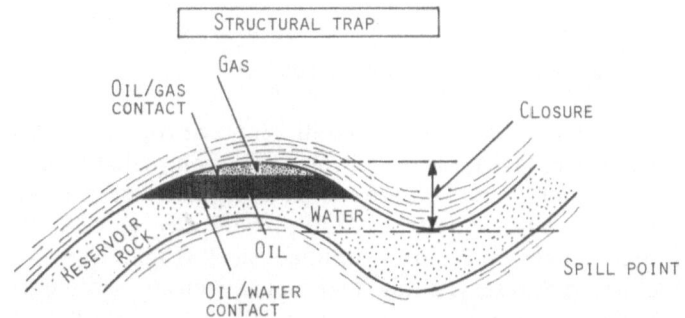

Fig. 13.16. Cross-section through a single anticlinal trap. The structure, with closure in all directions, and the spill point define the maximum volume of rock which can be oil-saturated

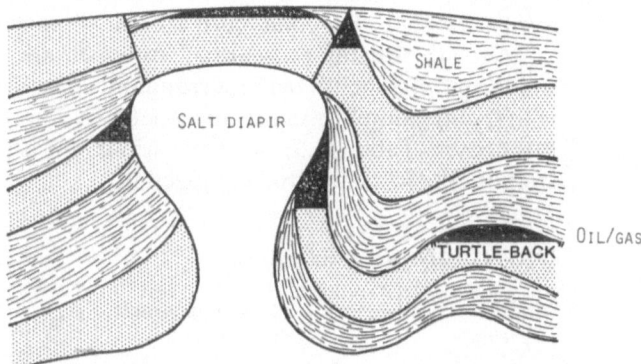

Fig. 13.17. Salt diapirs provide many potential environments for oil and gas traps. This applies to fault traps and anticlinal traps connected with the diapir. The upward flow of salt produces a rim syncline and then an anticline (turtle-back) further away. Salt is also a very good cap rock

in the strata which curve up against the salt structure, (4) due to stratigraphic pinch-out of strata around the salt dome (Fig. 13.17).

Cap rock reservoirs are caused by solution and brecciation at the top of salt domes.

Salt tectonics is of very great significance in many oil-bearing areas where we have thick salt deposits. In the eastern USA we find large areas with Silurian salt, and in the Gulf of Mexico there is Permian and Jurassic salt.

Salt deposits are particularly common in Permian-Jurassic sediments around the Atlantic Ocean. This is because in the period prior to the opening of the Atlantic Ocean, there were large areas with basins resulting from faulting (rifting) in the middle of a large super-continent (America + Europe, Asia, Africa) with little precipitation. Today we find a similar situation around the Red Sea and the Dead Sea. The Permian Zechstein salt in Germany and Denmark continues outwards into the North Sea and halokinetic movements have formed dome structures which are very effective traps, as for example in the Ekofisk area.

Evaporite basins also formed during early spreading of the Atlantic Ocean in Jurassic and Lower Cretaceous times, when it was still narrow, and salt formed at that stage is found extensively along the margins of the present Atlantic Ocean.

Growth Faults and Growth Anticlines

Growth faults are typical of sediment series deposited relatively rapidly, such as in deltas (see p. 84). The fault plane is often sealed (not always), and may prevent further migration upwards of oil and gas. Fault traps may form when sandstone beds are offset against the fault plane. However, oil traps are formed just as often in anticlines on the down-faulted block (Fig. 5.15b). These are "roll-over" anticlines. Because faulting takes place during sedimentation, the strata on the down-fault side will be thickest, hence the term "growth faulting". The throw between corresponding strata declines upwards along the fault plane. There is often formation of minor fault planes with an opposite throw (antithetic faulting) in the beds which curve in towards the main fault plane. Growth faults contribute greatly to reducing porewater circulation in sedimentary basins, and we often find under-compacted clay, which may turn into clay diapirs in connection with growth faults.

Growth anticlines are anticlines which are formed by one area in the basin sinking more slowly than the surrounding areas. Their formation is thus concurrent with sedimentation, and not due to later folding. Growth anticlines may form over salt domes (Table 13.3) or over various upward projecting features in the substratum due to differential compaction.

Table 13.3. Traps for hydrocarbons

Anticlinal dome traps:	Fault trap against sealing fault	Stratigraphic traps
a) Anticline with a culmination of the fold axis (fold fault axis dipping in both directions)	a) Normal fault including growth b) Strike-slip fault c) Upthrust fault	a) Primary pinch-pout of strata b) Isolated fluvial channels
b) Salt domes — dome produced by flow of salt		c) Submarine channels
c) Growth dome — sedimentary dome structure produced by differential subsidence		d) Reefs and other bioherms e) Beds cut by unconformities

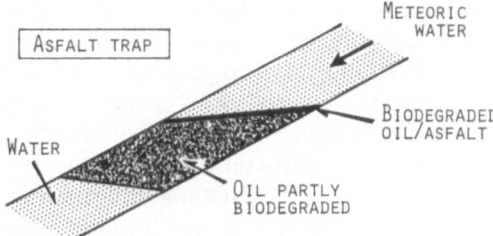

Fig. 13.18. Hydrodynamic trap. The oil water contact slopes because of the drag resulting from water flow

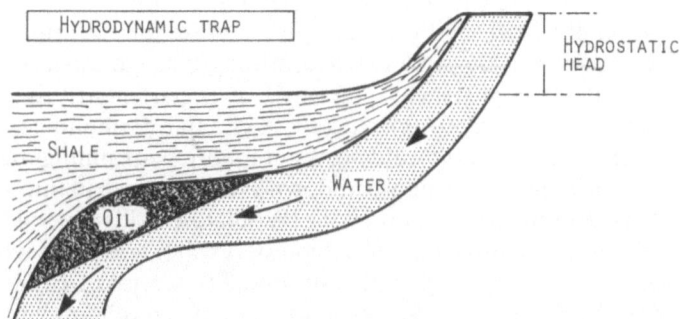

Fig. 13.19. Trap where asphalt formed by bacterial breakdown of oil is a cap rock

Other Types of Traps

There are also a number of less common types of faults. If porewater circulation in a sedimentary basin is sufficiently vigorous, the oil-water contact may deviate significantly from the horizontal because of the hydrodynamic shear stress which is set up. This has consequences for assessing the quantity of oil in structures, and in some cases oil may accumulate in the absence of closure, in a so-called hydrodynamic trap (Fig. 13.18). Flow of fresh (meteoric) water down through oil-bearing rocks will lead to biodegradation and formation of asphalt, which may then form a cap rock for oil (Fig. 13.19).

Drilling for Oil

Drilling for oil is a costly process. The object of a well is to prove the presence of or to produce oil or gas. Sometimes wells are also drilled to inject water, chemicals or steam into the reservoir during production. Even a well which fails to find hydrocarbons (a dry well) is of great value, however, because of the information it provides about the rocks in the area. This information forms part of the basis for the geological maps which are used in further exploration for oil and can be sold or exchanged for data from other companies. This is the reason that oil companies wish to keep the geological results of oil drilling confidential for some years after drilling has been completed. Wells drilled mainly to obtain stratigraphic information are called *stratigraphic wells.*

Oil drilling used to proceed largely on land, but now takes place to a steadily increasing extent in shallow ocean areas on our continental shelves. The costs of this type of drilling are many times greater than those associated with drilling on land. This has led to increased efforts to gain maximum information from wells. The costs of analysing samples and logs are small in relation to the costs of drilling the well. We shall not go very deeply into the technical aspects of drilling for oil here, but merely look briefly at some of the most important principles.

When drilling commences at the surface, the diameter of the well may be 20″–30″ (50–90 cm), but it decreases downwards to 3″–6″ (7–15cm) at great depths. Normally a roller bit is used, which crushes the rock into small pieces (about 2–5 mm) called *cuttings.* Only when drilling through especially important rock strata (usually reservoir rocks), of which it is particularly interesting to have whole samples, are core samples taken. A circular diamond core drill bit must then be used. This takes time, and costs a lot more per running metre.

Drilling mud is pumped down through the drill into the well during drilling. This mud has several functions. When one drills several hundred or a thousand metres down into rock, one encounters water, gas or oil which may be under high pressure. The drilling mud is supposed to serve as a counter-measure to this, and prevent the uncontrolled gush of water or petroleum into the well and up to the surface in a blow-out. The pressure in the well is determined by the columns of drilling mud, and can be regulated by pumping down more mud and by increasing its relative density. The pressure of the drilling mud must exceed the pressure of oil and water in the surrounding formation. Heavy minerals such as barytes are frequently used as an additive to drilling mud, which otherwise consists of montmorillonite-containing clays with a large number of different additives. The drilling mud also serves to cool off the drill bit, and cuttings are brought back to the surface suspended in the circulating mud. The cuttings are then washed out from the drilling mud onto a sieve (shale shaker) and the mud can be used again.

The cuttings are continuously analysed on the drilling platform by a geologist who describes the rocks of which the cuttings consist and their mineral composition. Samples of cuttings are usually taken every 10–30 feet drilled and these samples can be analysed in more detail in the laboratory. The cuttings are often poorly washed and need extra cleaning to get rid of the drilling mud. In poorly indurated mudstones it may be difficult to separate the drilling mud from the soft cuttings. Organic additions to drilling mud may cause problems when analysing cuttings, and sometimes oil-based rather than water-based drilling mud is used. The fossil content of the rock fragments, largely microfossils, is used to determine the age of the strata drilled through (see biostratigraphy). All the cuttings which come up with the drilling mud do not, however, necessarily come from precisely the strata being drilled through at that time. There is also a certain mixture due to "cave in" of overlying strata. This means that we find material of a different composition mixed in, and fossils from another age than the one being drilled through. Care has therefore to be taken with the stratigraphic interpretation of microfossils from cuttings. The surest way is to register the first occurences of a species when one is proceeding downwards from the top in the well. The last occurrence of a fossil may be the result of cave-ins from younger strata.

Since the pressure of the drilling mud may prevent oil and gas from penetrating into the well, one may drill through significant oil and gas occurrences without registering them. It is therefore advisable to carry out special tests in the most promising strata to find out if there is petroleum present, and in what quantities.

As drilling proceeds, the well is lined witn steel casing to prevent rocks falling into it, but prior to casing each section of the well it has to be logged with different logging tools which require contact with the rocks. Radioactivity logs, however, can also be run after the casing has been installed. It is useful to note when the different casings are installed, because that limits the strata which could have "caved in" and contaminated the cuttings. If the well is going to produce oil, a production pipe is used and installed running through the petroleum-bearing strata. It is then per-forated by shooting holes in the steel casing (the oil column) so that petroleum can flow into the well.

For an oil field to be capable of full production, several wells are normally required. A well drilled to recover oil from a known oil field is called a production well. A well drilled to estimate the extension of an oil field is called an appraisal well. The first well in a new area is called a wildcat well.

Calculating the Size of Reservoirs

The first thing which is done is to estimate the volume of the rock in the structures above the oil/water contact (Fig. 13.16). The estimations are based on seismic sections and wells through the construction of a *structure-contour* map, a kind of topographic map of the structure. Then contours are drawn through points on the same level in relation to a *datum*, which is a reference level like sea level or the contours on a map. The volume of rock can then be calculated for the whole structure, and also for that part which corresponds to the oil/water contact, or the gas/oil contact if there is gas. Usually the oil/water contact is horizontal, but if there is dynamic flow of water through the structure, the fluid drag may produce a sloping oil/water contact.

The average porosity must then be calculated for the oil-bearing part of the rock. We often divide up the whole structure and work out the porosity values for the different beds in it. We must also estimate how much of the pore space is occupied by oil and how much water is still left in the smaller pores. This is called the *oil saturation*. Some beds are too fine-grained and have too low a permeability to yield much of their oil. The beds that we think will produce are called *pay*. The whole oil column to the top of the reservoir is called the *gross pay* and the total thickness of the beds that will actually produce significant quantities of oil is called *net pay*. The oil found in the structure is called "oil in place". This is then:

Volume oil = volume oil-bearing reservoir rock × porosity × oil saturation.

Only part of this oil can be recovered, however, and recoverable oil often only makes up 20–60% of the oil in place.

Composition of Porewater in an Oil Field

Analyses of porewater occurring in the oil field along with the oil provide important information about the diagenetic history of the reservoir rock, and the migration of oil. The total salt content of the porewater and the concentration of individual elements are results of the origin of the porewater and processes which have later modified the composition. In a marine sediment basin there are four main sources of porewater:

1. Marine water which is buried in the basin as porewater in the sediments.
2. Fresh porewater which flows into the basin from meteoric sources (groundwater, see p. 248).
3. Water which is released through dehydration of water-containing minerals, e.g. smectite, kaolinite, gypsum etc. or through diagenesis and maturation of organic matter.
4. Juvenile water. Water which comes from magmatic sources.

Porewater which is buried with sediments and which has not later been in contact with the atmosphere is called "connate water". In marine basins it is characterised by a high chloride content and a low oxygen, bicarbonate and sulphate content. Sodium is the most important cation. Pure water of meteoric origin derived from fresh groundwater usually has less than 10,000 ppm dissolved material. It contains more bicarbonate and small amounts of magnesium, sodium and calcium. Because meteoric water comes from the surface, it brings with it oxygen and bacteria which can break down hydrocarbons (see "Biodegradation"). The composition of meteoric water alters as it reacts with the more readily soluble minerals which it passes through. Carbonates will act as a buffer, and evaporites in particular will rapidly increase the salt content of meteoric water.

Drilling in recent years on the continental shelves, for example off the east coast of the USA, shows that meteoric water is very widespread, and that water which is virtually fresh is sometimes found below the sea bed as far out as 100 km from the coast (Mannheim and Paul, 1981). In the North Sea there is also isotopic evidence that the formation water in shallow reservoirs is partly of meteoric origin.

Marine porewater, which is present in sediments when they are deposited, will initially have an approximately normal salinity. Clay minerals function as ion exchangers. Marine water has a composition which is not far from being in equilibrium with the most common clay minerals. Potassium is easily adsorbed onto clay minerals, so that the potassium content of porewater is usually less than in ordinary marine water. When clayey sediments are compacted, the pores will shrink and compacted clay and mud may function as a membrane. Clay minerals are normally negatively charged on the surface and particularly at the ends, where there are ruptured bonds. When the pores become small enough, it becomes easier for the metallic cations to move through them than the anion complexes, which are repulsed by the negative charges of the clay minerals. In order for the charges on both sides to equalize, the small ions, particularly H^+, must move in the opposite direction, and as a result there is a higher H^+ concentration (lower pH) on the lower side of the membrane. This process, which will concentrate salt, is called *salt sieving*. Membranes also discriminate selectively amongst different cations, depending on their size and charge. Those alkali ions which are strongly hydrated (Na^+, Li^+) will be adsorbed onto the surface of the clay to a lesser degree, and will be more mobile than, for example, K^+, Rb^+.

At higher temperatures, 80–100°C, smectite will dissolve and form illite + water. At 120–140°C kaolinite will become unstable and form illite, quartz + water. This process binds cations, particularly K^+, and releases pure water, so that the salinity of the porewater decreases. As a result the porewater in shales often has a lower salinity than that in sandstones at the same depth. The composition of porewater in sandstones varies greatly, depending on whether they contain meteoric water or not.

The quantity of dissolved matter in porewater increases as a function of depth in most cases. Concentrations of 100,000 to 300,000 ppm of dissolved matter (total dissolved solids, TDS) are typical for about 1 km to 3 km overburden, e.g. in the oil and gas fields around the Gulf of Mexico. In the North Sea the salt content is often very much lower, particularly outside the extension of the Permian evaporites.

Porewater in reservoir rocks is often called formation water. If it has a high salt content it is also called "oil-field brine". Characteristic of its composition is the fact that Cl^- is by far the dominant anion, so that the Na^+:Cl^- ratio is less than 1. The rest of the positive charge is made up of Mg^{2+} and Ca^{2+}, such that $Cl^- - Na^+ = Ca^{2+} + Mg^{2+}$.

The composition of connate porewater is a function of the mineral composition of the sediments, the temperature and the quantity of dissolved gases, particularly CO_2. With increasing temperature, the kinetic obstacles to mineral solution and precipitation reactions are reduced. At temperatures above about 80° the porewater will tend to be in equilibrium with most of the minerals present. In sedimentary basins with evaporites, these will greatly influence the composition of the porewater in overlying formations. Dissolved salts move upwards through flow of porewater due to compaction, and diffusion due to high concentration gradients also plays a major role. Sedimentary basins along the Atlantic Coast have porewater compositions in areas where Mesozoic evaporites are deposited which are essentially different from those in areas north of this palaeoclimatic belt. The South American continental shelf, the Gulf Coast and the area off the East Coast of the USA are characterised by Jurassic/Cretaceous evaporites. In the North Sea the extension of Zechstein evaporites forms an important boundary which is reflected in the porewater composition of overlying Mesozoic sediments.

In addition to chemical analysis, isotope composition ($^{13}C/^{12}C$ and $^{18}O/^{16}O$) can provide important information concerning the origin of porewater.

Formation of Reservoir Rocks

The most essential requirements for a good reservoir rock are that it has high porosity and permeability and a thickness and volume sufficient to enable it to hold large quantities of oil. In order for a reservoir rock to actually contain large volumes of petroleum, a number of different conditions must be fulfilled.

1. The reservoir rock must have preserved much of its primary porosity (more than 10–20%) or developed secondary porosity prior to oil migration.

2. The reservoir rock must be overlain by a cap rock with low permeability which forms a structure with closure upwards.
3. The reservoir must be within reach of hydrocarbons migrating from mature source rocks.

Primary Depositional Environment of Reservoir Rocks

1. A knowledge of the distribution of potential reservoir rocks and their properties is one of the most important preconditions for oil exploration. Mapping the primary sedimentary facies is therefore vital for delimiting potential prospecting targets.

 Primary sedimentary facies determine the geometry of the reservoir rock and provide models for prospecting and also for planning production on an oil field.
2. The environment of deposition very largely determines the initial vertical and horizontal variations in porosity and permeability of the reservoir rock.
3. Diagenetic processes, and the accompanying development of porosity and permeability following removal of the overburden, are as a rule also controlled by primary sedimentary facies.

The combination of high porosity and permeability is most often found in well sorted sandstones, i.e. a large percentage of the grains are over 63 μ, and the clay and silt content is low. We must be aware that grain-size distribution is not the same as it was at the time of deposition, and we must distinguish between primary clastic minerals and new minerals which are formed through diagenesis. Some of the clastic minerals may also have gone partially or wholly into solution. Consequently we cannot just use grain-size distribution to interpret the depositional environment, even if the error is not so great for the larger grain sizes. It is also often difficult to disintegrate sandstones for grain-size analyses, and analyses are difficult to do on thin sections.

 Relatively well-sorted sandstones are deposited in different environments:

1. On beaches and shorefaces in shallow marine areas, due to washout of fine material through wave activity. This gives greatest porosity and permeability, since sediments have little fine material and a high proportion of larger sediment grains (negative skewness). Into this group fall a number of important types of reservoir rocks.

 a) Beach sand
 b) Barrier island sand
 c) Washover sand deposited in fans landward of the barrier
 d) Sand deposited in shallow marine environments below the fair-weather wave base during storms.

2. *The Tidal Water Environment:* Tidal flats are characterised by strong currents alternating with still-standing water, and typically produce thin lenses of sand with

current bedding in a clay-rich matrix formed through deposition from suspension at high tide. Sediments of this type will have low porosity and permeability. Tidal channels, on the other hand, will represent well-sorted sand fining upwards. Sand ridges deposited by tidal currents at great depths (sub-tidal) will be characterised by cross-bedding and current ripples and will show increased sorting upwards.

3. *Delta Deposition:* Here we find a series of facies within the delta with very different depositional environments. Delta-front sands will be increasingly sorted and more coarse-grained upwards. The same wil be true of sand which is deposited just outside the river mouth (distributary mouth bars). Sand which is deposited in the channel itself, conversely, will characteristically fine upwards from the bottom of the channel. Channel sand may provide good reservoir rocks, but it is important to remember that they have an almost one-dimensional extension, and that the reservoir volume cannot be very great. Fluvial sandstones will also have a primary clastic clay content.

4. *Bed-Load Fluvial Channel Sandstones:* The geometry of sandstones, and how much of the various facies is represented, will vary greatly according to the type of depositional system. In constructive deltas, the fluvial facies will dominate, and the sand/clay ratio will to a large extent determine whether we have thin, isolated sandstones surrounded by clay, or whether we have a more continuous system of channel facies. Sand which is deposited where channels break through onto the delta (crevasse splays) will produce a more continuous but local thin sand bed.

Destructive deltas will consist of wave-dominated or tide dominated facies, and we will find more continuous sand beds of delta-front sand or sand ridges, and hence larger sand bodies.

Sandstones in delta facies have a great potential for becoming reservoir rocks, additionally so because they are often associated with clay sediments deposited under reducing conditions, both between the outlets of rivers (interdistributary facies) and in delta slope facies.

Fluvial sandstones in more continental facies, which are deposited by meandering rivers, will form continuous horizons with cross-bedded point bar sequences, separated by fine-grained silt and clay sediments ("overbank" sediments), through lateral migration.

Branched or "braided" rivers will, however, give rise to cross-bedded sandstone beds with more vertical continuity, without such frequent occurrence of fine-grained clay and silt layers, which reduce the vertical permeability.

Sediments which are deposited in alluvial fans will also have very high porosity and permeability. However, these are deposited in an oxidising continental environment, and are relatively seldom associated with good source rocks and cap rocks.

Eolian sandstones are extremely well-sorted and the grains are about 0.2 mm. The sandstones are good reservoir rocks. What is crucial is their association with source rocks and cap rocks. Cap rocks may, for example, be formed through deposition of evaporites, as is the case in the North Sea, where Zechstein salt overlies Permian eolian and fluvial sandstones.

Marine Sandstones Deposited on Submarine Slopes

Below the wave base most sedimentation will occur through transport in suspension, and the sediments deposited are mainly fine-grained and poorly sorted. Organic material is more easily retained, and we primarily find source rocks formed.

In a number of environments, however, there will also be strong currents and transport along the bottom. This applies particularly to submarine canyons, where relatively pure sandstones may be deposited. These could become very good reservoir rocks and have a high potential for hydrocarbons because they will be surrounded by fine-grained clay sediments which could become source rocks and cap rocks. Sand deposits of this nature are deposited on a slope and may pinch out upwards towards the shelf, forming stratigraphic traps.

Sediments which have been transported in suspension may also become reasonably well-sorted. This is particularly true of proximal turbidites. Turbidites tend to have low primary porosity, but on the other hand often thickness, volume and extension, so that they may form large reservoirs.

It is important to interpret the environment in which reservoir rocks are deposited. Their internal properties and geometrical form are a direct function of the processes which deposited the sediments.

Carbonate Reservoirs

About 40% of all oil and gas produced is found in carbonate rocks, including most of the large oil fields around the Persian Gulf.

Carbonate reservoirs distinguish themselves from sandstone reservoirs in a number of important respects:

1. Biological build-up of bioherms (reefs) gives rise to upward projecting structures which are themselves hydrocarbon traps.
2. Carbonate minerals are more soluble than silicate minerals, and solution and formation of secondary porosity is even more important than in sandstones.
3. Primary mineralogical composition, and consequently also diagenetic development, is largely a function of biological conditions during deposition.
4. Carbonate rocks which otherwise have low porosity and permeability often form fracture reservoirs.
5. Carbonate minerals have essentially different surface properties from silicate minerals and generally tend to be more oil-wetting than sandstones.

The extension of carbonate reservoirs is largely a function of the ecological conitions in the sea during deposition. It is therefore important to try and reconstruct the sedimentary basin as precisely as possible. Areas with a clastic supply source must be mapped because even small amounts of suspended material prevent a number of carbonate-secreting organisms from growing, particularly corals. It is important to map shores, shelves and the margin abutting on the ocean basins. This will determine the orientation of the carbonate in the beach facies, and reef structures also often lie out on the edge of the shelf.

In many areas of the southern USA, oil exploration concentrates on finding old reef structures under younger sediments (so-called "reef trends").

A palaeogeographic and palaeobathymetric reconstruction of the basin will also be necessary in order to understand the diagenetic development of carbonate reservoirs, which to a large extent is controlled by circulation of fresh (meteoric) porewater. Carbonate reservoirs consist largely of shallow marine carbonate deposits, and are therefore very sensitive to changes in sea level. Calcareous debris precipitated by pelagic organisms seldom form oil and gas reservoirs. The Ekofisk Field is one of the few examples of such a chalk reservoir.

Porosity in Carbonate Reservoirs

Carbonate reservoirs can only be understood against a background of general carbonate sedimentology and diagenesis, as discussed in Chapter 8.

Primary porosity in carbonate rocks consists of:

1. Interparticle porosity in grainstones, e.g. between ooids, pellets, lithoclasts and fossils.
2. Intraparticle porosity in fossils, e.g. snails (gastropods).
3. Protected cavities under fossils (shelter porosity).
4. Cavities formed in carbonate mud due to gas bubbles (fenestral porosity).
5. Primary cavities in reefs or coralline algae beds (growth framework porosity) (See Fig. 13.20).

Secondary porosity can be formed through:

1. Biological breakdown — cavities formed by boring organisms, e.g. living mussels.
2. Chemical breakdown of minerals which are unstable in relation to the composition of the porewater.

Most boring organisms, like boring algae and sponges, produce holes too small to be of significance for porosity. On the other hand, they form micrite, which is more stable during later diagenetic transformation, so that a micritic envelope may be formed (see p. 160). Large boring organisms, particularly boring mussels, make large cavities which easily become cemented.

The most important form of secondary porosity is due to solution of aragonite through percolation of freshwater. If calcite is precipitated at the same time as aragonite is dissolved, so that a neomorphic transformation takes place, little secondary porosity develops. This happens if the porewater is undersaturated with respect to aragonite and saturated with respect to calcite. If the porewater is undersaturated with respect to both aragonite and calcite, cavities will be formed, e.g. after aragonite fossils, which may give rise to secondary porosity. The decisive factor is whether oil migrates into the pore before it refills with drusy cement (Fig. 8.16). The development of this type of porosity is very largely controlled by the flow of meteoric water along the margin of and under sedimentary basins. Periods of uplift of land or lowering of sea level cause extensive solution by meteoric water,

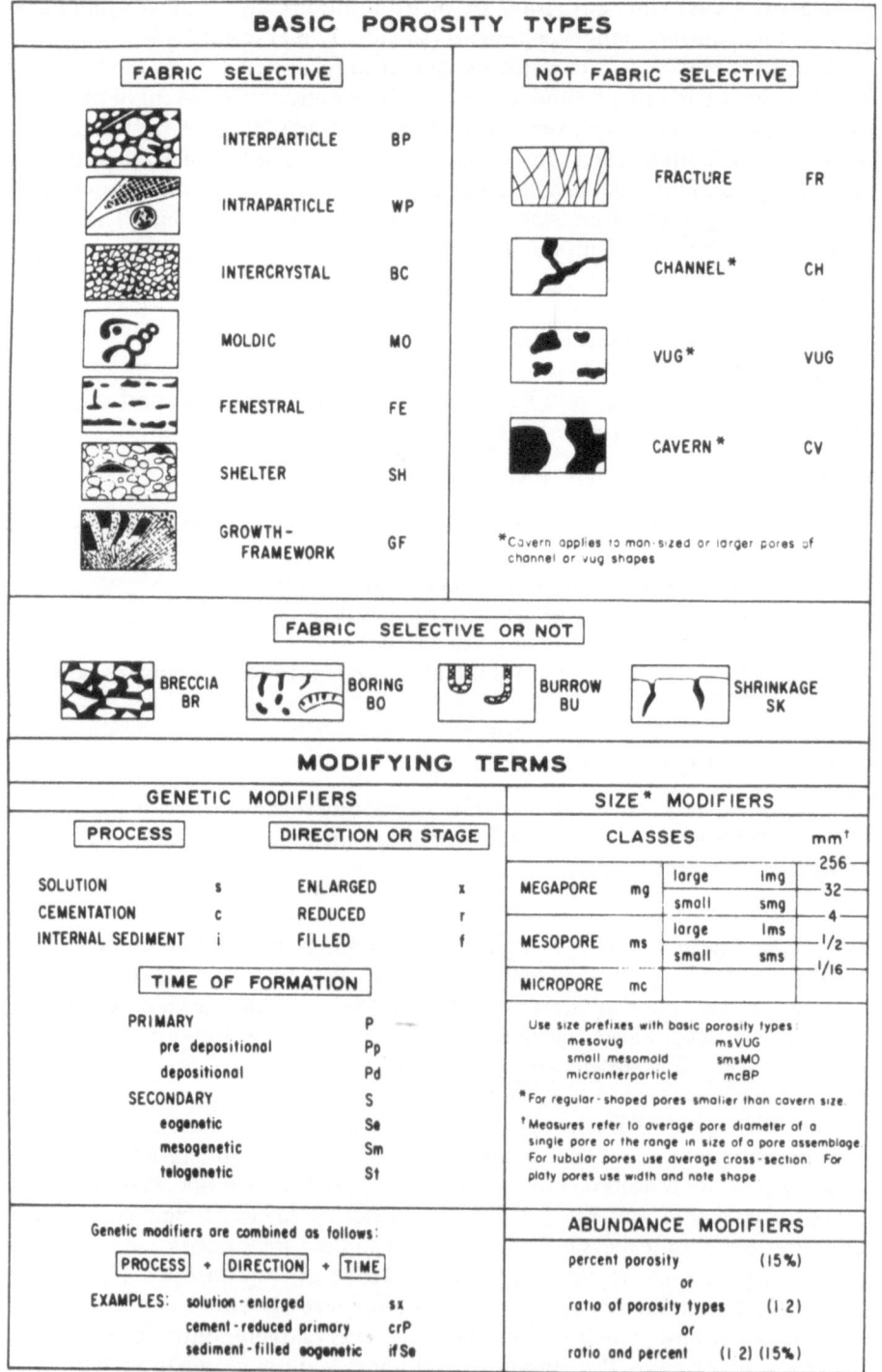

Fig. 13.20. Different types of porosity in limestones. (After Choquette and Pray 1970)

often with formation of karst holes, which may be an important type of porosity in carbonate reservoirs.

The other main type of secondary porosity is due to dolomitisation. During dolomitisation the amount of dolomite precipitated is often less than that corresponding to the dissolved calcite (or aragonite), the result being a net increase in porosity. Since dolomitisation is often associated with zones where fresh and marine porewater meet and mix, there is a tendency for reservoirs with this sort of porosity to lie near the margin of the continental shelf, where meteoric porewater flows down into the basin. In many basins, (e.g. the Alberta Basin in Canada), oil exploration is very largely a question of finding dolomite zones.

Reduction of Porosity

The most important cause of reduction of both primary and secondary porosity in carbonate rocks is pressure solution. The same principles apply here as for sandstone diagenesis, but carbonate minerals are more soluble under pressure as well. Well-sorted carbonate sand will be subjected to pressure solution at the contact points between grains, and will precipitate cement between the grains. Typical examples of this are often found in oolitic sand. In the same way as in sandstones, overpressure contributes to reducing pressure solution. Precipitation of limited amounts of early carbonate sediment, e.g. in beach rock, causes the pressure to be better distributed, and reduces pressure solution with burial. Consequently we often find that carbonate sandstones which have been subjected to rather early diagenesis retain their remaining porosity best with depth. Pressure solution often also leads to formation of stylolites, which are continuous zones where solution is in progress. Because of different rates of solution, the stylolite surface will be very irregular, with a relief which represents minimum solution (Fig. 8.18).

When carbonates are dissolved from a surface, silicates and other minerals with low solubility remain behind and form a thin membrane. This membrane consists largely of clay minerals, and may be almost impermeable to water and especially to oil. In the Middle East there are oil reservoirs which are limited by stylolites, and communication within reservoirs is often greatly reduced because of stylolites. This is true, for example, of the Ekofisk reservoir in the North Sea.

Carbonate reservoirs also often tend to be more oil-wet than clastic reservoir rocks (higher wetting angle), but this depends very much on the composition of the hydrocarbons and on the pressure.

Further reading: Dickey 1979; Mobson and Tiratsoo 1981; Link 1982; Selley 1982; Perrodon 1983; Tiratsoo 1984; North 1985.

14 Well Logs

Logging

Logging is a way of recording the physical properties of a rock by taking continuous measurements in a well.

One of the advantages of this sort of record is that it is continuous, so that one obtains a picture of the gradual change in physical properties from one bed to the next. Taking samples from the rock itself from cuttings for analysis will not give the same continuous profile, even if numerous samples are taken. Through logging one also measures properties which cannot be measured in a laboratory.

A borehole is logged by sending a probe with measuring instruments down a well. The measurements from the instruments in the logging tool are recorded digitally at intervals of between 3 and 15 cm, and the data is processed near the well on land, or on the platform in the case of offshore wells.

Well logging is carried out by special logging companies which work under contract for the oil companies.

Most logs (except radioactivity logs) are dependent on direct contact with the rock via the walls of the well, and have to be run at regular intervals while drilling, before the steel casing is installed in the well.

Modern logging tools make several types of records at the same time, and the instruments are built into a long steel pipe which is only about 10 cm in diameter.

The following types of log are the most important:

1. Electric logs – self-potential, resistivity and conductivity logs. Electric logs were the first type of log to be implemented in exploration for petroleum, and involve measurement of specific resistivity and self-potential (S.P.).
2. Radioactivity logs – gamma ray, neutron logs and density logs. Gamma ray logs measure the natural emission of gamma rays from rocks in the well. A neutron log is obtained by using a neutron source which sends radiation into the rocks. The absorption, mostly by hydrogen atoms, is then measured.
3. Acoustic (sonic) logs. These are measurements which record how sound travels through rocks, and provide information about porosity in particular.
4. Dipmeter logs. A dipmeter log is an electric log which measures the slope of beds and laminations in rocks.
5. Caliper logs – record the diameter of the borehole.
6. Temperature logs – record borehole temperature and can be used to calculate the true formation temperature.

Well logs are used both qualitatively and quantitatively. Qualitatively, the characteristic reactions from different types of rocks are used for stratigraphic correlation, identification of sedimentary facies etc. Quantitatively it is possible to determine porosity and, if relevant, the water and oil saturation of the rock on the basis of logs. Logs are the most important basis for correlating sequences in a sedimentary basin and for evaluating the properties of reservoir rocks for production purposes.

Electric Logs

Resistivity Logs

Resistivity is a relatively simple property to measure, and resistivity logs were one of the first types to be used by the oil industry.

Resistivity logs are the result of measuring the resistance between two and four electrodes which are in contact with the rocks of the well walls. Resistivity is measured as a function of the cross-section between the rock and the distance between the electrodes (Ohm/m^2/m). Most minerals are very good insulators and it is only the clay minerals and salts such as KCl, NaCl which have a fair degree of conductivity. Almost all conduction takes place through the liquid phase, and the resistance therefore depends first and foremost on pore liquid. Rocks containing porewater with a high salt concentration have a lower resistivity than rocks with fresh porewater. Conductivity is also a function of the amount of porewater relative to rock volume (hence porosity) and the distribution of pores in the rock (permeability). Each of the electric log measurements depends on the degree to which drilling mud invades the formation, because this will influence the electrical properties.

Since it is not the conductivity of the porewater we are interested in, but the properties of the formations, we measure what is called the *formation factor* (F).

$$F = \frac{R_o}{R_w} .$$

R_o is the overall resistivity of the rock (formation) saturated with formation fluid. R_w is then the resistivity of the formation water or other fluids (oil or gas). The formation factor is a function of the porosity and permeability of the rock and is an expression of rock properties independent of the conductivity of the porewater. For sediments with a high primary porosity the formation factor will be an expression of the diagenetic alteration of the rock. The relationship can be expressed:

$$F = \emptyset^{-m},$$

where $\emptyset$ is the porosity and m is the cementing exponent which varies from 1 for porous rock to 3 for very well-cemented rock.

Sediments with porewater with low salinity (meteoric water) have a higher resistivity than sediments with normal marine porewater. Oil, and particularly gas, will greatly increase the resistivity. Well-cemented limestones (with a low porosity and permeability) have a very high resistivity. This is a characteristic feature which helps to identify thin limestones which may be important for correlations. Coal beds have an even higher resistivity, because coal has virtually zero conductivity. Even relatively thin coal beds show up strongly on the resistivity log. The thickness which can be recorded depends on the distance between the coils. Pure, well-cemented sandstones have a higher resistivity than impure, clay-rich sandstones. Evaporites are characterised by very good conductivity and low resistivity.

An induction tool is a means of focussing the induced current used for resistivity measurements. With a distance of 1 m between the coils, resistivity is measured 1–5 m into the formation away from the well.

Spontaneous Self-Potential Logs (SP Logs)

Self-potential logs measure the electric potential which develops between drilling fluid and porewater. A current is formed due to the difference in the concentration of electrolytes in the liquid phases. This is called the spontaneous self-potential. What one is chiefly interested in is the relative variations in the spontaneous self-potential. The SP log is therefore calibrated so that the self-potential in shale in the sedimentary sequences in question becomes the baseline. Readings normally cover a range of about 100 mV, and sandstones have more negative values than shale. The readings for porewater with a lower resistivity than the drilling mud will be negative (the curve will swing to the left). The readings are a good indication of how pure a sandstone is, i.e. the degree of sorting, or clay content. SP logs are well suited for use in interpreting sedimentary facies because coarsening and fining-upward sequences appear very distinctly on the logs. SP logs are often used as an expression of sorting and permeability in clastic sediments. Sandstones with relatively fresh porewater will give less negative readings than sandstones with marine porewater and if the formation water is fresher than the mud filtrate, the values will be positive.

In most cases readings from limestones lie between those which correspond to a shale and to a well-sorted sandstone, depending on porosity.

Self-potential is generated from porewater over a certain depth range, and SP logs do not generally give the best bed resolution. Partly for this reason, SP logs are often replaced by gamma-ray logs, which give very much the same information with better bed resolution.

Radioactive Logs

Gamma-Ray Logs

Gamma-ray logs measure the natural radioactivity which is produced in the rock. The elements which produce gamma radiation of significance in ordinary sedimentary rocks are potassium, thorium and uranium. The relative contribution to the total recorded radiation is such that 1 ppm uranium corresponds to 3.65 ppm thorium and 2.70% potassium. Shale normally contains most of these elements and the gamma reading of shales is almost always higher than that of sandstones. Although the potassium content of clay minerals varies a good deal, and sandstones also contain potassium, the gamma log will give readings which are basically a function of the sand/shale ratio. In consequence it is very similar to the SP log.

Sandstones with a high content of felspar and mica will have a proportionately greater gamma intensity than purer, quartz-rich sandstones. The Jurassic sandstones from the North Sea have rather unusually high gamma-ray readings due to their high mica content. Glauconite will also produce an intense gamma-ray response. Black shales in particular, with their substantial organic content, produce marked reactions on the gamma log because they normally have a higher uranium content than other shales. Kimmeridge (Upper Jurassic) shale in the North Sea contains 2–10 ppm uranium, and the contribution to gamma radiation from the uranium is therefore very great. Normally shales contain less than 1 ppm/uranium, but 10–12 ppm thorium, which represents a very significant percentage of the total radioactivity (49–50%). For this reason Kimmeridge shale is called the "hot shale". Gamma radiation is measured in API units from 1–200 and this scale can be calibrated with a standard radiation intensity (micrograms of radium per tonne).

Limestones have very low concentrations of U, Th and K, and give very low gamma-ray responses. In evaporite sequences, however, gamma logs are very sensitive indicators of potassium salts.

In recent years the gamma-ray logging tool has been equipped with a scintillation counter capable of distinguishing different energy levels (expressed in MeV of gamma radiation from rock, so that it can distinguish between the relative contributions of K (1.46 MeV), U (1.76 MeV) and Th (2.62 MeV). Such logs are called *spectral gamma-ray logs* and by using only the Th log, for example, it is often possible to obtain a better estimate of the shale content, because the concentration of thorium in shales varies less than that of uranium and potassium.

Neutron Logs

The neutron log method is based on a probe which emits neutrons at high velocity. The neutron rays are absorbed by rock and particularly the water in rock. This is due to collisions with atomic nuclei, and the absorption of the neutron radiation is primarily a function of hydrogen atom concentrations (the Hydrogen Index). The reduction in neutron radiation at a certain distance from the neutron source can then be measured and is an indication of hydrogen concentration. The frequency of collision of the neutrons can also be recorded by measuring the secondary X-ray radiation which is created by absorbing the neutron radiation. Since most of the hydrogen in rocks is present as water, neutron logs provide an expression of water content and thereby the porosity of a sediment. Neutron logs are particularly useful for determining the porosity of shales. As opposed to SP logs and resistivity logs, neutron-log response is not dependent on permeability. Gas is less dense and has fewer hydrogen atoms per unit volume than water and oil, and therefore has a lower hydrogen index (Fig. 14.1). Neutron logs can therefore be used to detect gas and distinguish it from oil.

Organic matter such as coal and other kerogens also has a high hydrogen index.

The hydrogen index of limestones and sandstones can be converted into neutron porosity units, and neutron logs are the best logging tool for determining the porosities of reservoir rocks.

Neutron logs often bear the symbol "CNL" (Compensated Neutron Tool) or simply "CN".

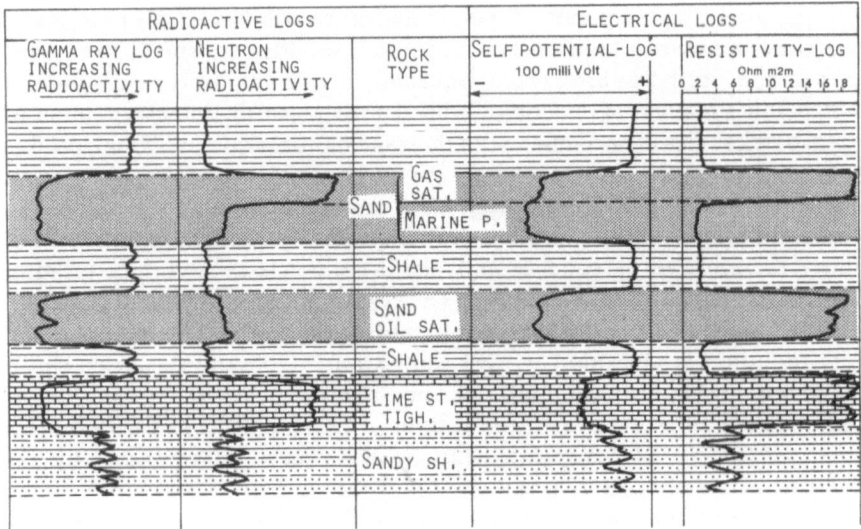

Fig. 14.1. Theoretical log patterns for different rock types. (Hobson and Tiratsoo 1981)

Density Logs

Density logs record the electrodensity of a rock and thereby its density.

Medium-energy gamma rays are focused on the formation, and their attenuation is detected. The attenuation is a function of electron density (Compton scattering), which is very closely related to rock density, expressed in g/cm^3.

Density logs produce important information which helps to identify different lithologies as a function of their densities. If the rock type and its matrix density are known, the porosity can be calculated from the bulk density.

$$\text{Porosity} = \frac{\rho \text{ matrix} - \rho \text{ bulk}}{\rho \text{ matrix} - \rho \text{ fluid}} \, .$$

The bulk density of a rock also depends on the pore fluid and in gas reserves the density of gas must be used when calculating porosity. At the same time the gas/oil or gas/water contact can be detected as a change in bulk density if it occurs in a homogeneous part reservoir rock. Figure 14.2 shows a log heading for a neutron density log used to determine porosity.

Sonic Logs or Acoustic Logs

With this method a probe sends out acoustic pulses which travel through the rock surrounding the well to the other end of the logging tool, and the velocity of sound in the rock is recorded. The velocity is usually presented in microseconds per foot even if the log is otherwise metric. This is called "interval transit time", and is presented on the logs on a scale of 40–140 $\mu s/ft$ ($\mu s = 10^{-6}$). 100 $\mu s/ft$ corresponds to 10,000 ft/s, or 3048 m/s. The interval transit time is the reciprocal of the sonic

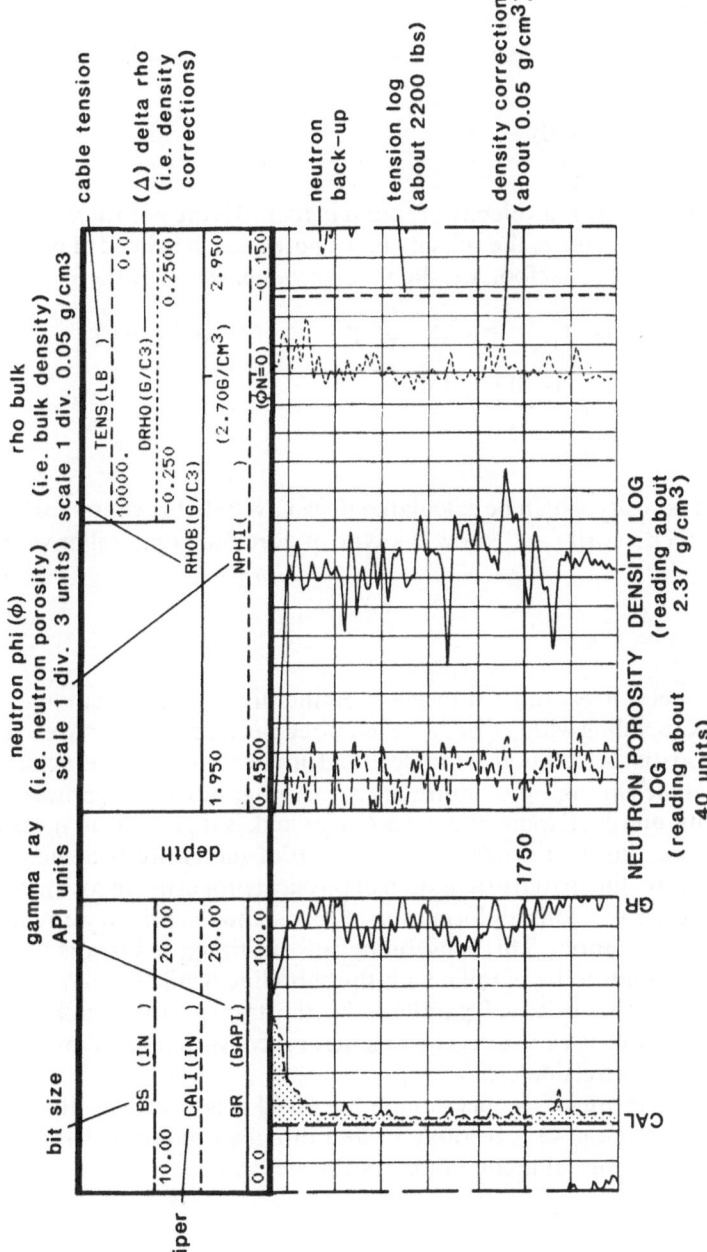

Fig. 14.2. Log heading used to determine neutron porosity. (Rider 1986)

transit velocity. Since the velocity of sound in water is considerably lower than it is in minerals and rocks, it will be more or less inversely proportional to rock porosity.

Wylli's equation expresses the relationship between the velocity of sound and porosity:

$$\frac{1}{V_r} = \frac{(1-\emptyset)}{V_s} + \frac{\emptyset}{V_f} .$$

V is the measured velocity of sound in rock. $\emptyset$ is the porosity. V_s is the velocity in the rock matrix and V_f the velocity in the pore fluid. If instead of measuring the velocity in m/s we use μs/foot, i.e. the inverse value, we obtain:

$$\frac{1}{V} = T_r, \qquad \frac{1}{V_s} = T_s \qquad \text{and} \frac{1}{V_f} = T_f$$

$$\text{and } T_r = (1-\emptyset) \cdot T_s + \emptyset T_f$$

$$\text{and } \emptyset = \frac{T_r - T_s}{T_f - T_s}$$

The porosity, $\emptyset$, can be calculated if we have reliable values for T_s (interval time of the rock matrix) and know the type of pore fluid (gas, oil or water).

Dipmeter Logs

Dipmeter logs are used to measure the dip of beds or laminations, for example cross-bedding within a bed. We use three electrodes at 120° to each other to measure the spatial orientation of the electrical properties of rocks. Previously this could only be done through discontinuous measuring, but now dipmeters measure curves continuously. Dipmeter measurement makes it possible to measure the dip of the laminae inside a sandstone or the dip of the entire bed. Because the dipmeter measures the spatial orientation of physical properties in rocks, it could also be used to measure tectonic deformation of rocks, and the primary orientation of sediment particles (fabric). This is useful because in beach sediments the long axis of sand grains tends to be parallel with the shoreline, while in fluvial sandstone the sand grain will be oriented parallel with the transport direction. Dipmeter logs can therefore be used to a certain extent for reconstructing sedimentary environments, transport direction etc.

There are many uncertainties involved in interpreting dipmeter logs, however, and it is not always possible to find dipmeter patterns that are compatible with sedimentological models of cross-bedded strata.

Temperature Logs

Geothermal gradients in sedimentary basins are extremely important for the calculations necessary for predicting kerogen maturation, and also for the general modelling of basin subsidence. When logging a well, the temperature of the mud may be recorded as one of the physical variables. Because the mud is circulating, the drilling mud will not be in thermal equilibrium with the formation water. When the circulation of drilling mud is stopped, it will have risen in temperature, slowly approaching that of the formation water. By plotting the increase in temperature against time after circulation stopped, the correct formation temperature can be estimated. However, it takes several days or even weeks before the temperature of the drilling mud approaches that of the formation water, and there is therefore considerable uncertainty involved in such calculations. The temperatures measured during logging are called "bottom-hole temperature" (BHT). Before using such data one should see if the proper corrections have been made.

If the total heat flow in the basin is known from other measurements, the conductivity of the sediment can be calculated.

$$\text{Conductivity (k)} = \frac{\text{Heat flow (Q)}}{\text{temperature gradient } (\Delta T/\Delta Z)}.$$

The SI unit of thermal conductivity is 1 W/m°C, and heat flow is expressed as mW/m². These parameters are also often expressed as cal/cm.s.°C for thermal conductivity, and μ cal/cm.s.°C for heat flow. A heat flow of 50 mW/m² corresponds to 1.2 μ cal/cm²/s.

If we have a constant flow of heat vertically through a sedimentary sequence in a basin, which is usually a reasonable assumption, the geothermal gradient will be inversely proportional to the conductivity of the local lithological types. This means a relatively high temperature gradient over formations that are poor conductors, such as poorly compacted shales, porous limestones and coal beds' and low gradients over good conductors.

Sandstones are normally better conductors, and salt deposits have a rather high thermal conductivity. Thermal conductivity decreases with increasing temperature, and the difference between the conductivity of the different lithological types decreases with increasing overburden.

Caliper Logs

Caliper logs measure the diameter of the borehole. Some also record the three-dimensional shape of the borehole walls. Normally caliper logs are presented as hole diameter in inches, and thus record cavities where the well has caved in, and also the hardness of the rocks cut during drilling. Holes in clay and poorly compacted shale will have a larger diameter than the drill bit due to caving in after drilling and erosion by the drilling mud. Where we have porous sandstone, a mud cake may develop, causing the hole diameter to become smaller.

Variations in hole diameter and cave-ins do influence the records of the different logs described above. Therefore it is important to consult the caliper log to identify any artefacts due to variation in the geometry of the hole which might appear on the other types of logs.

Interpretation of Environment of Deposition by Means of Well Logs

In addition to providing information about the porosity and permeability of rock types, well logs can be used directly for interpreting environments of deposition. Gamma logs and self-potential logs record characteristic coarsening-upward and fining upward sequences very efficiently. We can therefore often recognise fluvial channels (fining-upward) and shallow marine deposits (coarsening-upward). Turbidites give a characteristic alternation between coarse and fine material.

In delta sequences coal or lignite beds give characteristically strong deflections on the resistivity log.

Figure 14.3b shows some typical logs from rocks deposited in different environments of deposition.

It is important, however, to bear in mind that the response of the logs is not due to sedimentological parameters such as grain size and sorting, but can be indirectly correlated with these parameters. The gamma-ray log does not measure the clay content, but the content of radioactive gamma-ray emitters in the sediments. Sand

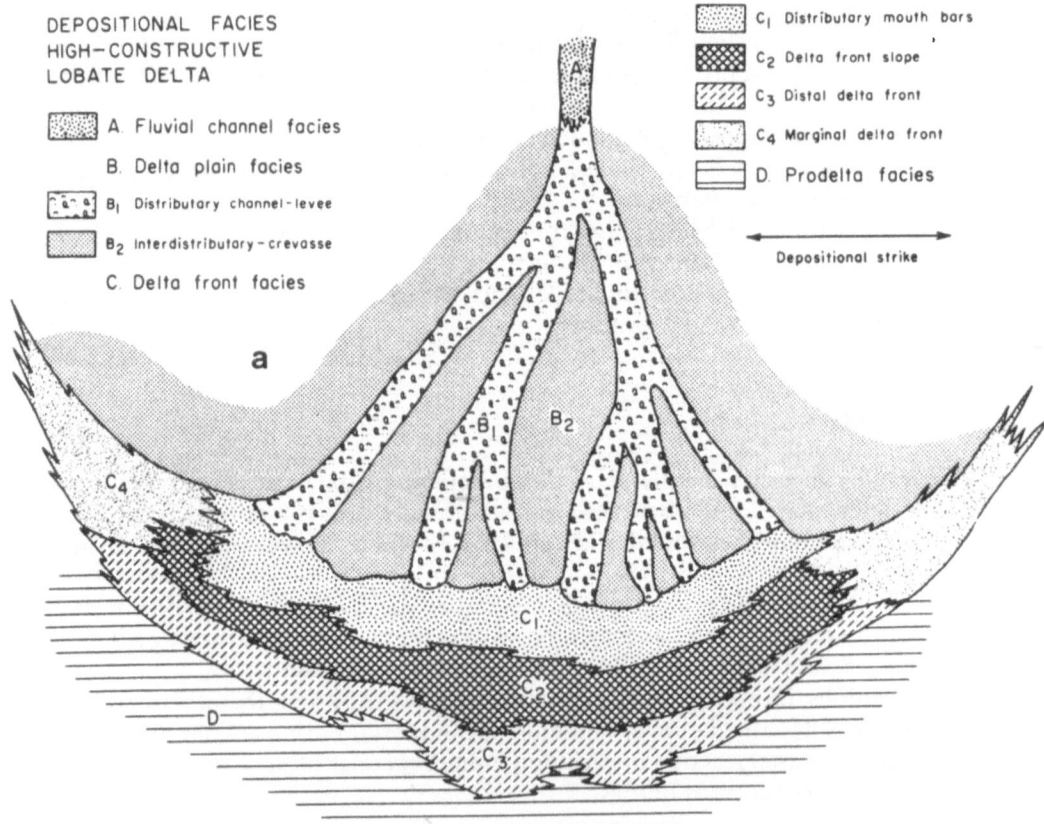

Fig. 14.3. a Enrivonmental of deposition in a lobate delta. (Fisher et al. 1974). **b** Logs from various environments shown in **a**. Sp logs on the *left* and resistivity logs on the *right*. (Fisher et al. 1974)

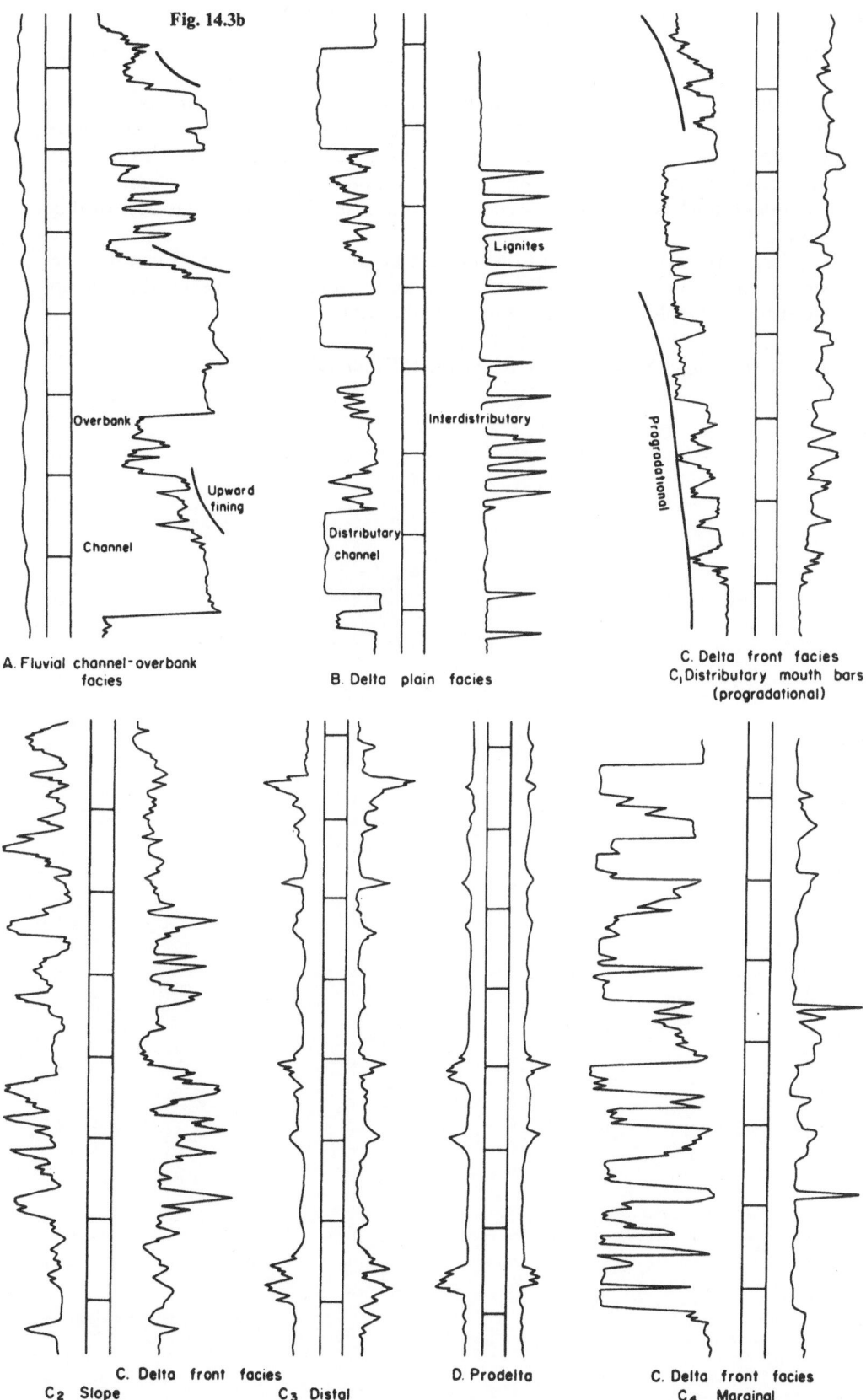

Fig. 14.3b

A. Fluvial channel-overbank facies

Overbank

Upward fining

Channel

B. Delta plain facies

Lignites

Interdistributary

Distributary channel

C. Delta front facies
C₁ Distributary mouth bars (prograradational)

progradational

C. Delta front facies
C₂ Slope

C. Delta front facies
C₃ Distal

D. Prodelta

C. Delta front facies
C₄ Marginal

frequently also has a high concentration of such elements (uranium, potassium, thorium). Fluvial channels should have a nice "bell-shaped" curve due to a sharp erosional base and a fining-upward trend which results in increasing clay content and consequently an increasing gamma response upwards. However, fluvial channels often have clay or shale clasts near the base which will tend to destroy this simple pattern.

Logs are very convenient to work with because all the log data can now be stored on computers and easily retrieved and compared. However, there is no substitute for looking at cores in the core laboratory.

Further reading: Asquith and Gibson 1982; Rider 1986.

15 Plate Tectonics and Oil Prospecting

The formation of sedimentary basins is a result of large-scale crustal movements like sea-floor spreading and crustal thinning. The size of sedimentary basins, thickness of sedimentary strata and development of facies depends very largely on tectonic factors which can only be understood in a plate-tectonic context. Plate tectonic classification of sedimentary basins therefore provides a very good basis for prospecting for petroleum.

We cannot go into the whole background for plate tectonics here, but must simply refer to one of the many textbooks which gives an introduction to the subject.

The basic facts are that the crust of the earth can be regarded as consisting of a relatively small number of rigid plates, and that relative movement very largely takes place at the boundaries between these plates.

We have three different types of relative movement (Table 15.1).

1. Converging plates move towards one another, and the heavier plate (the one with the greatest density) is bent down and subducted under the other, resulting in volcanicity and formation of mountain chains with a pronounced topography because we have two crusts stacked on top of one another.
2. Diverging plates drift away from one another, so that a rift basin is created by faulting and subsidence due to tension and crustal thinning. If the diverging movements continue, large quantities of basalt flow up between the plates, and a new oceanic floor of basalt is formed (sea-floor spreading).
3. With conservative plate boundaries we have movement of plates relative to one another without subduction or ocean-floor spreading. Plate movements are then parallel to the plate boundary, and we get strike-slip faults like the San Andreas fault in California.

Table 15.1. Types of sedimentary basin as a function of relative plate movements and plate contacts

| Relative plate movements | Simplified classification of plate boundaries | | |
	Oceanic-oceanic	Continental-oceanic	Continental-continental
Diverging plates	Atlantic-type spreading		Rift valley
Converging plates	Subduction trench island arc	Andes-type subduction	Alpine-type (Himalayan)
Conservative boundaries	Oceanic transform	Nansen-ridge type strike-slip	San Andreas type strike-slip

Basins Formed During Diverging Plate Movements

1. Rift basins in the continental crust. These are basins which form during the initial phase of tension in the crust of the earth. This leads to thinning of the continental crust and to faulting along normal faults and rotation of pieces of the crust along listric faults. The thinning of the continental crust will cause subsidence, but the thinning will also cause increased geothermal gradients, resulting in uplift. On the side of the rift structure, where the thickness of the continental crust is normal, we find isostatic elevation due to a higher geothermal gradient. This leads to repeated faulting against the central part of the graben where there is maximum thinning of the crust and therefore greatest subsidence. We then have very rapid erosion along the edge of the rift and high rates of sedimentation in the basin. Erosion on the margin of the rift will represent isostatic unloading so that much of the erosion is compensated for by isostatic uplift, while sediments in the bottom of the graben will cause increased loading, which in turn will result in further subsidence. Instead of a symmetrical graben structure with two main faults, we often find the development of a half-graben, with one main fault plane.

Because of the high relief around these basins and the short transport distance from the time of erosion to deposition, these basins will be characterised by very immature sediments, largely arkoses and conglomerates deposited in fan deltas along the active faults around the basins. In the central and deeper parts of the basins we find deposition of finer-grained sandstone and clay sediments. Rift valley basins may be continental lacustrine basins, as in East Africa, or marine like the Jurassic basins of the North Sea. Their elevation will depend on the thickness of the continental crust in the area and on the heat flow. Both marine and lacustrine rift valley basins will tend to have reducing conditions in the deeper part due to limited circulation of oxygenated water. We will therefore often find black, organically rich shales which are good source rocks being deposited in these basins. Lacustrine basins often have an even better potential for producing source rocks than marine rift basins, because there is usually more marked water stratification (density stratification) in lakes. Sandstones and conglomerates in the marginal facies (fan deltas) and shallow marine sandstones are good potential reservoir rocks, and very high oil potential is therefore associated with rift basins. Lacustrine rift valleys often provide a less favourable environment for the development of suitable reservoir rocks because they have low wave and tidal energy. The high geothermal gradients in rift basins cause organic matter (kerogen) to mature and expel hydrocarbons earlier than this would take place in basins on passive margins, where the geothermal gradients are normally lower (cold basins). The crucial features, however, are cap rocks and sealing structures. When volcanic activity declines we often find fine-grained marine clay sediments deposited over the entire area, onlapping unconformities produced by earlier uplifts, forming a low permeability layer which seals off the underlying sandstones (cap rocks). In the North Sea, for example, we find examples of Jurassic rifting being followed by subsidence and transgression in the Cretaceous and Tertiary, with deposition of fine-grained clayey sediments which function as cap rocks. Block-faulting and rotation along listric faults produces almost ideal hydrocarbon traps. This is why rift basins have a very high petroleum potential.

   Evaporite deposits are also sediments which are characteristic of rift basins formed in an arid zone. Block faulting will easily form isolated basins or horsts, which cut off contact with the open sea. Evaporites are typical of rift deposits today, e.g. in East Africa, and were widespread in Europe and North America during the Permian-Triassic, before the ocean-floor spreading which created the Atlantic Ocean started in the Mid-Jurassic. The Zechstein Salt deposits in Germany and the North Sea are typical examples of this. During the Jurassic and early Cretaceous there was also rifting, and during the early phase of the opening of the Atlantic Ocean evaporite basins formed in the arid regions of the time. There are two aspects of these salt deposits which give them special significance in the context of petroleum reservoir formation:

a) They may be cap rocks, as the Zechstein Salt is for gas pockets in the southern art of the North Sea.
b) If they are thick enough they may form salt domes which produce structural traps in overlying rocks, as in the Ekofisk area in the North Sea and in the Gulf Coast basin.
c) Rising salt domes also strongly influence the clastic sediment distribution in a basin.

Associations of evaporites and rift basins are therefore of vital interest for prospecting.
   Mature source rocks will be found mainly in the central part of grabens, where they have been subjected to the deepest burial and the highest geothermal gradients. Petroleum, however, will often migrate from source rocks in the central part of a graben up into traps in shallower sediments along the edge of the graben, if the central part of the rift system continues to subside, as is the case in the North Sea. The source rock in the deepest part will be overmature and mainly produce gas.

2. Proto-oceanic rift basins. These are basins formed where ocean-floor spreading has started, but is still limited in width, so that the whole basin has a high geothermal gradient, and the continental crust on both sides still has a relatively high relief. An example of such a situation today is the Red Sea. The oil potential in such basins is good because of early maturation of the source rocks and its association with reservoir rocks as in rift basins.

3. Sediment basins along passive continental margins (Atlantic type) (Fig. 15.1). As ocean-bottom spreading continues and the ocean widens, the geothermal gradient along the continental margin will decrease. We will then find subsidence of the earth's crust. At the transition to oceanic crust, the continental crust is stretched and thinned out so that we have isostatic subsidence and formation of large sedimentary basins. Along the continental margin sediments will be deposited by building-out (progradation) of coastlines onto what is known as the continental shelf. Where drainage from the continents collects in large rivers, thick delta deposits may be partly reworked into shallow marine sediments on the boundary between the continental and oceanic crust. The oceanic crust continues to subside as the spreading ridge moves further away from the edge of the continent and the oceanic and thin continental crusts became colder. Sediment loading, particularly due to

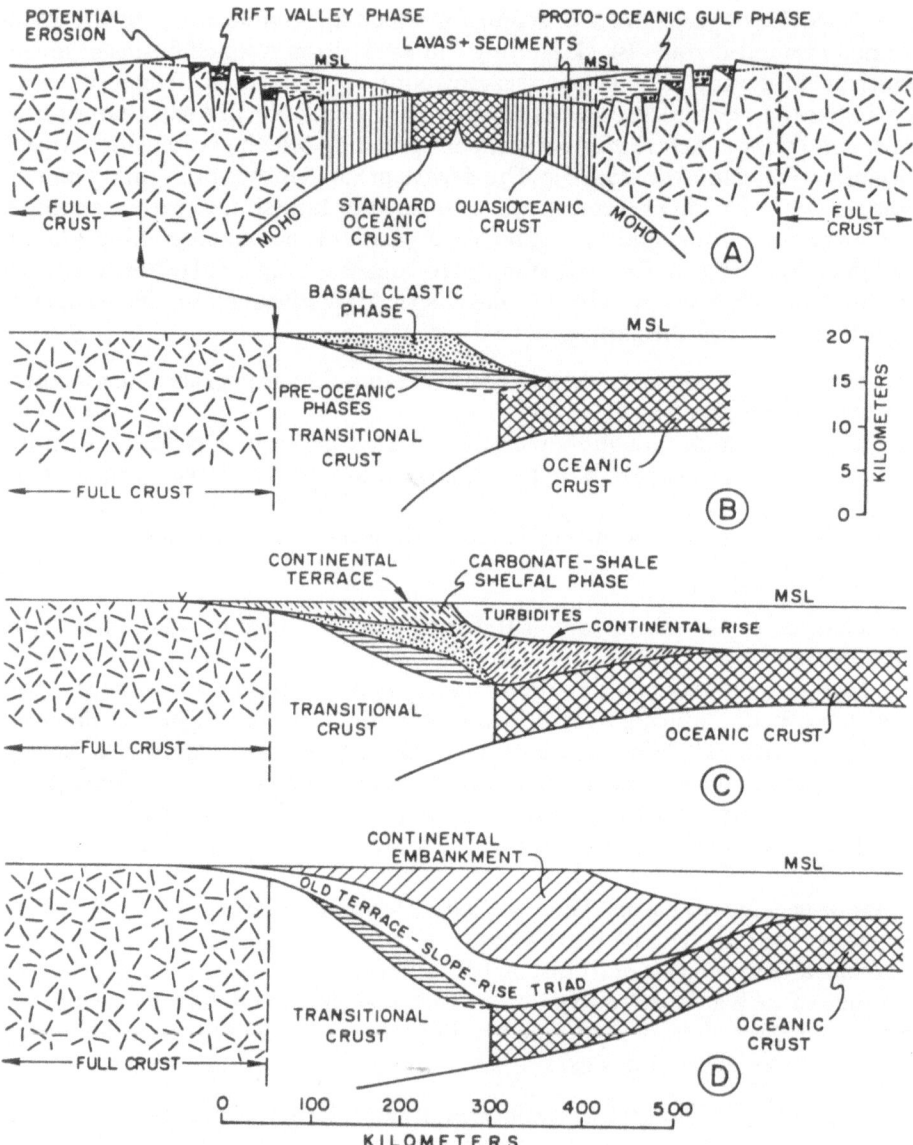

Fig. 15.1. Development from continental rifting to sea-floor spreading and formation of a passive margin. (Dickinson and Yarborough 1977). Note the development of thick clastic wedges at the boundary between oceanic and continental crust

deltas, will cause the oceanic crust to subside isostatically and make room for up to 17 km of sediment. We find such a thick sediment sequence underneath the Mississippi Delta in the Gulf of Mexico, where sediments from large expanses of the North American continent have been depositing since Mesozoic times. Sediment basins along passive margins are normally characterised by low geothermal gradients, and we must therefore have a relatively great thickness of sediment (>3 km) to achieve maturation. We call sedimentary basins along passive margins "cold basins".

Converging Plate Boundaries

Converging plates produce three principle types of boundary:

1. Oceanic crust against continental crust, such as along the west side of South America (Andes type).
2. Continental crust against continental crust (Alpine or Himalayan type mountain chain).
3. Oceanic crust against oceanic crust, with formation of volcanic islands.

Where we have subduction of oceanic crust beneath continents we find the following types of basin (Fig. 15.2):

1. Trench basin and accretionary prism.
2. Fore-arc basin (relatively shallow basin in front of island arcs).
3. Inter-arc basin (sedimentary basin within island arcs of continental crust).
4. Back-arc basin (sedimentary basin on the oceanic crust behind the island arcs).
5. Retro-arc basin forms a secondary basin in the innermost part of the continental crust (Fig. 15.3).

1. Trench Basins. Trench basins seldom contain large amounts of sediment if separated from the continent by an island arc. The ocean floor under trench basins is a dynamic system, and part of it is constantly being consumed through subduction. The downward movement of the oceanic crust helps to maintain a low geothermal gradient in the trench since heat flow will operate counter to the movement of the crust. Because the floor is in constant motion under the trenches, the sediment thickness will be limited.

Most of the sediment coming from the continents will be trapped in back-arc basins. The sediments in trench basins will consist of deep-water conglomerates

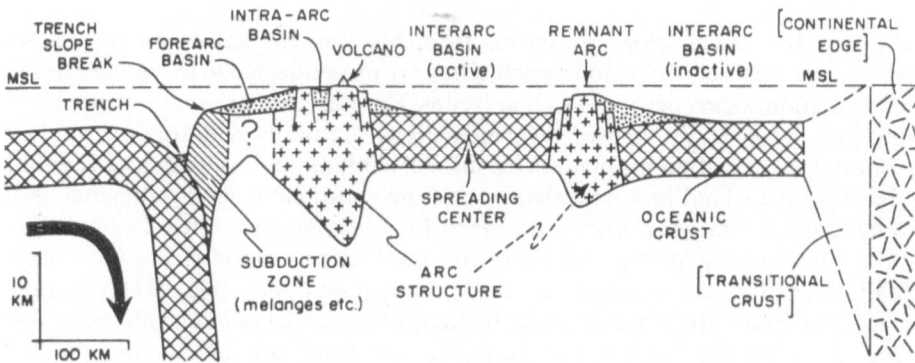

Fig. 15.2. Sediment basins associated with converging plate boundaries. (Dickinson and Yarborough 1977)

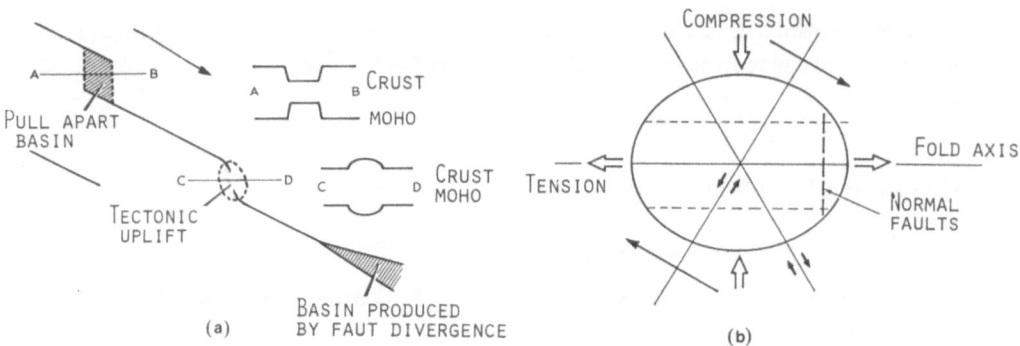

Fig. 15.3. a Tension and compression along conservative strike-slip plate boundaries (Crowell 1974). b Tectonic deformation associated with strike-slip movements presented in a strain ellipsoid. (Crowell 1974)

turbidites and pelagic sediments. Trench basins lie mostly below the carbonate compensation depth and therefore contain little carbonate. Sedimentary transport of very fine-grained pelagic sediments also tends to occur along the long axis of trenches. The inner slope up to the island arcs is steeper than the outer one, and at the foot of this slope we have the outgoing part of the subduction zone (Benioff zone).

When the oceanic plate passes down into the Benioff zone, sediments deposited on the sea bed may be scraped off, resulting in a complicated pattern of folds (accretionary prism) which exhibit overthrusts and underthrusts along listric faults, with the opposite relative movement to what we find in an extensional rift basin.

On the inside of the trench slope, sediment is supplied by the island arcs. The sediments will be folded, deformed and reworked shortly after sedimentation during the subduction processes.

Although sediments in trench basins may be rich in organic material, their oil potential is poor due to lack of reservoir rocks and the intense tectonic deformation. The geothermal gradient will also be lower than in other basins, at least at the time of deposition.

2. Fore-Arc Basins. Fore-arc basins form between the actual island arcs with volcanoes and the subduction trench. They are not subjected to the intense tectonic deformation which occurs in trench basins. Fore-arc basins often transgress onto the island arcs as the belt of volcanic activity gradually withdraws towards the continent. The sediments in these basins will contain fluvial and deltaic sediments closest to the island arcs, and shallow marine continental shelf sediments. In the outer parts sediments are deposited as turbidites in deeper water. There are good possibilities of organically rich sediments being deposited here. The San Joaquin Valley in California, with its many oil fields (around Bakersfield), is an example of a fore-arc basin formed in front of the Sierra Nevada island arc with its batholith belt, when the volcanic crust of the Pacific was being subducted along the Franciscan complex, under the North American continent, in Cretaceous and early Tertiary times.

3. Intra-Arc Basins. Intra-arc basins are a result of tension tectonics in island arcs. Sediments deposited in these basins will for the most part be immature and volcanic material will be common. The geothermal gradient will be high, but conditions for formation of source and reservoir rocks are not very good. Volcanic sandstones will often lose their primary porosity rapidly due to unstable minerals and diagenetic transformation. Intra-arc-basin sediments will frequently be subjected to intense tectonic deformation.

4. Back-Arc Basins. Back-arc basins develop on the oceanic crust due to ocean-floor spreading between island arcs and the continent. There is likely to be an abundant supply of sediment from the continent, and back-arc basins may fill with deltaic and shallow marine sediments.

Intra-arc basins form as a result of back-arc spreading behind island arcs of continental crust surrounded by oceanic crust.

The Jurassic and Cretaceous sea way through North America east, of the Rocky Mountains, was a back-arc basin during subduction of the Pacific plate. The area is a good oil region (Rice and Gautier 1983) but Tertiary uplift resulted in a very strong meteoric water drive which flushed out many of the oil fields.

5. Retro-Arc Basins. Retro-arc basins form a continental crust behind a continental arc with a fold thrust belt, and are a type of foreland basin.

In foreland basins the load of the nappes on the crust contributes to basin subsidence and also provides a source of sediment.

Conservative Plate Boundaries

When two plates move parallel to one another, we have *strike-slip* faults like the San Andreas fault in California. We distinguish between "right-lateral" or "dextral" faults where the opposite side of the fault has moved to the right, and "left-lateral", or "sinistral" faults where the movement has been the reverse. The San Andreas fault is a dextral fault, and the western part of California (Salina block) has moved northward in relation to the North American continent. If the fault plane follows a completely straight structure in the rock parallel with the direction of movement we will get neither tension nor compression along the fault plane. On the other hand, if the fault plane is a little irregular due to inhomogeneity of the crust, there may be both compression and tension along the fault plane (Fig. 15.3). Compression will lead to folding and to formation of small mountain chains, and tension will produce small, deep sedimentary basins.

We find the most typical examples of strike-slip faulting in California. When a fault branches we may also see both compression and tension, with elevation and subsidence respectively of blocks, depending on their orientation.

If the relative movement shifts from one fault to another, parallel fault plane, we have crustal tension in the area in between, and "pull-apart" basins of the Salten Trough type will be formed, often associated with volcanicity.

A general feature of basins formed by strike-slip fault systems is that they are a result of tension and thinning of the continental crust. The basins are very deep holes in the continental crust, and are sometimes floored with volcanic rocks. Because of the limited size of the basins and the elevation of the source areas around the basin, we get a very high rate of sedimentation. In Miocene-Pliocene basins in California, 5–8 km or more of sediments have been deposited in roughly as many million years. This corresponds to an average of 1 mm per year, which is a very high figure in this context. The sediments consist largely of turbidites, proximal turbidites along the edge of the basin, and distal turbidites and fine-grained clay sediments in the central, deeper parts. In the Ventura Basin, which continues into the Santa Barbara Channel, there are at least 15 km of sediment (Fig. 15.4).

Basins of this type provide almost ideal conditions for oil accumulation. The Californian basins are amongst the richest oil regions in the USA. The Los Angeles basin is a pull-apart basin and a very rich oil region. Very few basins can compete with the California strike-slip-related basins in terms of barrels of oil per square kilometre. This is because these deep basins often have reducing conditions in their bottom water, due to limited circulation of oxidised water. We therefore get sediments rich in organic matter deposited. Turbidite sandstones, particularly the proximal parts, often have a porosity sufficiently high to make them a suitable reservoir rocks. They may be quite large.

Sediments deposited in basins of this type tend to undergo weak folding, which will result in good structural traps. Geothermal gradients will be relatively high, and allow thorough maturation. Figure 15.5 illustrates the structural setting of such basins and its consequence for sedimentation, faulting and folding.

Fig. 15.4. Anticlinal structure in the Ventura basin in California. Oil is produced from deeper anticlinal structures similar to those we see on the surface. The Ventura basin is one of the most oil-rich in the world in proportion to its size and continues out onto the continental shelf (Santa Barbara Channel)

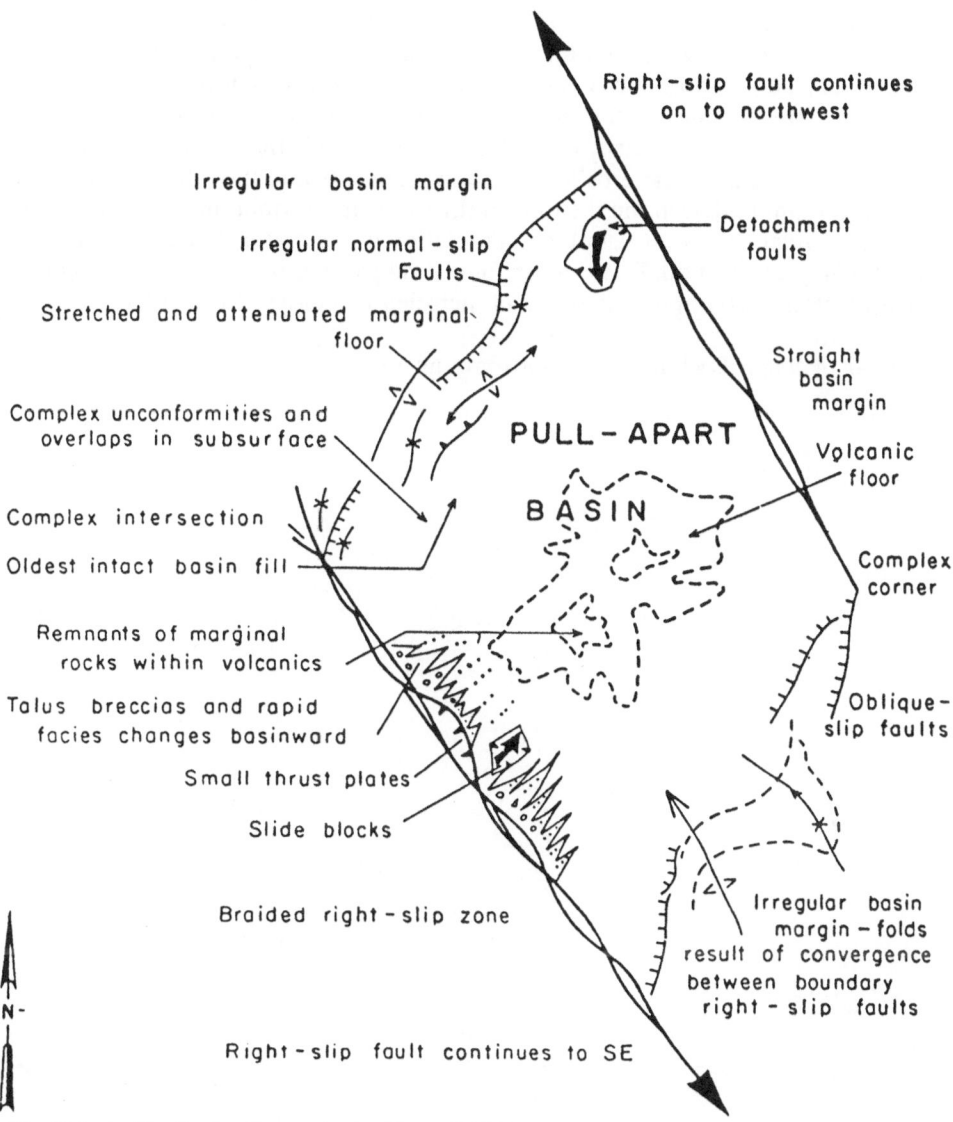

Fig. 15.5. An idealized pull-apart basin. (Crowell 1974)

Oil Prospecting in Mountain Chains

Until just a few years ago, areas with intense tectonic deformation and imbricated nappes were regarded as holding little promise in the search for oil and gas. For this reason the imbricated plates of the Rocky Mountains were not subjected to systematic oil exploration earlier. In recent years, however, there has been intense exploration in these areas, with very good results. Other as yet unexplored mountain chains may also contain considerable quantities of oil and particularly gas. It turns out, as one should perhaps have realised earlier, that tectonic thrusting is no obstacle to the formation of hydrocarbon traps. On the contrary, the thrust plane often forms a good seal for oil and gas. Arched and imbrication duplexes create very good traps. What is crucial is maturation, which in this setting is a function of the thickness of the pile of nappes rather than of the stratigraphic depth. Temperatures may very easily exceed the "oil window" and gas is more common in older mountains where erosion has cut more deeply into the pile of nappes. It is also important to determine the time of thrusting and formation of structures, for example under curved thrust planes, in relation to the maturation of source rocks and the migration of oil. Exploration for oil and gas in mountain chains has made thrust tectonics an important aspect of petroleum geology (Mitra 1986).

Further reading: Dickinson and Yarborough 1977.

16 Production Geology

Reservoir Energy

In most cases oil and gas in reservoirs is under pressure. When a well penetrates a reservoir, petroleum will flow up the well to the surface. Petroleum in the reservoir is capable of doing work, so the reservoir is self-producing. The energy in a reservoir is used to overcome surface tension in the porous rock and to overcome viscous resistance to the flow of oil and gas. As oil is produced from the reservoir, the pressure falls. Oil and water have a relatively low coefficient of expansion, and the pressure will thus drop rapidly as a result of production. The smaller the reservoir, the more rapid the fall in pressure. The drop in pressure during production therefore provides very important information about the size of the reservoir and communication within it. However, gas has a high coefficient of expansion, and fluid gases may become gaseous and expand very considerably when the pressure is released (Fig. 16.1). This will help to maintain pressure longer during production, and we have so-called "gas drive". Gas which is dissolved in oil will come out of solution as the pressure falls and thereby help to reduce the fall in pressure. It is important to maintain the pressure in reservoirs as long as possible, and not to reduce gas pressure unnecessarily.

In large reservoirs water pressure can be maintained somewhat longer because of the large volumes. However, the optimum situation is when the water in the reservoir is under pressure from an aquifer. We then have almost constant pressure during production because new water keeps flowing in to replace the oil produced. This is called "water drive" (Fig. 16.1).

Production from a reservoir without water drive is greatest just after production has begun, and then falls asymptotically towards zero. In order to counteract the fall

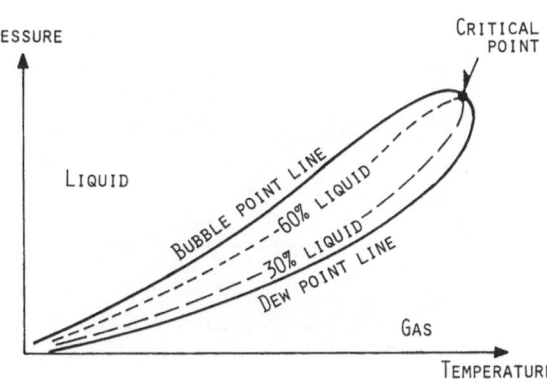

Fig. 16.1. Phase diagram for hydrocarbons which can occur both as liquid and as gas. Condensate can be formed through reduction of pressure and temperature near the critical point, where liquid and gas have the same properties

in pressure in the reservoir, it is usual to inject water or gas into the reservoir through special injection wells. Injected water may, however, react with the minerals in the reservoir and alter its properties. When there is no more energy in terms of pressure in the reservoir the oil will have to be pumped. This is only economical in onshore wells.

If there are several liquid and/or gas phases present in the reservoir rock, the permeability is not merely a function of the size of the pores, but will also depend on the relative amounts of the other phases. The term *effective permeability* expresses the permeability of a phase in combination with other phases. If, for example, we have 40% water present together with 60% oil occupying the pores of a rock, the permeability to oil will be less than if all the pores were filled with oil (100% oil saturation). The relationship between the effective permeability at a particular degree of saturation and the effective permeability at 100% saturation is called the *relative permeability*. These relationships are very important concepts, and apply not only during oil production, but also during migration of petroleum.

Figure 16.2 shows relative permeability as a function of the percentage of water and oil saturation. We see that when we have less than 40–50% water (50–60% oil saturation) a reservoir only produces oil. With a water saturation of about 45–85%, both oil and water are produced, and at high water saturations (85–100%), only water is produced. The reason for this is that in most cases water wets the mineral surface more easily than oil. We may have up to 30–40% water around mineral grains, while only oil flows through the pore throats. When the water content becomes 40–50% or more, we no longer have a continuous oil phase, but oil drops which flow together with water. When there is little oil, these will remain as small drops in the pores and water will flow past. This applies to water-wet systems. Under certain pressure and temperature conditions, however, reservoir rocks may be more easily wet by oil than by water. This is particularly true of carbonate reservoirs.

Reservoir rocks with oil and gas will produce gas only until the oil saturation is greater than about 30–40%. This is because gas is less viscous and easily flows past water. At higher oil-saturation levels, the gas will carry oil with it, and at about 55%

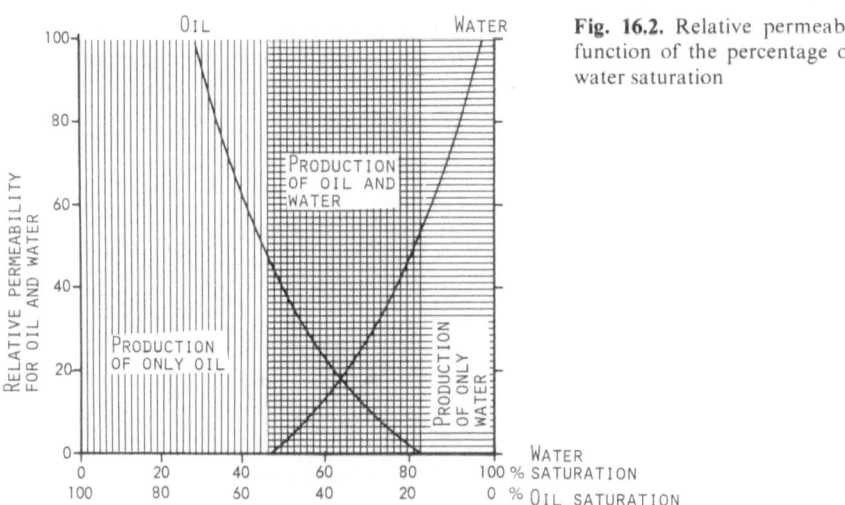

Fig. 16.2. Relative permeability as a function of the percentage of oil and water saturation

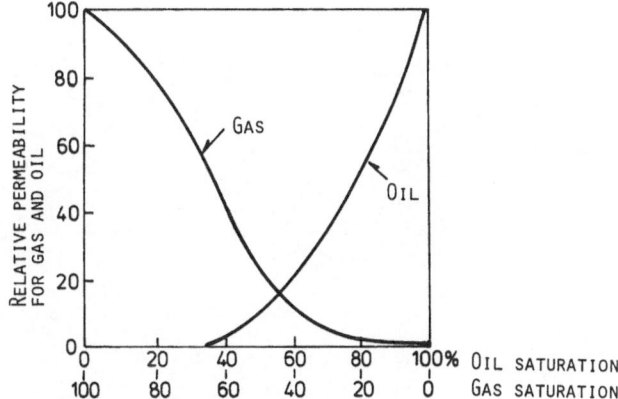

Fig. 16.3. Relative permeability as a function of oil and gas saturation

oil saturation oil and gas will have the same relative permeability. At higher gas-saturation levels gas occurs as separate bubbles and we will therefore find a certain amount of gas production, even with 90% oil saturation (Fig. 16.3).

Production from Sandstone Reservoirs

With production from sandstone reservoirs, the physical properties of sandstone must be taken into account, e.g. distribution of porosity and permeability. The actual production, however, will lead to changes in the reservoir rock which may have an adverse effect on the properties of the reservoir, i.e. cause reservoir damage. This may happen in two ways:

1. Chemical reactions between minerals and liquids which are used in drilling or injection of water.
2. Mechanical damage through relatively loose clay minerals or other small grains being carried by the fluid flowing towards the well, and obstructing the pore throats.

As oil flows into a well, the flux (cm^3/cm^2) is inversely proportional to the distance to the well. If we can improve the permeability of the reservoir nearest the well, we may be able to step up the production rate considerably. For this reason chemicals are sometimes injected into the reservoir to improve the permeability. We then need to understand some of the chemical properties of minerals in order to be able to predict the reactions which will result from treatment with acids and other chemicals.

As far as chemical reactions are concerned, clay minerals are important because they have a large specific surface and ion exchange capacity. The specific surface varies with the size and shape of the mineral grains. Kaolinite will typically have a specific surface of 5–30 m^2/g, chlorite 10–50 m^2/g, illite and smectite (montmorillonite) 50–120 m^2/g. The high specific surface of illite and smectite is due to the fact that they often occur as very thin sheets or fibres (Fig. 16.4).

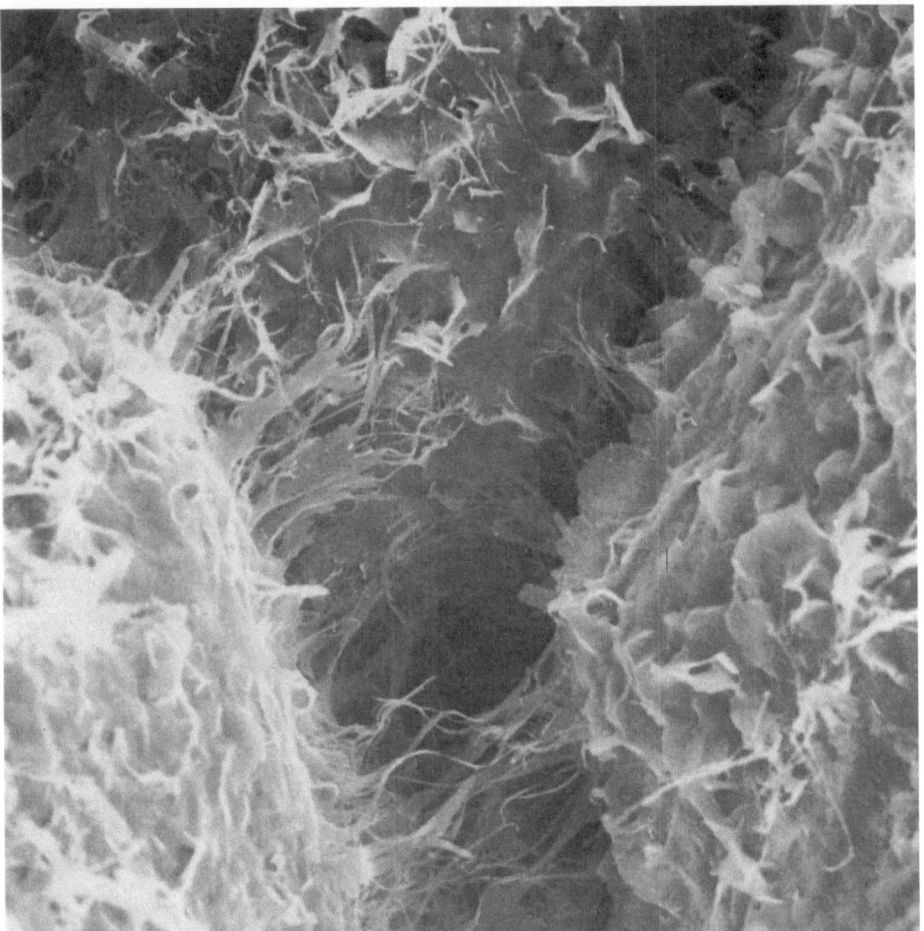

Fig. 16.4. Pores with fibrous illite formed diagenetically among sand grains in Triassic sandstone from the North Sea. The thin illite fibres greatly reduce the permeability but do not reduce the porosity very much. The cross-section in the SEM picture is about 0.1 mm square. Similar developments are found in Permian gas reservoir sands from the southern part of the North Sea

Kaolinite has a very low solubility in water, and strong acid (HF) is required to dissolve it. Kaolinite is therefore chemically stable and its ion exchange capacity is lower than that of other minerals. Chlorite, on the other hand, is soluble in acid (e.g. HCl). If acid is used to dissolve minerals like chlorides in order to increase the permeability around the well, it is important to avoid iron being precipitated as iron hydroxide ($Fe(OH)_3$). This could reduce the porosity very severely because it forms large crystals. To avoid this, one can add chemicals which form complexes with iron (e.g. citrate).

Illite and smectite are not very soluble in HCl, and fluoric acid (HF) must be added to dissolve these minerals. They are sensitive to variations in salinity and in particular to the K^+ content. K^+ and other alkali ions will be held in the interlayer position in smectite so that water is expelled and the mineral contracts in volume. On the other hand, injection of fresh or slightly saline water into a reservoir may lead to uptake of water and expansion of smectite and mixed layer minerals, which may greatly reduce the porosity. Sodium smectites can swell to five to ten times their original volume. Swelling can be reduced by injecting dilute acid (HF, HCl) or water with a high salinity (e.g. KCl). Water is also often injected to maintain reservoir pressure. Offshore drilling platforms in particular depend on using sea water. However, sea water may also have adverse effects on the reservoir. The high SO_4^{2-} concentration in sea water may cause precipitation of baryte ($BaSO_4$) which is the least soluble of common sulphate minerals. Sea water also contains sulphate-reducing bacteria which will start to reduce SO_4^{2-} to H_2S in the reservoir. This is most undesirable since H_2S is a very poisonous gas which also has a highly corrosive effect on steel during production.

To avoid unforeseen chemical damage to the reservoir rock, it is necessary to carry out very thorough mineralogical analyses so as to be able to predict the effect of various types of chemical stimulation.

Physical damage to the reservoir is as previously mentioned due to mineral grains being loosened and carried by oil and water and obstructing the passage. If the reservoir is water-wet, fine-grained minerals will tend to remain in the aqueous phase, and will not come into the oil phase very easily because of surface tension. When considerable quantities of water are produced together with oil, therefore, the aqueous phase is particularly likely to carry clay minerals and other small mineral particles, for example of quartz and felspar, which may block the pore throats. This effect can be reduced by lowering the production rate. In the laboratory this type of mechanical formation damage can be tested in an experimental flow rig. If the measured permeability increases temporarily when the flow direction is reversed, formation damage has been caused by "a moving clay and silt fraction". Sandstones with a high percentage of secondary porosity will have relatively large pores with small pore throats. The ratio between the pore diameter and the pore throat is often referred to as the *aspect ratio*. Pores with high aspect ratios will be particularly vulnerable to mechanical formation damage. Pore geometry is also important with respect to two-phase flow (oil and water) because of the capillary forces in the pore throats that have to be overcome. This is even more the case with condensate reservoirs, where we may have three phases (oil, gas, water).

The planning of production from a reservoir involves a detailed production strategy and optimal positioning of production and injection wells. The trend production will take with time is simulated on very powerful computers which are fed with a very large number of parameters relating to the reservoir. Simulations depend very heavily on the geometry of the reservoir and internal communication, e.g. between sandstone bodies, the exact position of the sealing fault etc. The internal properties of the reservoir can be measured on cores from the wells, but a three-dimensional representation of the distribution of porosity and permeability and the pore geometry must be constructed, using models for the detailed depositional environment and diagenetic alteration.

Carbonate Reservoirs

Many of the same principles apply to carbonate reservoirs as apply to sandstones. There is a tendency for carbonate reservoirs to be less water-wet than sandstone reservoirs, but this depends on the composition of the oil and on the temperature and pressure. Pores often have a high aspect ratio, particularly in limestones, where we often have intragranular porosity inside fossils or vuggy moldic porosity after selective dissolution of organic fossils. The reservoir properties of bioclastic and reef limestones are largely controlled by the palaeoenvironment, which controls the distribution of fauna.

Fractures are generally much more important in carbonate than in sandstone reservoirs. Stylolites are sometimes found in sandstone reservoirs that have been buried to at least 3.4–4 km. In limestones stylolites develop at much shallower depths, and often form quite an impermeable, thin clay surface, which may divide up a reservoir into compartments or even serve as a cap rock.

Fractured Reservoirs

In fractured reservoirs there is a high percentage of porosity due to fractures rather than more uniformly distributed porosity. Fractures are most extensively developed in brittle rocks like well cemented limestones and chert. They may also form in metamorphic and igneous rocks. Fractures are often invaded by drilling mud, causing lost circulation, and this may damage the reservoir near the well. The flow of petroleum through a fractured reservoir during production is very complex, and it is difficult to map out the fracture system of the subsurface in sufficient detail.

As a result of reduced pressure during production, fractures may start to close due to increased net stress. The introduction of some coarse-grained material may help to wedge the fractures and prevent them from closing. By injecting water at very high pressures (higher than the geostatic pressure) fractures may be widened (hydrofracturing) and this method is also sometimes used in other reservoirs to create or widen fractures close to wells and improve the permeability of the critical area around the well. A detailed treatment of fractured reservoirs is given by Van Golf-Racht (1982).

Enhanced Oil Recovery Methods (Tertiary Oil Recovery)

Only a fraction of the oil present in a reservoir is recovered. This can constitute from 10% to 60% of the total amount of oil in the reservoir rock (oil in place) and averages only 30–40%. How much of the oil is recovered depends on a number of factors. The most important are:

1. Location of the oil wells and the distance between them.
2. Regulation of rate of production from the various wells in relation to the pressure in the reservoir and injection of water to maintain reservoir pressure. Gas injection may also increase production. These are called *secondary recovery methods*.

Methods which involve changes in the *internal properties* of the reservoir are often called *tertiary recovery methods*. However, since these methods are not always used after the secondary recovery methods, the term *enhanced recovery* is to be preferred. There are three main enhanced recovery methods. They employ:

I.   *Thermal processes:* These are processes which increase the temperature of the oil and thus lower its viscosity. The most common are:
     a) Steam injection
     b) In situ combustion.
II. *Chemical processes:* These involve injection of chemicals into the reservoir. The most important types are:
     a) Chemicals which reduce surface tension (surfactants)
     b) Polymers, which are injected into water to increase its viscosity
     c) Caustic (alkaline) chemicals which may be added (caustic flooding) to reduce the surface tension.
III. Injection of various types of gas to increase the miscibility of hydrocarbon phases. These may include carbon dioxide, neutral gases (e.g. nitrogen) or hydrocarbon gases.

I. Thermal Processes: These are only used with reservoirs with highly viscous oi, where an increase in temperature will greatly reduce the viscosity. This will normally be heavy, asphaltic oil which as been subjected to bacterial breakdown and evaporation. Oil reservoirs of this type are normally fairly shallow (less than 1500 m). Deeper reservoirs will have a high temperature anyway, so there is less to be gained by artificially heating the oil.

In situ combustion can be induced by pumping air down into the reservoir, and the combustion can be regulated by addition of oxygen. The hydrocarbons which do not burn will then become hotter and less viscous. Burning also produces gases, which may increase production. Special wells can be used for injecting air, and combustion spreads from these to the production wells, forcing a zone of hot water, gas and light oil ahead of it to the production wells. Combustion may also increase the porosity of the reservoir rock, making it possible to produce later on from the wells in which combustion has taken place.

Steam Injection

Steam injection is responsible for 90% of all production through secondary recovery methods apart from water injection. One of the major reasons for the low recovery percentage from oil reservoirs is that the oil is too viscous. This is particularly true of oil which has been subjected to bacterial degradation (biodegradation), leaving the viscous asphaltic part of the oil behind as heavy oil (low gravity 16–17 API). The Cretaceous Athabasca tar sand in Alberta, Canada, is the most famous example, and contains huge oil reserves. When oil reservoirs with such heavy biodegraded oil are exposed at the surface, the oil is almost solid at normal temperatures, and is called tar sand. For it to be possible to produce oil, the pressure difference between the oil in the reservoir rock and the well must exceed the resistance to flow exerted by capillary forces (is water-wet reservoirs) and viscosity forces (friction).

Biodegraded oil is normally found at relatively shallow depths, and the temperature of the oil is low and the viscosity consequently high. By injecting steam the temperature can be raised and the viscosity lowered considerably. There are examples of the viscosity of oil being reduced from 300 centipoise to 10 centipoise after steam injection, and heavy biodegraded oil is often too viscous to flow at all without steam injection. Steam injection is particularly important in California, where 300,000 barrels per day are produced using this method. Steam injection must be repeated at regular intervals because the viscosity increases again as the reservoir cools off. Sandstones which are to be subjected to steam injection must also have high oil saturation, because otherwise water and steam can flow past the oil. Steam is recovered from the produced oil, and in some fields one out of every three barrels produced is used for steam production.

II. Chemical Processes: When oil flows out of a reservoir rock into a producing well, it must overcome the capillary forces involved in two-phase flow. By reducing the surface tension between oil and water which is injected into the wells, the capillary forces can be reduced. The addition of surfactants, a sort of soap, will reduce the surface tension so that the *wettability* changes. The wettability can be measured by means of the contact angle, i.e. the angle which the liquid phases make with a mineral surface, for example (Fig. 13.9).

If an oil reservoir is *oil wet*, it means that oil is more strongly bound to the mineral surface than water. By reducing the surface tension in the water phase, the *wetting angle* is altered and the system may become *water wet*, and the oil will become more mobile so that the relative permeability to oil is increased. A mixture of different chemicals (*slug*) is injected into oil reservoirs to reduce the surface tension. It includes microemulsions, soluble hydrocarbons, micelles and frequently also alcohol and various salts.

One serious drawback of this method is that chemicals can easily be adsorbed onto the reservoir rock. Clay minerals in particular have high ion exchange and adsorption capacities. The specific surface (m^2/g) of rocks is also of significance. Consequently chemicals are often pumped down in advance (*preflush*) to reduce the adsorption of chemicals which are intended to reduce surface tension. This treatment ought also to take into account which clay minerals occur in the rock.

Polymers. In oil reservoirs with an undesirably high relative water permeability, large amounts of oil will remain in the reservoir rock. Water will flow where there is least resistance to its movement, e.g. along cracks, and injection of water will have little effect. This is particularly relevant with oil reservoirs with low oil saturation (high water saturation). If the viscosity is increased so that the pressure gradient in the water phase increases, the gradient in the oil phase will also increase. Polymers reduce the mobility of the water, causing a piston-like replacement mechanism. With increased viscosity, the polymer-bearing water will exert greater shear forces on the oil phase, so that oil drops are more easily carried along. Water containing polymers is no longer a Newtonian liquid, but a pseudoplastic liquid in which the apparent viscosity is reduced by the rate of deformation.

Polymers will also have a tendency to be adsorbed onto the surface of minerals so that their effect decreases with increasing distance from the injection well.

Addition of Alkaline (Caustic) Chemicals. Strong alkaline solutions, for example NaOH, will also reduce the surface tension of aqueous phases. Injection of a mixture of these alkaline solutions may therefore increase both the water-wetness of the reservoir and oil production. The most common chemicals are sodium and potassium hydroxide, sodium orthosilicate, sodium carbonate, sodium phosphate etc. The method is used largely in sandstone reservoirs. Small amounts of gypsum or anhydrite will cause precipitation of $Ca(OH)_2$ and neutralise caustic soda (NaOH).

Gas Injection to Increase the Miscibility of the Hydrocarbon Phase

The object of gas injection is to produce a fluid phase which will dissolve the reservoir oil. The flow of such a fluid through reservoirs with low oil saturation may dissolve the oil and carry it along to the production well. Hydrocarbon gases which are in the liquid phase at reservoir pressure (*Liquid Petroleum Gas - LPG*) will be fully miscible with oil. This requires that the reservoir temperature be below the critical temperature for the gas. At higher temperatures the gas is in gaseous form regardless of pressure, and is not fully miscible with oil. Propane is often used in the oil-expelling mixture, and gas and water are injected subsequently.

CO_2 has a critical temperature of about 31°C, and is therefore gaseous at all reservoir pressures. Carbon dioxide is very soluble in oil, which increases in volume when it is saturated with respect to CO_2. This creates a sort of gas drive.

Increasing the CO_2 content also reduces the viscosity and density of oil, making it very mobile. The method is used for oil reservoirs with oil saturation of 25–54%.

Neutral gases such as nitrogen may also be injected into reservoirs, but far greater pressure is required to make the gas soluble in oil.

Oil shale is a source rock exposed at the surface. If the source rock (shale) is mature it will have a characteristic smell of hydrocarbons. Since the hydrocarbons are thoroughly disseminated in the fine-grained sediment, oil cannot be produced in the same way as from a sandstone reservoir. The hydrocarbons can only be obtained by breaking and crushing the shale and heating it to distill off the interspersed hydrocarbons.

Oil shales are found in Precambrian, Palaeozoic, Mesozoic and Tertiary sequences. In Scandinavia the Cambrian Oil Shale is a good example. The Tertiary Green River Shale in Colorado, Utah and Wyoming represents one of the largest petroleum reservoirs in the world. This is a lake deposit, and the organic matter consists mainly of algae, so the kerogen is mainly Type I.

Although very large quantities of petroleum can be produced from oil shale, production costs are at present too high compared to conventional oil. There are also serious environmental problems involved in production from oil shale, and the process requires very large quantities of water, a resource which is not always plentiful.

Further reading: Dake 1978; van Poolen 1980; Dikkers 1985; Donaldson et al. 1985.

Bibliography

Aagaard P, Helgeson HC (1982) Thermodynamic and kinetic constraints on reaction rates among minerals and aqueous I. Theoretical considerations. Am J Sci 282:237–285

Allen JRL (1970) Physical processes of sedimentation – an introduction. Allen & Unwin, London, pp 248

Allen JRL (1982) Sedimentary structures, their character and physical basis. Developments in Sedimentology 30 A,B. Elsevier, Amsterdam

Andersen A, Bjerrum L (1968) Slides in subaqueous slopes in loose sand and silt. Norges Geotekniske Institutt Publ 81:1–9

Andersland OB, Anderson DM (1978) Geotechnical engineering for cold regions. McGraw-Hill, New York, pp 556

Anderton R, Bridges PH, Leeder MR, Sellwood BW (1979) A dynamic stratigraphy of the British Isles. A study in crustal evolution. Allen & Unwin, London, pp 301

Anikouchine W, Sternberg RW (1973) The World Ocean. An introduction to oceanography. Prentice-Hall, Englewood Cliffs, New Jersey, pp 338

Anstey NA (1982) Simple seismic for the petroleum geologist, the reservoir engineer, the well-log analysist, the processing technician and the man in the field. International Human Resources Development Corporation, Boston, pp 168

Arthur MA, Jenkins HC (1981) Phosphorites and paleooceanography. Oceanologica Acta, 26th Geological Congress Paris 1980, Colloquia no 4, pp 83–86

Arthur MA, Anderson TF, Kaplan IR, Veizer J (1983) Stable isotopes in sedimentary geology. S.E.P.M. Short Course no. 10, Dallas

Asquith G, Gibson CR (1982) Basic well-log analysis for geologists. Methods in exploration series. Am Assoc Pet Geol, pp 215

Badiozomani K (1973) The Dorag dolomitization model – application to the Middle Ordovician of Wisconsin. J Sediment Petrol 43:965–984

Badley ME, Egeberg T, Nipen O (1984) The role of faulting in the development of the structural evolution of the Oseberg Feature. Block 30/6 Offshore Norway. Geol Soc Pap Lond 141:639–649

Badley ME (1985) Practical seismic interpretation. Int Human resources Development Corporation, Boston, USA, pp 266

Bagnold RA (1956) The flow of cohesionless grains in fluids. R Soc Lond Philos Trans Ser A 249:235–297

Baker PA, Kastner M (1981) Constraints of the formation of sedimentary dolomite. Science 213:214–216

Bally AW, Watts AB, Grow JA, Manspeizer W, Bernoulli D, Schreiber C, Hunt JM (1981) Geology of passive continental margins: history structure and sedimentology record (with special emphasis on the Atlantic Margin). For the AAPG Eastern Section Meeting and Atlantic Margin Energy Conference, Education Course Note Series no 19. Am Assoc Pet Geol 1–1/7–31

Barker C (1979) Organic geochemistry in petroleum exploration. Continuing Education Course Note Series no 10. Am Assoc Pet Geol, pp 159

Barker C, Hinch H, Jones RW, Mcauliffe C, Momper J, Price L (1978) Physical and chemical constraints on petroleum migration. Continuing Education Course Note Series no 8. Am Assoc Pet Geol A-2/F-13

Barnes RSK (ed) (1977) The coastline. Wiley, New York, pp 355

Bathurst RGC (1975) Carbonate sediments and their diagenesis, 2nd edn. Development in Sedimentology No 12. Elsevier, Amsterdam, pp 658

Bebout D, Davies G, Moore CH, Scholle PS, Wardlaw NC (1980) Geology of carbonate porosity. Continuing Education Course Note Series no 11. Am Assoc Pet Geol A1/E23

Beckmann H (1974) Applied geophysics. Geology of petroleum, Vol. 1. Enke, Stuttgart, pp 272

Beckman H (1976) Geological prospecting of petroleum. Enke, Stuttgart, pp 183

Bell FG (1981) Engineering properties of soil and rocks. Butterworth, London, pp 149

Bentor YK (1980) Marine phosphorites — geochemistry, occurrence, genesis. A symposium. S.E.P.M. Special Publication No. 29, pp 249

Berg R (1979) Exploration for sandstone stratigraphic traps. Continuing Education Course Note Series No. 3. Am Assoc Pet Geol 1-2/9-4

Berg OR, Woolverton DG (eds) (1985) Seismic stratigraphy II: An integrated approach to hydrocarbon exploration. Am Assoc Pet Geol Memoir No. 39, pp 276

Berner A (1971) Principles of chemical sedimentology. McGraw-Hill, New York, pp 240

Berner A (1980) Early diagenesis. A theoretical approach. Princeton University Press, NJ, pp 241

Birkenmajer K (1981) The geology of Svalbard, the western part of the Barents Sea, and the Continental margin of Scandinavia. In: Narn AEM, Churkin M Jr, Stehli FG (eds) The ocean basins and margins, Vol. 5. Plenum, New York

Bjørlykke K (1973) Origin of limestone nodules in the Lower Paleozoic of the Oslo Region. Nor Geol Tidsskr 53:419-431

Bjørlykke K (1975) Mineralogical and chemical changes during weathering of acid and basic rocks in Uganda. Nor Geol Tidsskr 55:81-89

Bjørlykke K (1983) Diagenetic reactions in sandstones. In: Parker A, Sellwood (eds) NATO Advanced Study Institute, Reading, UK. Reidel, Dortrecht, pp 169-213

Bjørlykke K, Palm E (1988) Convection in sedimentary basins and its relevance to diagenetic reactions. Marine and Petroleum Geology, 5:332-351

Bjørlykke K, Sangster DF (1981) An overview of sandstone lead deposits and their relation to red-bed copper and carbonate-hosted lead-zinc deposits. Econ Geol 75th Anniversary Volume, pp 179-213

Bjørlykke K, Elvsborg A, Høy T (1976) Late Precambrian sedimentation in the Central Sparagmite Basin of South Norway. Nor Geol Tidsskr 56:233-290

Bjørlykke K, Bue B, Elverhøi A (1978) Quaternary sediments of the Northwestern Part of the Barents Sea and their relation to the underlying Mesozoic bedrocks. Sedimentology 25:227-246

Bjorøy M (1983) Advances in organic geochemistry 1981. Proceedings of the 10th Int Meeting on Organic Geochemistry, University of Bergen, Norway, 14-18 Sept 1981. Wiley, New York, pp 880

Bjorøy M, Bue B, Elvsborg A (1981a) Organic geochemical analysis of the first two wells in the Troms (Barents Sea). In: Bjorøy M (ed) Advances in Organic Geochemistry. Proceedings of the 10th Int Meeting on Organic Geochemistry, University of Bergen, Norway, 14-18 Sept 1981. Wiley, New York

Bjorøy M, Hall PB, Solli H, Vigran JO (1981b) Organic geochemistry in search of petroleum. Methods in interpretations as used at IKU. Institutt for Kontinentalundersøkelser, Trondheim

Bjorøy M et al. (1981c) Advances in organic geochemistry. Proceedings of the 10th International Meeting on Organic Geochemistry, Bergen, Norway, Sept 1981. Wiley, Chichester, pp 879

Blatt H (1982) Sedimentary petrology. Freeman, San Francisco, pp 551

Blatt H, Middleton GV, Murray RC (1980) Origin of sedimentary rocks. Prentice-Hall, Englewood Cliffs, New Jersey, pp 782

Boillot G (1981) Geology of the continental margins. Longman, London, pp 314

Bott MHP (1976) Sedimentary basins of continental margins and cratons. Developments in geotectonics No 12. Elsevier, Amsterdam, pp 314

Bouma AH (1962) Sedimentology of some flysh deposits. A graphic approach to facies interpretation. Elsevier, Amsterdam, pp 168

Bouma AH, Hampton MA, Hollister D, Kulm VD, Middleton GV, Mutti E, Nelson HC, Walker RG (1973) Turbidites and deep water sedimentation. SEPM. Pacific Section Short Course Anaheim 1973. Los Angeles, pp 157

Bouma AH, Moore GT, Coleman JM (1978) Framework, facies, and oil-trapping characteristics of the upper continental margin. Studies in Geology No 7. Am Assoc Pet Geol, pp 326

Bouma AH, Normark WR, Barnes NE (1985) Submarine fans and related turbidite systems. Springer, Berlin Heidelberg New York Tokyo, pp 351

Boyer SE, Elliott D (1982) Thrust systems. Am Assoc Pet Geol Bull, vol 66, pp 1196-1230

Brenchley PJ, Williams BPJ (ed) (1985) Sedimentology, recent developments and applied aspects. Geol Soc Spec Publ, vol 18, Blackwell, pp 342

Brindley GW, Brown G (1980) Crystal structures of clay minerals and their x-ray identification. Monograph no. 5, Mineralogical Society, London, pp 495

Brooks J, Glennie KW (1987) Petroleum geology of North West Europe. Graham & Trotman, London, pp 982

Brown LF, Fisher WL (1977) Seismic-stratigraphic interpretation of depositional systems. Examples from Brazilian rift and pull-apart basins. In: Payton CE (ed) Seismic stratigraphy — application to hydrocarbon Exploration. Am Assoc Petr Geol Mem 26, Tulsa, Ok, pp 213-248

Brown LF, Fisher WL (1979a) Seismic stratigraphic interpretation and petroleum exploration. Am Assoc Pet Geol Continuing Education Course Note Series, no. 16

Brown FL, Fisher WL (1979b) Principles of seismic stratigraphic interpretation. AAPG Course Note, series 16, pp 104

Bull WB (1977) The alluvial fan environments. Prog Phys Geogr 1:222-270

Carlson WD (1983) Aragonite-calcite nucleation kinetic: An application and extension of avrami transformation theory. J Geol 91:57-72

Carrol D (1970) Clay minerals: a guide to their x-ray identification. Geol Soc Am Spec Pap 126:80

Chilingarian GV, Wolf KH (1975) Compaction of coarse-grained sediments. Developments in sedimentology 18A. Elsevier, Amsterdam, pp 552

Chilingarian GV, Yen TF (1978) Bitumens, asphalts and tar sands. Developments in petroleum science 7. Elsevier, Amsterdam, pp 332

Choquette PW, Pray LC (1970) Geological nomenclature and classification of porosity and sedimentary carbonates. Am Assoc Pet Geol Bull 54:207-250

Cluff RM, Barrows MH (1982) Hydrocarbon generation and source rock evaluation. Am Assoc Pet Geol, reprint series no 24, pp 215

Cohee GV, Glaessner MF, Hedberg HD (1978) The geologic time scale. Studies in geology No. 6. Am Assoc Pet Geol, Oklahoma, pp 388

Coleman JM, Prior DB (1980) Deltaic sand bodies. Continuing Education Course Note Series No 15. Am Assoc Pet Geol, Tulsa, pp 171

Collins AG (1975) Geochemistry of oilfield waters. Developments in petroleum science 1. Elsevier, Amsterdam, pp 496

Collinson JD, Levin J (1983) Modern and ancient fluvial systems. Spec Publ, no 6, Int Assoc Sedimentologists, Blackwell, London, pp 575

Collinson JD, Thompson DB (1982) Sedimentary structures. Allen & Unwin, New York, pp 194

Conybeare CEB (1979) Lithostratigraphic analyses of sedimentary basins. Academic Press, New York, pp 555

Cook HE, Enos P (1977) Deep-water carbonate environments. S.E.P.M. Special Publication No. 25, pp 336

Coustau M (1977) Formation waters and hydrodynamics. J Geochem Explor 7:213-241

Crowell JC (1974) Origin of late cenozoic basins in Southern California. In: Dickinson WR (ed) Tectonics and sedimentation. S.E.P.M. Publication No. 22

Crowell JC (1982) Continental glaciations through geologic times. I: Studies in geophysics, climate and earth history. National Academic Press, Washington, DC, pp 77-82

Curran HA (1985) Biogenic structures: their use in interpreting depositional environments. S.E.P.M. Special Publication no. 35, Tulsa, pp 347

Currey JR (1964) Transgressions and regressions. In: Miller RL (ed) Papers in marine geology. Commemorative vol. Macmillian, New York, pp 175-203

Dahlberg EC (1982) Applied hydrodynamics in petroleum exploration. Springer, Berlin Heidelberg New York, pp 161

Dake LP (1978) Fundamentals of reservoir engineering. Developments in petroleum science No. 8. Elsevier, Amsterdam, pp 443

Daly RA (1936) Origin of submarine canyons. Am J Sci Ser 5,33:401-420

Davies SN, de Wiest RJM (1966) Hydrogeology. Wiley, New York, pp 463

Davis, Jr RA (1978) Coastal sedimentary environment. Springer, Berlin Heidelberg New York, pp 420

Dean WE, Schreiber BC (1978) Marine evaporites. Lecture Notes for Short Course. S.E.P.M. Oklahoma

Dean WE, Friedman GM, Hite RJ, Nurmi RD, Raup OB, Schreiber CB, Shearman DJ (1978) Marine evaporites. S.E.P.M. Short Course No. 4, Oklahoma City, pp 188

Demaison GJ, Moore GT (1980) Anoxic environment and source oil genesis. Org Geochem 2:9-31

Demaison G, Murris RJ (1984) Petroleum geochemistry and basin evaluation. Am Assoc Petrol Geol, Tulsa, memoir 35, pp 426

D'Heur M (1980) Chalk reservoir of the West Ekofisk field. The sedimentation of the North Sea Reservoir rocks. Norsk Petroleums Forening (NPF), Geilo

Dickey PA (1979) Petroleum development geology. Division of Petroleum Publishing Co., Tulsa, pp 398

Dickinson WR, Yarborough H (1977) Plate tectonics and hydrocarbon accumulation. Continuing Education Course Note Series No. 1. Am Assoc Pet Geol, pp 62

Dikkers AJ (1985) Geology in Petroleum production. Developments in petroleum science 26, Elsevier, Amsterdam, pp 239

Dingman LS (1984) Fluvial hydrology. Freeman, San Francisco, pp 383

Donaldson EC, Chillingarian GV, Yen TF (1985) Enhanced oil recovery. In: Fundamentals and analyses developments in petroleum science, 17. Elsevier, Amsterdam, pp 357

Dott RH (1964) Wacke, greywacke and matrix — what approach to immature sandstone classification? J Sediment Petrol 34:625–632

Dott RH, Bourgeois J (1982) Hummocky stratification. Significance of its bedding sequences. Geol Soc Am Bull 93:663–680

Doyle LJ, Pilkey OH (1979) Geology of continental slopes. S.E.P.M. Special Publication Nr. 27, pp 374

Drever JI (1982) The geochemistry of natural waters. Prentice-Hall, Englewood Cliffs, pp 254

Dunham RJ (1962) Classification of carbonates rock according to depositional texture. In: Ham WE (ed) Classification of carbonate rocks. Am Assoc Pet Geol Mem No. 1, Tulsa, OK

Dyer KR (1973) Estuaries. A physical introduction. Wiley, New York, pp 140

Dzulynski S, Ksiazkiewics M, Kuenen PH (1959) Turbidities in flysh of the Polish Carpathian Mountains. Geol Soc Am Bull 70:1089–1118

Eglinton G et al. (1985) Geochemistry of buried sediments. Proceedings of a Royal Soc discussion meeting held on 22–28 June 1984, London, pp 233

Ekdale AA, Bromley RG, Plemberton SG (1984) Ichnology, trace fossils in sedimentology and stratigraphy. S.E.P.M., pp 317

Elverhøi A, Roaldseth E (1983) Glaciomarine sediments and suspended particulate matter, Weddell Sea, Antarctica. Polar Research 1:1–21

Elverhøi A, Liestøl O, Nagy J (1980) Glacial erosion, sedimentation and microfauna in the inner part of Kongsfjorden, Spitsbergen. Norsk Polarinstitutt - Skrifter 172:33–58

Elverhøi A, Lonne Ø, Seland R (1983) Glaciomarine sedimentation in a modern fjord environment, Spitsbergen. Polar Research 4

England WA, Mackenzie AS, Mann DM, Quigley TM (1987) The movement and entrapment of petroleum fluids in the subsurface. J Geol Soc 144:327–347

Englund JO (1980) Generell Hydro geologi. Landbruksbokhandelen, AS-NLH

Ernst WG, Morin JG (1982) The environment of the deep sea. Rubey, vol II. Prentice-Hall, Englewood Cliffs, pp 371

Erxleben AW (1975) Depositional systems in the Canyon Group (Pennsylvanian System), North-Central Texas. Report of Investigations No. 82, Bureau of Economic Geology. The University of Texas at Austin

Eslinger E, Pevear D (1988) Clay minerals for petroleum geologists. S.E.P.M. Short course note no 22, Tulsa, USA

Eugster HP, Jones BF (1979) Behaviour of major solutes during closed-basin brine evolution. Am J Sci 279:609–631

Fairbridge RW, Bourgeois J (1978) The encyclopedia of sedimentology. Dowden, Hutchinson & Ross, Stroudsburg, PA, pp 901

Faure G (1986) Principles of isotope geology. Wiley, New York, pp 589

Fayers JF (ed) (1981) Enhanced oil recovery. Proceedings 3rd European symposium on enhanced oil recovery. Bournemouth, UK, Developments in petroleum science 13, pp 596

Fertl WH (1976) Abnormal formation pressures. Developments in petroleum science 2, Elsevier, Amsterdam, pp 382

Fischer G, Judson S (1975) Petroleum and global tectonics. Princeton University Press, pp 322

Fischer JS, Dolan R (eds) (1977) Beach Processes and coastal hydrodynamics. Benchmark papers in geology No. 39, Dowden, Hutchinson & Ross, Stroudsburg, PA, pp 382

Fisher WL, McGowen JH, Brown Jr LF, Groat CG (1972) Environmental geological atlas of the Texas coastal zone. 6 volumes, Bureau of Economic Geology of the University of Texas at Austin, Texas 78712

Fisher WL, Brown LF, Scott AJ, McGowen JD (1974) Delta systems in the exploration for oil and gas. A Research Colloquium, Geology Building University of Texas campus, pp 78

Flint FR (1971) Glacial and Quaternary geology. Wiley, New York, pp 892

Flügel E (1978) Mikrofazielle Untersuchungsmethoden von Kalken. Springer, Berlin Heidelberg New York, pp 454

Folk RL (1959) Practical classification of limestones. Am Assoc Pet Geol Bull 42:1-38

Folk RL (1974) Petrology of sedimentary rocks. Hemphill, Austin, Texas, pp 182

Folk RL, Ward WC (1957) Brazos River bar: a study in the significance of grain size parameters. J Sediment Petrol, 27:3-27

Frazier DE (1974) Depositional episodes: their relationship to the Quaternary stratigraphic framework in the northwestern portion of the Gulf Basin. Geological Circular 74-1, pp 28

Freach MM (1976) The periglacial environment. Longman, London, pp 308

Freeze RA, Cherry JA (1979) Groundwater. Prentice-Hall, Englewood Cliffs, pp 604

Friedman GM, Johnson KG (1982) Exercises in sedimentology. Wiley, New York, pp 208

Friedman GM, Sanders JE (1978) Principles of sedimentology. Wiley, New York, pp 792

Füchtbauer H (1974) Sediments and sedimentary rocks 1. Halsted Press Division. Wiley, New York, pp 464

Füchtbauer H (1978) Zur Herkunft des Quarzsements. Abschätzung der Quarzauflösung im Silt und Sandsteinen. Geol Rundsch 67:991-1009

Füchtbauer H, Müller G (1977) Sediment und Sedimentsgesteine. Schwlizerbart, Stuttgart, pp 784

Galloway WE, Brown Jr LF (1972) Depositional systems and shelf-slope relationships in Upper Pennsylvania rocks. North-Central Texas Bureau of Economic Geology, The University of Texas at Austin, pp 62

Galloway WE, Hobday DK (1983) Terrigenous clastic depositional systems. Springer, Berlin Heidelberg New York, pp 420

Galloway W, Kreitler W, McGowen JH (1979) Depositional and groundwater flow systems in the exploration for uranium. Bureau of Economic Geology, Austin, Texas, pp 267

Garrels RM, Christ L (1965) Solutions, minerals and equilibria. Harper & Row, New York, pp 450

Garrels RM, MacKenzie FT (1971) Evolution of sedimentary rocks. Norton, New York, pp 397

Garrels RM, MacKenzie FT (1974) Chemical history of the ocean deduced for post depositional changes in sedimentary rocks. In: Hay WW (ed) Studies in paleo-oceanography. Special Publication S.E.P.M. 20:193-204

Gautier DL (1986) Rates of organic matter in sediment diagenesis. S.E.P.M. Special Publication no 38, Tulsa, USA, pp 203

Gautier DL, Claypool GE (1984) Interpretation of methanic diagenesis in ancient sediments by analogy with processes in modern diagenetic environments. In: McDonald DA, Surdam RC (eds) Clastic diagenesis. Am Assoc Pet Geol, memb 37:111-126

Gautier DL, Kharaka YK, Surdam RC (1985) Relationship of organic matter and mineral diagenesis. S.E.P.M. Short Course no 17, Tulsa, USA, pp 279

Geoff JC (1983) Hydrocarbon generation and migration from Jurassic source rocks in the E. Shetland Basin and Viking Graben at the North Sea. Geol Soc Lond 140:445-474

Ginsburg RN (ed) (1971) Geology of calcareous algae. Miami, FL. University of Miami Comparative Sedimentology Laboratory, Short Course Note

Ginsburg RN (1975) Tidal ore deposits. A casebook of recent examples and fossil counterparts. Springer, Berlin Heidelberg New York, pp 428

Ginsburg RN, Rezak R, Wray JL (1971) Geology of calcareous algae. Comparative Sedimentology Lab., Fisher Island, Miami Beach

Goldschmidt VM (1954) Geochemistry. Clarendon, Oxford, pp 730

Gouidie AS, Pye K (eds) (1983) Chemical sediments and geomorphology, precipitates and residues in the nearsurface environments. Academic Press, New York, pp 43

Gradstein FM, Agterberg FP, Brower JC, Schwarzaeher WS (1985) Quantitative stratigraphy. Reidel, Dortrecht, pp 598

Graf DL (1982) Chemical osmosis, reverse chemical osmosis, and the origin of subsurface brines. Geochim Cosmochim Acta 4:1431-1448

Granat L, Rohde R, Hallberg RO (1976) The global sulphur cycle. In: Svensson BH, Söderlund R (eds) Nitrogen, phosphorus and sulphur — global cycles. SCOPE Report 7, Ecol Bull (Stockholm) 22:89-134

Gray H (1977) Field guide to some carbonate rock environments. Florida Keys and Western Bahamas, Kendall/Hunt, Iowa, pp 415

Gregory KJ (1977) River channel changes. Wiley, New York, pp 448

Gretener PE (1978) Pore pressure: fundamentals, general ramifications and implications for structural geology. Continuing Education Course Note Series No. 4. Am Assoc Pet Geol, pp 87

Gretener PE (1982) Geothermics: Using temperature in hydrocarbon exploration. Education Course Note Series No. 17. Am Assoc Pet Geol Oklahoma, pp 170

Gries R (1983) Oil and gas prospection beneath Precambrian of foreland thrust plates in rocky Mountains. Am Assoc Pet Geol Bull 67:1-28

Griffiths DH, King RF (1965) Applied geophysics for engineers and geologists. Pergamon, New York, pp 223

Grim RE (1968) Clay mineralogy. Mc-Graw-Hill, New York, pp 595

Grow JA, Bowin CO, Hutchinson (1979) The gravity field of the US Atlantic Continental Margin. Tectonophysics 59:27-52

Habicht JKA (1979) Paleoclimate, paleomagnetism and continental drift. Studies in Geology No. 9. Am Assoc Pet Geol

Hakanson A, Jansson M (1983) Principles of lake sedimentology. Springer, Berlin Heidelberg New York Tokyo, pp 320

Hallam A (1981) Facies interpretation and the stratigraphic record. Freeman, San Francisco, pp 290

Hallberg RO (1973) The microbiological C-N-S cycles in sediments and their effect of the sedimentwater surface. OIKOS Supplementum 15:51-62

Ham WE (1962) Classification of carbonate rocks. A symposium. Am Assoc Pet Geol, Oklahoma, pp 279

Hambrey MJ, Harland WB (1981) Earth's pre-Pleistocene glacial record. Cambridge University Press, pp 1004

Hardenbol J, Vail PR, Ferrer J (1982) Interpreting paleoenvironments, subsidence history, and sea-level changes of passive margins from seismic and biostratigraphy. In: Climate in earth history, Studies in Geophysics. National Academy Press, Washington DC, pp 139-153

Harland WB, Cox AV, Llewellyn PG, Pickerton CAG, Smith AG, Walters R (1982) A geologic time scale. Cambridge University Press, pp 131

Harms JC, Southard JB, Walker RG (1975) Depositional environments as interpreted from primary sedimentary structures and stratification sequences. S.E.P.M., Short Course Notes No 2, pp 161

Harms JC, Southard JB, Walker RG (1982) Structures and sequences in clastic rocks. Lecture notes for Short Course No. 9. S.E.P.M., Tulsa, IK, USA, pp 242

Harris PM (1983) Carbonate buildups in a core workshop. S.E.P.M. Core workshop No. 4, Dallas, April 1983

Harvie CE, Eugster HP, Weare JH (1982) Mineral equilibria in six-component sea water system $Na-K-Mg-Ca-SO_4-Cl-H_2O$ at 25° II. Compositions of the saturated solutions. Geochim Cosmochim Acta 46:1603-1618

Hay WW, Holser WT, Peterson WH (1982) Ocean circulation, plate tectonics and climate. In: Climate in Earth History, Studies in Geophysics. National Academy Press, Washington DC, pp 83-96

Heath GR (1974) Dissolved silica and deep sea sediments. S.E.P.M. Special publication 20:77-93

Hedberg HD (ed) (1972) An international guide to stratigraphic classification, terminology and usage. Introduction and summary. Lethaia 5:283-295

Hedberg HD (1976) International stratigraphic guide. Wiley, New York

Heezen BC, Hollister D (1971) The face of the deep. Oxford University Press, pp 659

Hobson GD, Tiratsoo EN (1981) Introduction to petroleum geology. Blackwell Scientific Press, London, pp 352

Hobson GD, Tiratsoo EN (1983) Introduction to petroleum geology. Gulf Publishing, Houston, TX, pp 384

Holland CM et al. (1978) A guide to strategraphic procedure. Geol Soc Lond Spec Rep, No 11, pp 18

Holtedahl H (1965) Recent turbidites in Hardangerfjord, Norway. In: Whittard WF, Brandshaw R (eds) Submarine geology and geophysics. Colston paper 17, Butterworth, London

Howard JD, Scott RM (1983) Comparison of Pleistocene and Holocene barrier island beach-to-offshore sequences, Georgia and northeast Florida, USA. Sediment Geol 34:167-183

Hower J, Eslinger EV, Hower ME, Perry EA (1976) Mechanism of burial metamorphism of argillacious sediments. 1) Mineralogical and Chemical Evidence. Geol Soc Am Bull 87:725-737

Hsu KJ, Jenkyns HC (eds) (1974) Pelagic sediments: on land and under the sea. Special publication No. 1, International Association of Sedimentologists. Blackwell, London, pp 447

Hubbard RJ, Pape J, Robertz DZ (1985) Depositional sequence mapping as a technique to establish tectonic and stratigraphic framework and evaluate hydrocarbon potential on a passive continental margin. APG memoir 39, pp 79-91

Hudson JD (1977) Oxygen isotope studies on Cenozoic temperatures, ocean and ice accumulations. Scott J Geol 13:313–326

Hudson JD (1978) Concretions isotopes and the diagenetic history of the Oxford clay (Jurassic) of Central England. Sedimentology 25:361–387

Hunt CB (1972) Geology of soils, their evolution, classifications and uses. Freeman, San Francisco, pp 344

Hunt JM (1979) Petroleum geochemistry and geology. Freeman, San Francisco, pp 617

Hurst A, Irwin H (1983) Application of geochemistry to sandstone reservoir studies. In: Brooks J (ed) Petroleum geochemistry and exploration of Europe. Geol Soc Spec Publ 12, Blackwell, Oxford, pp 127–146

Illing LV, Hobson CD (eds) (1981) Petroleum geology of the continental shelf of North-West Europe. Heyden on behalf of Institute of Petroleum, London, pp 518

Inderbitzen AL (1974) Deep-sea sediments. Plenum, New York, pp 497

Ion DC (1980) Availability of world energy resources. Graham & Trotman, London, pp 345

Irwin H (1980) Early diagenetic carbonate precipitations and pore fluid migration in the Kimmeridge Clay of Dorset, England. Sedimentology 27:577–591

Irwin H, Curtis CD, Coleman ML (1977) Isotopic evidence for source of diagenetic carbonates formed during burial of organic-rich sediments. Nature 269:209, 213

Irwin H, Curtis C, Coleman M (1979) Isotopic evidence for source of diagenetic carbonates formed during burial of organic-rich sediments. Nature (Lond) 269:209–213

Jeynon MK, Fitch AA (1985): Seismic reflection interpretation. (Geoexploration monographs, series 1, no 8). Boerntraeger, Berlin, pp 318

Kenneth JP (1982) Marine geology. Prentice Hall, Englewood Cliffs, pp 813

King RE (1979) Stratigraphic oil and gas fields – classification, exploration methods and case histories. Edwards, Michigan, pp 687

Kinsman DJJ (1969) Modes of formation, sedimentary associations and diagnostic features of shallow-water and supratidal evaporites. Am Assoc Pet Geol Bull 53:830–840

Kinsman DJJ (1975a) Rift valley basins and sedimentary history of trailing continental margins. In: Fischer AG, Judson S. Princeton University Press, pp 83–126

Kinsman DJJ (1975b) Salt floors to geosynclines. Nature 255:275–378

Kinsman DJJ (1976) Evaporites: relative humidity control of primary mineral facies. J Sediment Petrol 46:273–279 (describes effect of lowered $^a H_2 O$ of brines on evaporite mineral sequence). Sediment Petrol 48:1357–1359

Kinsman DJJ, Park RK (1976) Algal belt and coastal sabkha evolution,Trucial coast. Persian Gulf. In: Walter MR (ed) Development in sedimentology. 20, Stromatolites. Elsevier, Amsterdam, pp 421–433

Klein Gde V (1985) Sandstone depositional models for exploration of fossil fuels. 3rd eds, International Human Resources Development Cooperation, Boston, pp 209

Koster EM, Steel RJ (1984) Sedimentology of gravel and conglomerates. Can Soc Pet Geol, Memoir no 10, pp 441

Kraft JC (1971) Sedimentary facies patterns and geologic history of a Holocene marine transgression. Geol Soc Am Bull 82–2131–2158

Kraft JC, John CJ (1979) Lateral and vertical facies relations of transgressive barriers. Am Assoc Pet Geol Bull 63:2145–2163

Krauskopf KB (1979) Introduction to geochemistry. McGraw-Hill, New York, pp 617

Kuenen PH, Miglionini CI (1950) Turbidity currents as a cause of graded bedding. J Geol 58:91–127

Kumar N, Sanders JE (1974) Inlet sequence, a vertical succession of sedimentary structures and textures created by the lateral migration of tidal inlets. Sedimentology 21:491–532

Kupsch WO (1967) Postglacial uplift – a review. In: Andrews I (ed) Glacial isostasy. Geneharm papers in geology, No. 10. Dowden, Hutchinson & Ross, Stroudsburg, pp 125–151

Larsen G, Chilingarian GV (1967) Developments in sedimentology 8. Diagenesis in sediments. Elsevier, Amsterdam, pp 551

Larter SR (1985) Integrated kerogen typing in the recognition and quantitative assessment of petroleum source rocks. In: Petroleum geochemistry in exploration of the Norwegian Shelf. Graham & Trotman, London, pp 269–286

Leatherman SP (ed) (1979) Barrier Island from the Gulf of St. Lawrence to the Gulf of Mexico. Academic Press, New York, pp 375

Leeder MR (1982) Sedimentology process and product. Allen & Unwin, London, pp 344

Leeds A, Buller AT (1979) Modern temperate water and warm water shelf carbonate sedimentation. Mar Geol 19:159–198

Leggett JK (1982) Trench-forearc geology. Geological Society, Special Publication No. 10. Blackwell Scientific, London, pp 576

Leroy LW, Leroy DO, Raese JW (1977) Subsurface geology, petroleum mining construction. Colorado School of Mines, Colorado

Levorsen AI (1967) Geology of petroleum. Freeman, San Francisco, pp 724

Leythaeuser D, Mackenzie A, Schaeffer RG, Bjorøy M (1984) A novel approach for recognition and quantification of hydrocarbon migration effects in shale-sandstone sequences. Am Assoc Pet Geol Bull 68:196–219

Leythaeuser D, Rullkötter J (1986) Advances in organic geochemistry 1985, part 1. Petroleum geochemistry. Pergamon, Oxford, pp 647

Link PK (1982) Basic petroleum geology. Oil and Gas Consultants International, Tulsa, pp 235

Lisitzyn AP (1972) Sedimentation in the world ocean. S.E.P.M. Special Publication 17, pp 218

Logan BW, Davies GR, Read JF, Cebuski DE (1970) Carbonate sedimentation and environments, Shard Bay, Western Australia. Am Assoc Pet Geol, Memoir, Tulsa, Oklahoma

Longman W (1981) Carbonate diagenesis as a control on stratigraphic traps. Education Course Note Series No. 21. Am Assoc Pet Geol, pp 159

Loudon TV (1979) Computer methods in geology. Academic Press, New York, pp 269

Lowe DR (1976) Subaqueous liquified and fluidized sediment flows and their deposits. Sedimentology 23:285–308

Lowe DR (1979) Sediment gravity flows, their classification and some problems of application to natural flow deposits. S.E.P.M. Special Publication 27:75–82

Lowe DR (1982) Sediment gravity flow: II Depositional models with special reference to the deposits of high density turbidity currents. J Sediment Petrol 52:279–297

Lowe DR, Lopeceole RD (1974) The characteristics and origin of dish and pillar structures. J Sediment Petrol 44:484–501

Lowell JD (1985) Structural styles in petroleum exploration. OGCI Publ Oil and Gas Consultants International, Tulsa, pp 460

Macenzie AS, Leythaeuser D, Muller P, Radke M, Schaeffer RG (1987) The expulsion of petroleum from Kimmeridge Clay Fm. source rocks in the area of the Brae Oilfield, UK continental shelf. In: Brooks J, Glennie KW (eds) Proc of the 3rd Conference on the Petroleum Geology of NW Europe Graham and Trotman, London, pp 865–878

Mannheim FT, Paul Ch (1981) Patterns of groundwater salinity changes in a deep continental oceanic transect off the Southern Atlantic coast in the USA. J Hydrol 54:95–105

Margara K (1978) Compaction and fluid migration. Practical petroleum geology. Developments in Petroleum Science No. 9. Elsevier, Amsterdam, pp 319

Marshall CE (1977) The physical chemistry and mineralogy of soils. Vol. II, Soil in place. Wiley, New York, pp 313

Marshall JD (1987) Diagenesis of sedimentary sequences. Geol Soc Spec Publ no 36:0–360

Martin JM, Burton JD, Eisma D (1980) River input to ocean systems. Proceedings of SCOR/ACMRR/FCOR/IAHS/UNESCO/CMG/IABO/IAPSO, Review at FAO (ROME), United Nations Environmental Programme, pp 383

Mason B, Moore CB (1982) Principles of geochemistry. Wiley, New York, pp 344

Matter A, Tucker MF (eds) (1978) Modern and ancient lake sediments. Int Assoc Sedimentol Spec Publ 2

Mayer-Gurr A (1976) Petroleum engineering geology of petroleum. Vol. 3. Enke, Stuttgart, pp 208

McDonald DA, Surdam RC (1984) Clastic diagenesis. Am Assoc Pet Geol Memoir 37, pp 434

McKee DE (1979) A study of global sand seas. US Geol Survey Professional paper 1052, US Government printing office, Washington, pp 429

McKenzie JA, Hsu KJ, Schneider JF (1980) Movements of subsurface waters under the sabkha, Abu Dhabi, UAE and its relation to evaporite dolomite genesis. S.E.P.M. Special Publication No. 28:11–30

McKerrow WS (1978) The ecology of fossils, an illustrated guide. Duckworth, London, pp 383

McMullan JT, Morgan TR, Murray RB (1977) Energy resources. Arnold, London, pp 177

McQuillin R, Ardus DA (1977) Exploring the geology of shelf seas. Graham and Trotman, London, pp 234

Menard HW (1974) Geology, resources and society, an introduction to earth science. Freeman, San Francisco, pp 621

Merkel RM (1979) Well-log formation evaluation. Cont. Education Course Note 14. Am Assoc Pet Geol, Tulsa, Oklahoma, USA

Merriam DF (ed) (1975) Quantitative techniques for the analysis of sediment. Pergamon, New York, pp 174

Miall AD (1978) Fluvial sedimentology. Can Soc Pet Geol Memoir No. 5, pp 859

Miall AD (1981a) Analysis of fluvial depositional systems. Am Assoc Pet Geol, Education Course Note Series No. 20, pp 75

Miall AD (1981b) Sedimentation and tectonics in alluvial basins. Geol. Society of Canada, Special Paper 23, pp 212

Miall AD (1984) Principles of sedimentary basin analysis. Springer, Berlin Heidelberg New York, pp 490

Miall AD (1986) Eustatic sea level changes interpreted from seismic stratigraphy: A critique of the methodology with particular reference to the North Sea Jurassic records. Am 70:131–137

Miall AD (1988) Reservoir heterogenities in fluvian sandstones, lessons from outcrop studies. Am Assoc Pet Geol Bull 72:682-697

Michael AA (1982) The carbon cycle-controls on atmospheric CO_2 and climate in geologic past. In: Climate in earth history, studies in geophysics. National Academy, Washington DC, pp 55–67

Middleton GV, Southard JB (1978) Mechanics of sediment movement. S.E.P.M. Short Course No 3, pp 1.1./10.1

Milliman JD (1974) Marine Carbonates. In: Milliman JD, Müller G, Förstner U (eds) Recent sedimentary carbonates. Part 1. Springer, Berlin Heidelberg New York, pp 375

Milliman JD (1983) World-wide delivery of river sediment to the oceans. J Geol 91:1–21

Millot G (1970) Geology of clay weathering, sedimentology, geochemistry. Springer, Berlin Heidelberg New York, pp 429

Mitra S (1986) Duplex structures and imbricate thrust systems, geometry structural position and hydrocarbon potential. Am Assoc Pet Geol Bull 70:1087–1112

Momper JA (1978) Oil migration limitations suggested by geological and geochemical considerations. AAPG Course Note Series 8, B1–B60

Monicard RP (1980) Properties of reservoir rocks: core analysis. Institut Français du Petrole Publications, Editions Technic, 27 Ginoux 75737, Paris

Moore CH (1980) Porosity in carbonate rock sequences. Am Assoc Pet Geol A 124

Muir M, Lock D, von der Borch C (1980) The Coorong model for penecontempoeraneous dolomite formation in the Middle Preterozoic Macarthur Group, Northern Territory, Australia. S.E.P.M. Special Publication No. 28, pp 51–67

Natland ML (1933) Depth temperature distribution of some recent and fossil foraminifera in the southern California region. Bull Scrips Inst Oceanogr, La Jolla, ser 3, pp 225–230

North FK (1985) Petroleum geology. Allen and Unwin, Boston, pp 607

Notholt AJG, Brasier MD (1986) Proterozoic and Cambrian phosphorities – regional review, Europe. In: Cook PJ, Shergold JH (eds) Phosphate deposits of the world. vol 1, Proterozoic and Cambrian phosphorites, pp 91–100

Nummedal D, Fischer IA (1978) Process-response models for depositional shorelines: the German and the Georgia bights. Proc. 16th Coastal Engineering Conference, Hamburg 2, pp 1215–1231

Ollier C (1969) Weathering. Oliver & Boyd, Edinburgh, pp 304

Palacas JG (1984) Petroleum geochemistry and source rock potential of carbonate rocks. Am Assoc Petrol Geol, Studies no 18, Tulsa, pp 207

Parasnis DS (1979) Principles of applied geophysics. Chapman & Hall, London, pp 275

Parker A, Sellwood BW (1983) Sediment diagenesis. Reidel, Dortrecht

Patterson RJ, Kinsman DJ (1981) Hydraulic framework at a sabhka along the Arabian Gulf. Am Assoc Pet Geol Bull 65:1457–75

Patton CC (1977) Oilfield water systems. 2nd edn, Champbell Petroleum Series, Oklahoma, USA, pp 252

Payton CE (1977) Seismic stratigraphy – application to hydrocarbon exploration. Am Assoc Pet Geol, Tulsa, Oklahoma, USA, pp 516

Penn IE, Cox BM, Gailos RW (1986) Towards presession in stratigraphy in geophysical log correlation of Upper Jurassic (including Callovian) strata of the eastern England Shelf. Geol Soc Lond 143:381-410

Perrodon A (1983) Dynamics of oil and gas accumulations. Bull des Centre de recherches exploration. Elf Aquitaine Documentation Micoulau, 64018 Pau, France, pp 368

Pettijohn FJ (1957) Sedimentary rocks. 2nd edn. Harper & Row, New York, pp 717

Pettijohn FJ, Potter PE, Siever R (1972) Sand and sandstone. Springer, Berlin Heidelberg New York, pp 618

Picard DM, High Jr LR (1973) Sedimentary structure of ephemeral streams. Developments in Sedimentology No. 17, Elsevier, Amsterdam, pp 223

Pirson SJ (1970) Geologic well-log analysis. Gulf Publishing, Houston, pp 370

Potter PE, Maynard BJ, Pryor WA (1980) Sedimentology of shale. Study guide and reference source. Springer, Berlin Heidelberg New York, pp 306

Powell TG (1986) Petroleum geochemistry and depositional setting of Lacustrine source rocks. Marine and Petroleum Geology 3:200–219

Press F, Siever R (1978) Earth. 2nd edn. Freeman, San Francisco, pp 649

Purdy EG (1963) Recent calcium carbonate facies of the great Bahamas bank, 2 sedimentary facies. J Geol 71:472–497

Purser BH (1973) The Persian Gulf Holocene carbonate and diagenesis in a shallow epicontinental sea. Springer, Berlin Heidelberg New York, pp 471

Purser BH, Evans GE (1973) Regional sedimentation along the trucial coast (E. Persian Gulf). In: Purser BH (ed) The Persian Gulf: Holocene carbonate sedimentation in a shallow epicontinental sea. Springer, Berlin Heidelberg New York, pp 471

Rachochi A (1981) Alluvial fans. Wiley, New York, pp 161

Raiswell R (1971) The growth of Cambrian and Liassic concretions. Sedimentology 17:147–171

Ramberg IB, Neumann ER (1978) Tectonics and geophysics of continental rifts. Reidel, Dortrecht, pp 444

Rawson PF (1982) Early cretaceous events and the "Late Kimmerian Unconformity" in the North Sea. Am Assoc Pet Geol 66:2628–2648

Reading HG (1978) Sedimentary environments and facies. Blackwell, London, pp 557

Reading HG (1986) Sedimentary environments and facies. Blackwell, Oxford, pp 615

Reeckmann A, Friedman GM (1982) Exploration for carbonate petroleum reservoirs. Wiley-Intern Sci, New York, pp 213

Reineck HE (1970) Das Watt. Ablagerungs- und Lebensraum. Kramer, Frankfurt, pp 142

Reineck HE, Singh IB (1975) Depositional sedimentary environments. Springer, Berlin Heidelberg New York, pp 439

Reusch HH (1878) Iagttagelser over isskuret fjeld of forvitret fjeld. Vid. Selsk. Forh. 1:1–27

Rice DD, Gautier DL (1983) Patterns of sedimentation, diagenesis and hydrocarbon accumulation in Cretaceous rocks of the Rocky Mountains. S.E.P.M. Short Course no 11

Richards AF (1981) Coastal upwelling. Coastal and Estuarine Monograph Series. Am Geophys Union

Rider MH (1986) The geological interpretation of well logs. Blackie, Glasgow, pp 274

Rieckmann A (1982) Exploration for carbonate petroleum reservoirs (Translated from Essai de Characterisation sédimentologique des dépots carbonates. Elf Aquitaine, Boussens et Pau, 1977). Wiley, New York, pp 213

Rieke III H, Chilingarian V (1974) Developments in sedimentology. 16, Compaction of argillaceous sediments. Elsevier, Amsterdam, pp 424

Rigby JK, Hamblin WK (1972) Recognition of ancient sedimentary environments. S.E.P.M. Special Publication No. 16

Riggs SR (1979) Phosphorite sedimentation in Florida. A model phosphogenic system. Econ Geol 74:285–314

Robert A (1977) Geotechnology. An introductory text for students and engineers. Pergamon, New York, pp 347

Roberts WH, Cordell RJ (1980) Problems of petroleum migration. Am Assoc Pet Geol. Studies in Geology No. 10, pp 273

Roehl PO, Choquette PW (1985) Carbonate petroleum reservoirs. Springer, Berlin Heidelberg New York Tokyo, pp 622

Rosenquist IT (1985) Mineralogical physical and chemical factors determining mechanical properties of remoulded clays and clay sediments. 5th meeting of the European Clay Groups, Prague 1983 (Charles University), pp 485–490

Rupke NA (1977) Growth of an ancient deep-sea fan. J Geol 85:725–744

Ryder RT, Lee MW, Smith GN (1981) Seismic models of sandstone stratigraphic traps in Rocky Mountain basins. Am Assoc Pet Geol, Oklahoma, pp 77

Saigal GC, Bjørlykke K (1987) Carbonate cements in clastic reservoir rocks from offshore Norway-relationships between isotopic composition, textural development and burial depth. In Marshall JD (ed) 1987: Diagenesis of sedimentary sequences. Geol Soc London Spec Publ No 36 pp 313–324

Savin SM (1982) Stable isotopes in climatic reconstructions. In: Studies in geophysics, climate in earth. National Academy Press, Washington DC, pp 164–171

Saxby JD (1982) A reassessment of the range of kerogen maturities in which hydrocarbons are generated. J Pet Geol 5:117–128

Saxov S, Nieuwenhuis JK (1982) Marine slides and their mass movements. Plenum, New York, pp 353

Schlee JS (1984) Unconformities and hydrocarbon accumulation. Am Assoc Pet Geol, Memoir no 35, pp 184

Schlumberger SP (1982) Well evaluation development. Published by SP Schlumberger

Schmidt V, McDonalds DA (1980) Secondary reservoir porosity in the course of sandstone diagenesis. Continuing Education Course Note Series No. 12, Tulsa. Am Assoc Pet Geol, pp 125

Schneidermann, Harris PM (1985) Carbonate cements. S.E.P.M. Special Publication no 36, Tulsa Ok, pp 397

Scholle PA (1971) Chalk diagenesis and its relation to the petroleum exploration – oil from chalks, a modern miracle. Am Assoc Pet Geol Bull 61:982–1109

Scholle PA (1978) Carbonate rock constituents textures, cements, and porosities. Am Assoc Pet Geol, memoir 27, pp 241

Scholle PA (1979a) A color illustrated guide to constituents, textures, cements and porosities of sandstones and associated rocks. Am Assoc Pet Geol, Memoir 27, pp 201

Scholle PA (1979b) Deposition, diagenesis and hydrocarbon potential of "deeper-water" limestones. Continuing Education Course Note Series No. 7. Am Assoc Pet Geol, pp 25

Scholle PA, Schluger PR (1979) Aspects of diagenesis. S.E.P.M., pp 443

Scholle PA, Spearing D (eds) (1982) Sandstone depositional environments, Am Assoc Pet Geol, memoir 31, pp 410

Scholle PA, Bebout DG, Moore CH (eds) (1983) Carbonate depositional environments. Am Assoc Pet Geol, Memoir No. 33, pp 708

Schwartz ML (ed) (1972) Spits and bars. Benchmark papers in geology. Dowden, Hutchinson & Ross, Stroudsburg, pp 452

Scoffin TP (1987) An introduction to carbonate sediments and rocks. Blackie, Glasgow, pp 274

Scrutton RA, Talwani M (eds) (1982) The ocean floor. Wiley, New York, pp 318

Seemann U (1982) Depositional facies, diagenetic clay minerals and reservoir quality of Rotlieground sediments in the Southern Permian Basin (North Sea). A review. Clay and clay minerals 17:55–67

Seibold E, Berger WH (1982a) An introduction to marine geology. Springer, Berlin Heidelberg New York, pp 288

Seibold E, Berger WH (1982b) The sea floor. Springer, Berlin Heidelberg New York, pp 288

Seibold E, Meulenkamp JD (1982) Stratigraphy quo vadis? Am Assoc Pet Geol, Studies in Geology no 16, IUGS Spec Publ no 14, pp 69

Selley RC (1978) Ancient sedimentary environments and their sub-surface diagnosis. Chapman & Hall, London, pp 288

Selley RC (1979) Concepts and methods of sub-surface facies analysis. Continuing Education Course Note Series No. 9. Am Assoc Pet Geol, pp 86

Selley RC (1982) An introduction to sedimentology. 2nd edn. Academic Press, New York, pp 417

Selley RC (1985a) Ancient Sedimentary environments. Chapman & Hall, London, pp 317

Selley RC (1985b) Elements of petroleum geology. Freeman, San Francisco, pp 449

Shah DO (1981) Surface phenomena in enhanced oil recovery. Plenum, New York, pp 874

Shanmugam G, Moiola RJ (1988) Submarine fans: characteristics, models, classification and reservoir potential. Earth Sci Rev 24:383–428

Shelton JW (1973) Models of sand and sandstone deposits. A methodology for determining sand genesis and trend. Okla Geol Surv Bull 118:122

Shepard FB (1967) Submarine geology. 2nd edn. Harper & Row, New York, pp 555

Shepard FP, Marshall NF, McLoughlin PA, Sullivan GG (1979) Currents in submarine canyons and other sea-valleys. Am Assoc Pet Geol. Studies in Geology No. 8, pp 173

Sheriff RE (1981) Structural interpretation of seismic data. A continuing education course presented for the Dallas Geological Society, February 1981. Education Course Note Series No. 23. Am Assoc Pet Geol, pp 72

Simone L (1981) Ooids. A review. Earth Sci Rev 16:319–355

Smith PJ (1973) Topics in geophysics. Open University Set Book. The Open University Press, pp 246

Sneidermann N, Harris PM (1985) Carbonate cements. S.E.P.M. Special Publication no 36, Tulsa Ok, pp 379

Spencer AM (ed) (1984) Petroleum geology of the North European Margin. Proceedings of the North European Margin Symposium (NEMS 83) organized by the Norwegian Petroleum Society in Trondheim, May 1983, pp 436

Spencer AM et al. (eds) (1986) Habitat of hydrocarbons on the Norwegian Continental Shelf. The Norwegian Petrol Soc, Graham & Trotman, London, pp 354

Spjeldnaes N (1981) Lower Palaeozoic Palaeoclimatology. In: Holland CH (ed) Lower Palaeozoic of the Middle East and Southern Africa, and Antarctica. Wiley, New York, pp 199–256

Stanley DJ, Moore GT (1983) The shelf break: critical interface on continental margins. S.E.P.M. Special Publication No. 33, pp 467

Stanley DJ, Wezel FC (1985) Geological evolution of the Mediterranean basin. Springer, Berlin Heidelberg New York, pp 669

Steckler MS, Watts AB (1978) Subsidence of the Atlantic-type continental margin off New York. Earth Panet Sci Lett 41:1–13

Steel RJ, Aasheim SJ (1978) Alluvial sand deposition in a rapidly subsiding basin (Devonian, Norway). In: Miall AD (ed) Fluvial sedimentology. Can Soc Pet Geol No. 5, Memoirs, pp 385–412

Stow DA, Piper DJW (1984) Fine-grained sediments: deepwater processes and facies. Geol Soc Lond, Blackwell, London, pp 659

Stowe KS (1979) Ocean science. Wiley, New York, pp 610

Stride AM (ed) (1982) Offshore tidal sands. Processes and deposits. Chapman & Hall, pp 222

Surlyk F (1978) Submarine fan sedimentation along fault scarps on tilted fault blocks (Jurassic-Cretaceous boundary, East Greenland). Grønland geologiske undersøkelse, Bull No 128, pp 108 and appendix

Surlyk F, Clemmensen LF, Larsen HC (1981) Post-paleozoic evolution of East Greenland continental margin. In: Kerr JW, Ferguson AJ (eds) Geology of the North Atlantic borderlands. Can Soc Pet Geol, Mem 7:611–645

Svensson BH, Öderlund (1976) Ecological Bulletins No. 22. Nitrogen, phosphorous and sulphur-global cycles. Scope Report 7, Ecological Bulletins, Swedish Natural Science Research Council, pp 192

Swart PK (1983) Carbon and oxygen isotope fractionation in sclerationian corals. A review. Earth Sci Rev 19:51–80

Taira A, Okao AH, Whitaker JHMO, Smith AJ (1982) The Shimants Belt at Japan: Cretaceous Lower Miocene active-margin sedimentation. In: Legget JK (ed) Trench-fore-arc geology: sedimentation and tectonics on modern and ancient active plate margin. Geol Soc Lond, Blackwell, Oxford, pp 5–26

Talbot MR (1973) Major sedimentary cycles in the corallian beds oxfordian of Southern England. Palaeogeogr Palaeoclimatol Palaeoecol 14:293–317

Taylor JCM (1980) Origin of the Werraanhydrit in the U.K. Southern North Sea – a reappraisal. Contrib Sedimentol 9:91–113

Thiede J, Suess E (1983) Coastal upwelling. Its sediment record. A: Responses of the sedimentary regime to present coastal upwelling, pp 603; B: Sedimentary records of ancient coastal upwelling, pp 610. NATO conference series IV, marine Sciences

Thomas BM et al. (1985) Petroleum geochemistry in exploration of the Norwegian Shelf. Proceedings at a Norwegian Petroleum Soc (NP) conference. Organic geochemistry in Exploration of the Norwegian Shelf. Stavanger Oct 1984, Graham & Trotman, London, pp 0–337

Thorez J (1975) Phyllosilicates and clay minerals. A laboratory handbook for their x-ray diffraction analysis. Lelotte, Belgia

Thornton EB, Guza RT (1982) Energy saturation and phase speeds measured on a natural beach. J Geophys Res 87:9499–9508

Tillman RW, Weber KJ (1987) Reservoir sedimentology. Soc Econ Paleont Mineral, Spec Publ no 40, Tulsa, USA, pp 357

Tiratso EN (1984) Oilfields of the world. Scientific Press Beacansfield, pp 392

Tissot BP, Welte DH (1978) Petroleum formation and occurrence. A new approach to oil and gas exploration. Springer, Berlin Heidelberg New York, pp 538

Tissot BP, Welte FC (1984) Petroleum formation and occurrence. Springer, Berlin Heidelberg New York, pp 669

Toomey DF (ed) (1981) European fossil reef models. S.E.P.M. Special Publication no 30, pp 546

Torrance JK (1983) Towards a general model of quick clay development. Sedimentology 30:547–556

Tucker ME (1981) Sedimentary petrology – an introduction. Blackwell, London, pp 252

Turner-Peterson CE (1980) Uranium in sedimentary rocks: application of facies concept to exploration. The Rocky Mountains Section of S.E.P.M., Denver, CO, pp 211

Urey HC, Lowenstam HA, Epstein S, Mckinney CR (1951) Measurements of paleotemperatures and temperatures of the upper cretaceous of England, Denmark and South Eastern United States. Bull Geol Soc Am, pp 62–399

Vail PR, Michum RM Jr, Todd RG, Widmier JM, Thompson III S, Sangree JB, Budd JB, Hatlied WG (1977) Seismic stratigraphy and global changes of sea level. In: Payton CE (ed) Seismic stratigraphy – applications to hydrocarbon exploration. Am Assoc Pet Geol memoir 26:47–212

Vail PR, Hardenbol J, Todd GG (1984) Jurassic unconformities, chronostratigraphy, and sea-level changes from seismic-stratigraphy. Am Assoc Memoir no 36:129–144

Van der Bark, Thomas OD (1982) Ekofisk: first of the giant oil fields in Western Europe. Am Assoc Pet Geol Bull 65:2341–2363

Van der Lingen J (1977) Diagenesis of deep-sea biogenic sediments. Dowden, Hutchinson & Ross, Stroudsburg, pp 384

Van Golf-Racht TD (1982) Fundamentals of fractured reservoir engineering. (Developments in petroleum science no 12), Elsevier, Amsterdam, pp 710

Van Olphen H, Fripiat JT (1979) Data handbook for clay materials and other non-metallic minerals. Pergamon, New York, pp 346

Van Poolen HK (1980) Fundamentals of enhanced oil recovery. Penn Well, Oklahoma, pp 155

Veizer J (1973) Sedimentation in geologic history recycling vs. evolution or recycling with evolution. Contrib Mineral Petrol 38:261–278

Veizer J (1980) Correlation of $^{13}C/^{12}C$ and $^{34}S/^{37}S$ secular variations. Geochim Cosmochim Acta 44:579–587

Veizer J (1982) Mantle buffering of the early oceans. Naturwissenschaften 69:173–180

Veizer J, Hoefs J (1976) The nature of O^{18}/O^{16} and C^{13}/C^{12} secular trend in sedimentary carbonate rocks. Geochim Cosmochim Acta, pp 1387–1395

Velde B (1977) Clay and clay minerals in natural and synthetic systems. Developments in sedimentology, No. 21. Elsevier, Amsterdam, pp 218

Visher GS, Howard JP (1974) Dynamic relationship between hydraulics and sedimentation in the Altamana Estuary. J Sediment Petrol 44:502–521

Vita-Finzi C (1973) Recent earth history. Macmillan, London, pp 138

Von Rad U, Hinz K, Sarntheim NM, Seibold E (1982) Geology of the Northwest African continental margin. Springer, Berlin Heidelberg New York, pp 703

Walker RG (1967) Turbidite sedimentary structures and their relationships to proximal and distal depositional environments. J Sediment Petrol 37:25–43

Walker RG (1979) Facies models. Geoscience Canada Reprint Series 1, pp 211

Walker RG (1982) Hummocky and Swaley cross stratification. In: Walker RG (ed) Clastic units of the front ranges foothills and plains in the area between Field BC and Drumheller, Alberta. Int Assoc of sedimentologists, 11th Int Congress on sedimentology, Hamilton, Canada, Guidebook to excursion 21:22–30

Walker RG (1984) Facies models. 2nd edn Geoscience Canada Reprint Series, Toronto, pp 317

Waples DW (1980) Time and temperature in petroleum formation: application of Lopatin's method to petroleum exploration. Am Assoc Pet Geol Bull 64:916–926

Waples DW (1985) Geochemistry in petroleum exploration. Reidel, Dortrecht, pp 223

Warme JE, Douglas RG, Winterer EL (1981) Deep sea drilling project: a decade of progress. S.E.P.M. Special Publication No. 32, pp 564

Watts AB (1981) The U.S. Continental margin: subsidence history, crustal structure and thermal evolution. In: Bally AW, Watts AB, Grow JA, Manspeizer W, Bernoulli D, Schreiber C, Hunt JM (eds) Geology of passive continental margins. Am Assoc Pet Geol Education Course Note Series No. 19, Tulsa, OK

Watts AB (1982) Tectonic subsidence, flexure and global changes of sea level. Nature 297:469–474

Watts NL (1983) Microfractures in chalk of Albuskjell field, Norwegian section, North Sea. Possible origin and distribution. Am Assoc Pet Geol Bull 67:201-234

Weaver CE, Pollard LD (1973) The chemistry of clay minerals. Developments in sedimentology No. 5. Elsevier, Amsterdam, pp 213

Wedepohl KH (1971) Geochemistry. Holt, Rinehart & Winston, New York, pp 231

Weimer RJ (1975) Deltaic and shallow marine sandstones: sedimentation, tectonics and petroleum occurrences. Continuing Education Course Note Series No. 2. Am Assoc Pet Geol, pp 94

Weimer RJ, Porter KW, Land CB (1985) Depositional modeling of detrital rocks, with emphasis on cored sequences of petroleum reservoirs. S.E.P.M. core workshop no 8, Golden, Colorado, Aug 1985

Weir AH, Ormsund EC, Mansey MI (1975) Clay mineralogy of sediments of the Western Nile Delta. Clay Minerals 10:369–386

Wilson MD, Pittman E (1977) Authigenic clays in sandstones. J Sediment Petrol 47:3–31

Wood BJ, Fraser DG (1976) Elementary thermodynamics for geologists. Oxford University Press, pp 303

Wood JR, Hewet TA (1982) Fluid convection and mass transfer in porous sandstones — a theoretical model. Geochim Cosmochim Acta 46:1707–1714

Woodrow DL, Sevon WD (1985) The Catskill Delta. Geol Soc Am Spec Pap 201, pp 254

Wright LD (1977) Sediment transport and deposition at river mouths, a synthesis. Bull Geol Soc Am 88:857–868

Wright LD, Coleman JM (1974) Mississippi River Mouth processes: effluent dynamics and morphology developments. J Geol 82:751–778

Wyllie PJ (1976) The way the earth works. An introduction to the new global geology and its revolutionary development. Wiley, New York, pp 296

Yalin MS (1977) Mechanics of sediment transport. Pergamon, New York, pp 295

Yarborough H, Emery KO, Dickinson WR, Seely DR, Dow DW, Curray JR, Vail PR (1977) Geology of continental margins. Continuing Education Course Note Series No. 5. Am Assoc Pet Geol A1/G10

Zenger DH, Dunham JB, Ethington RL (1980) Concepts and models of dolomitization. S.E.P.M. Special Publication No. 28, pp 320

Ziegler PA (1982) Geological atlas of Western and Central Europe. Encl. 37 — Maps, Shell International Petroleum, Maatchappij B.V., pp 130

Subject Index

Springer Books on Sedimentology

Springer-Verlag Berlin Heidelberg New York London Paris Tokyo Hong Kong

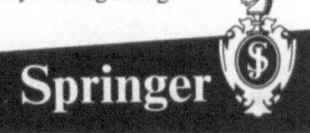

Springer

Springer Books on Sedimentology

K. L. Kleinspehn, Ch. Paola (Eds.)

New Perspectives in Basin Analysis

1988. XX, 453 pp. 225 figs. 23 tabs. (Frontiers in Sedimentary Geology) ISBN 3-540-96611-0

A. D. Miall

Principles of Sedimentary Basin Analysis

1984. XII, 490 pp. 387 figs. ISBN 3-540-90941-9

N. D. Naeser, T. H. McCulloh (Eds.)

Thermal History of Sedimentary Basins

1989. XIV, 319 pp. 197 figs. ISBN 3-540-96702-8

F. J. Pettijohn, P. E. Potter, R. Siever

Sand and Sandstone

2nd ed. 1987. XIX, 553 pp. 355 figs.
Hardcover ISBN 3-540-96355-3
Softcover ISBN 3-540-96350-2

P. E. Potter, J. B. Maynard, W. A. Pryor

Sedimentology of Shale

1984. X, 310 pp. 154 figs., 1 colored insert. ISBN 3-540-90430-1

H.-E. Reineck, I. B. Singh

Depositional Sedimentary Environments

2nd ed. 1986. XIX, 549 pp. 683 figs. 38 tabs. ISBN 3-540-10189-6

W. Ricken

Diagenetic Bedding

1986. X, 210 pp. (Lecture Notes in Earth Sciences, Vol. 6) ISBN 3-540-16494-4

P. O. Roehl, P. W. Choquette

Carbonate Petroleum Reservoirs

1985. XXII, 622 pp. 386 figs. (Casebooks Earth Sciences) ISBN 3-540-96012-0

W. Salomons, B. L. Bayne, E. K. Duursma, U. Förstner (Eds.)

Pollution of the North Sea

1988. XI, 687 pp. 238 figs. ISBN 3-540-19288-3

J. H. Schroeder, B. H. Purser (Eds.)

Reef Diagenesis

1986. IX, 455 pp. 187 figs. ISBN 3-540-16594-0

E. Seibold, W. H. Berger

The Sea Floor

1982. VII, 288 pp. 206 figs. ISBN 3-540-11256-1

P. G. Sly (Ed.)

Sediments and Water Interactions

1986. XXI, 521 pp. 276 figs. ISBN 3-540-96293-X

W. L. Steffen, O. T. Denmead (Eds.)

Flow and Transport in the Natural Environment: Advances and Applications

1988. VII, 384 pp. 138 figs. ISBN 3-540-19452-5

J. P. M. Syvitski, D. C. Burrell, J. M. Skei

Fjords

1987. X, 379 pp. 216 figs. ISBN 3-540-96342-1

O. H. Walliser (Ed.)

Global Bio-Events

1986. IX, 442 pp. (Lecture Notes in Earth Sciences, Vol. 8) ISBN 3-540-17180-0

J. L. Wilson

Carbonate Facies in Geologic History

1975. XIII, 471 pp. 183 figs. 30 plates. ISBN 3-540-90343-7

Springer-Verlag Berlin Heidelberg New York London Paris Tokyo Hong Kong

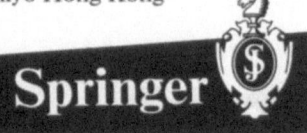